Pro SQL Server 2008 Analysis Services

Philo Janus
Guy Fouché

Apress®

Pro SQL Server 2008 Analysis Services

Copyright © 2010 by Philo Janus and Guy Fouché

ISBN-13 (pbk): 978-1-4302-1995-8

ISBN-13 (electronic): 978-1-4302-1996-5

Printed and bound in the United States of America 9 8 7 6 5 4 3 2 1

Publisher and President: Paul Manning
Lead Editor: Jonathan Gennick
Technical Reviewers: Dana Hoffman and Fabio Claudio Ferrachiatti
Editorial Board: Clay Andres, Steve Anglin, Mark Beckner, Ewan Buckingham, Gary Cornell, Jonathan Gennick, Jonathan Hassell, Michelle Lowman, Matthew Moodie, Duncan Parkes, Jeffrey Pepper, Frank Pohlmann, Douglas Pundick, Ben Renow-Clarke, Dominic Shakeshaft, Matt Wade, Tom Welsh
Coordinating Editors: Candace English and Fran Parnell
Copy Editors: Sharon Wilkey and Mary Ann Fugate
Compositor: Bytheway Publishing Services
Indexer: John Collin
Artist: April Milne
Cover Designer: Anna Ishchenko

Distributed to the book trade worldwide by Springer Science+Business Media, LLC., 233 Spring Street, 6th Floor, New York, NY 10013. Phone 1-800-SPRINGER, fax (201) 348-4505, e-mail orders-ny@springer-sbm.com, or visit www.springeronline.com.

For information on translations, please e-mail rights@apress.com, or visit www.apress.com.

Apress and friends of ED books may be purchased in bulk for academic, corporate, or promotional use. eBook versions and licenses are also available for most titles. For more information, reference our Special Bulk Sales–eBook Licensing web page at www.apress.com/info/bulksales.

The source code for this book is available to readers at www.apress.com. You will need to answer questions pertaining to this book in order to successfully download the code.

To Jodi Fouché: For her poetry, being my biggest fan, and unequivocal love

—*Guy Fouché*

Contents at a Glance

Contents

About the Authors

■ **Philo Janus** is a senior technology specialist with Microsoft. Over the years he has presented Microsoft Office InfoPath to thousands of users and developers, and assisted with enterprise implementations of InfoPath solutions. With that background, he is particularly sensitive to the difficulties users and developers have had with InfoPath.

He graduated from the US Naval Academy with a bachelor of science in electrical engineering in 1989 to face a challenging career in the US Navy. After driving an aircraft carrier around the Pacific Ocean and a guided-missile frigate through both the Suez and Panama Canals, and serving in the US Embassy in Cairo, a small altercation between his bicycle and an auto indicated a change of career (some would say that landing on his head in that accident would explain many things).

Philo's software development career started with building a training and budgeting application in Access 2.0 in 1995. Since then he's worked with Oracle, Visual Basic, SQL Server, and .NET, building applications for federal agencies, commercial firms, and conglomerates. In 2003 he joined Microsoft as a technology specialist, evangelizing Office as a development platform.

■ **Guy Fouché** is a business intelligence and decision support system consultant in the Dallas, Texas area. Guy spends his evenings playing one of his eight trumpets and expanding his composition skills by using the current generation of music technologies. On the weekend, he puts as many miles as he can on his bright yellow Honda F4i sport motorcycle. Guy and his wife Jodi enjoy taking nine-day trips in their Jeep 4×4, taking photographs and writing travelogs along the way. You can view their photography at http://photography.fouche.ws.

About the Technical Reviewers

■ **Fabio Claudio Ferrachiatti** is a senior consultant and a senior analyst/developer of Microsoft technologies. He works for Brain Force at its Italian branch (`www.brainforce.it`). He is a Microsoft Certified Solution Developer for .NET, a Microsoft Certified Application Developer for .NET, and a Microsoft Certified Professional, as well as a prolific author and technical reviewer. Over the past ten years, he's written articles for Italian and international magazines and coauthored more than ten books on a variety of computer topics.

■ Born in Brooklyn, New York, **Dana L. Hoffman** often jokes that her name should have been *Data*. She has always had a sharp eye for detail and an avid desire to create systems that are not just workable, but intuitive and easy to use. She always tries to see things from the user's point of view, and sees technical reviewing as an excellent opportunity to put her nitpicking skills to good use. With a background in programming and database development, Dana currently works as a data analyst. She lives in Connecticut and is nearly finished raising two sons.

Acknowledgments

I'd like to offer a huge thank-you to everyone at Apress who has had input into these pages!

Guy Fouché

Introduction

Pro SQL Server 2008 Analysis Services offers an in-depth look into the latest and greatest suite of analytic tools from Microsoft. This book will help you create business intelligence (BI) solutions that improve your company's analysis and decision making by focusing on practical, solution-oriented application of the technologies available in SQL Server 2008 Analysis Services (SSAS).

Using the examples and exercises in this book, you will further your understanding of online analytical processing (OLAP), BI, data mining, and SSAS itself. New SSAS features are also explained, including the Management Data Warehouse (MDW), dynamic management views (DMVs), and Aggregation Designer. Improvements to the Cube and Dimension Designers are also covered.

Chapters 1 and 2 introduce you to OLAP, and to the key concepts that are termed *cubes*, *dimensions*, and *measures*. With that foundation laid, Chapters 3 and 4 introduce you to what SQL Server provides. You'll get your first look at Analysis Services and its administration interface. Chapters 5 through 7 show you how to design and build a cube for analysis. The cube is the focal point of Analysis Services. Once you've created a cube, Chapter 8 shows how to deploy it for use.

After you've deployed a cube, it is available for you and other analyists to query. It is partly through queries that one examines and analyzes the data at one's disposal. To that end, Chapter 9 is devoted to Multidimensional Expressions (MDX), which is the query language underpinning Analysis Services solutions.

Key performance indicators (KPIs) are at the heart of every BI solution. In Chapter 10, you will learn how to define, create, and use these metrics. Chapter 10 also introduces you to perspectives, actions, and calculated members.

Data-mining algorithms enable you to sift through huge amounts of historical data, and create predictions based on trends and patterns. Working through Chapter 11, you will learn how to use data mining to create, execute, and validate a prediction model. Chapter 11 will also introduce you to Microsoft's Data Mining Extensions (DMX) language.

PowerPivot is an exciting set of technologies that provide powerful BI abilities to all business users. By integrating with Office 2010, your users can perform complex analysis and data mining on their workstations. Using SSAS language translation and automated currency conversions greatly enhances the usability of your company's data across the enterprise.

Finally, Chapter 13 offers important information for SSAS administrators. To effectively manage SSAS at the server level, you need to understand processing tasks and options, the SQL Server Profiler, the Performance Monitor, scheduling, and security.

CHAPTER 1

■ ■ ■

Introduction to OLAP

Online analytical processing (OLAP) is a technique for aggregating data to enable business users to dig into transactional data to solve business problems. You may be familiar with pivot tables from Microsoft Excel or other reporting solutions—for example, taking a list of order details (Figure 1-1) and creating a table that shows the total for each product ordered by month (Figure 1-2).

CompanyName	City	Country	OrderID	OrderDate	ProductName	UnitPrice	Quantity	ExtPrice
Vins et alcools Chevalier	Reims	France	10248	1996	Queso Cabrales	14	12	252
Vins et alcools Chevalier	Reims	France	10248	1996	Singaporean Hokkien Fried Mee	9.8	10	140
Vins et alcools Chevalier	Reims	France	10248	1996	Mozzarella di Giovanni	34.8	5	174
Toms Spezialitäten	Münster	Germany	10249	1996	Tofu	18.6	9	209.25
Toms Spezialitäten	Münster	Germany	10249	1996	Manjimup Dried Apples	42.4	40	2120
Hanari Carnes	Rio de Janeiro	Brazil	10250	1996	Jack's New England Clam Chowder	7.7	10	96.5
Hanari Carnes	Rio de Janeiro	Brazil	10250	1996	Manjimup Dried Apples	42.4	35	1855
Hanari Carnes	Rio de Janeiro	Brazil	10250	1996	Louisiana Fiery Hot Pepper Sauce	16.8	15	315.75
Victuailles en stock	Lyon	France	10251	1996	Gustaf's Knäckebröd	16.8	6	126
Victuailles en stock	Lyon	France	10251	1996	Ravioli Angelo	15.6	15	292.5
Victuailles en stock	Lyon	France	10251	1996	Louisiana Fiery Hot Pepper Sauce	16.8	20	421
Suprêmes délices	Charleroi	Belgium	10252	1996	Sir Rodney's Marmalade	64.8	40	3240
Suprêmes délices	Charleroi	Belgium	10252	1996	Geitost	2	25	62.5
Suprêmes délices	Charleroi	Belgium	10252	1996	Camembert Pierrot	27.2	40	1360
Hanari Carnes	Rio de Janeiro	Brazil	10253	1996	Gorgonzola Telino	10	20	250
Hanari Carnes	Rio de Janeiro	Brazil	10253	1996	Chartreuse verte	14.4	42	756
Hanari Carnes	Rio de Janeiro	Brazil	10253	1996	Maxilaku	16	40	800
Chop-suey Chinese	Bern	Switzerland	10254	1996	Guaraná Fantástica	3.6	15	67.5
Chop-suey Chinese	Bern	Switzerland	10254	1996	Pâté chinois	19.2	21	504
Chop-suey Chinese	Bern	Switzerland	10254	1996	Longlife Tofu	8	21	210
Richter Supermarkt	Genève	Switzerland	10255	1996	Chang	15.2	20	380
Richter Supermarkt	Genève	Switzerland	10255	1996	Pavlova	13.9	35	610.75
Richter Supermarkt	Genève	Switzerland	10255	1996	Inlagd Sill	15.2	25	475
Richter Supermarkt	Genève	Switzerland	10255	1996	Raclette Courdavault	44	30	1650
Wellington Importadora	Resende	Brazil	10256	1996	Perth Pasties	26.2	15	492
Wellington Importadora	Resende	Brazil	10256	1996	Original Frankfurter grüne Soße	10.4	12	156
HILARION-Abastos	San Cristóbal	Venezuela	10257	1996	Schoggi Schokolade	35.1	25	1097.5
HILARION-Abastos	San Cristóbal	Venezuela	10257	1996	Chartreuse verte	14.4	6	108

Figure 1-1. Tabular order data

Sum of Quantity	Column Labels ▼					
Row Labels ▼	April 1997	April 1998	August 1996	August 1997	December 1996	December 1997
Alice Mutton		12	30			108
Aniseed Syrup		25	30			20
Boston Crab Meat	30	55	100	60		1
Camembert Pierrot	106	174	76	30	111	15
Carnarvon Tigers	25	53	12	82	9	30
Chai	40	104	63	40	15	
Chang	12	201		100	25	85
Chartreuse verte	10			41	20	148
Chef Anton's Cajun Seasoning	50	25		35		
Chef Anton's Gumbo Mix		100	20	15	32	
Chocolade	15					
Côte de Blaye	15	25			40	15
Escargots de Bourgogne		40		6	95	
Filo Mix	6	32		5	20	23
Flotemysost		197	20	65	161	28
Geitost	16	55	44	20	60	
Genen Shouyu	12		20			
Gnocchi di nonna Alice	154	25		80	38	100
Gorgonzola Telino	20	15	55		164	149
Grandma's Boysenberry Spread				70		
Gravad lax						
Guaraná Fantástica	37	140	40	23	30	40
Gudbrandsdalsost	10	66		20	58	24
Gula Malacca	9	27	61	10		10
Gumbär Gummibärchen		81		35	70	21

Figure 1-2. Summarizing products by month ordered

This is an interesting report, but what if we want to see the breakdown by quarter? Or by year? Or by fiscal year? Perhaps we want to combine the products into groups—for example, if we don't care so much how a specific product is faring, but we do want to know how our condiments are selling overall. We might be able to create some of these reports from existing data, or we might be able to write a query to do so. Some will require modifications to the database.

However, we have another problem: in this database, we have only 2,155 order details and 77 products—a pretty easy group of data to deal with. What do we do when we have 500 products (or more—consider Amazon.com!) and tens of thousands of records? What about millions? We can't expect Microsoft Excel to create pivot tables from all those records for us. To solve this problem of analyzing large amounts of data, we turn to a server-based solution that can take large amounts of tabular data and create these aggregations for us.

From Pivot Tables to Dimensional Processing

Take another look at Figure 1-2, order totals of products by month ordered. We have two *dimensions* to our data: product and month. This is a pretty basic pivot table based on a tabular data source. Now let's say we want to see this same table, but broken down for each geographic region (Figure 1-3).

Sum of Quantity	Column Labels					
Row Labels	April 1997	April 1998	August 1996	August 1997	December 1996	December 1997
Alice Mutton		12	30			108
Aniseed Syrup		25	30			20
Boston Crab Meat	30	55	100	60		1
Camembert Pierrot	106	174	76	30	111	15
Carnarvon Tigers	25	53	12	82	9	30
Chai	40	104	63	40	15	
Chang	12	201		100	25	85
Chartreuse verte	10			41	20	148
Chef Anton's Cajun Seasoning	50	25		35		
Chef Anton's Gumbo Mix		100	20	15	32	
Chocolade	15					
Côte de Blaye	15	25			40	15
Escargots de Bourgogne		40		6	95	
Filo Mix	6	32		5	20	23
Flotemysost		197	20	65	161	28
Geitost	16	55	44	20	60	
Genen Shouyu	12		20			
Gnocchi di nonna Alice	154	25		80	38	100
Gorgonzola Telino	20	15	55		164	149

United States

Sum of Quantity	Column Labels					
Row Labels	April 1997	April 1998	August 1996	August 1997	December 1996	December 1997
Alice Mutton		12	30			108
Aniseed Syrup		25	30			20
Boston Crab Meat	30	55	100	60		1
Camembert Pierrot	106	174	76	30	111	15
Carnarvon Tigers	25	53	12	82	9	30
Chai	40	104	63	40	15	
Chang	12	201		100	25	85
Chartreuse verte	10			41	20	148
Chef Anton's Cajun Seasoning	50	25		35		
Chef Anton's Gumbo Mix		100	20	15	32	
Chocolade	15					
Côte de Blaye	15	25			40	15
Escargots de Bourgogne		40		6	95	
Filo Mix	6	32		5	20	23
Flotemysost		197	20	65	161	28
Geitost	16	55	44	20	60	
Genen Shouyu	12		20			
Gnocchi di nonna Alice	154	25		80	38	100
Gorgonzola Telino	20	15	55		164	149

Europe

Sum of Quantity	Column Labels					
Row Labels	April 1997	April 1998	August 1996	August 1997	December 1996	December 1997
Alice Mutton		12	30			108
Aniseed Syrup		25	30			20
Boston Crab Meat	30	55	100	60		1
Camembert Pierrot	106	174	76	30	111	15
Carnarvon Tigers	25	53	12	82	9	30
Chai	40	104	63	40	15	
Chang	12	201		100	25	85
Chartreuse verte	10			41	20	148
Chef Anton's Cajun Seasoning	50	25		35		
Chef Anton's Gumbo Mix		100	20	15	32	
Chocolade	15					
Côte de Blaye	15	25			40	15
Escargots de Bourgogne		40		6	95	
Filo Mix	6	32		5	20	23
Flotemysost		197	20	65	161	28
Geitost	16	55	44	20	60	
Genen Shouyu	12		20			
Gnocchi di nonna Alice	154	25		80	38	100
Gorgonzola Telino	20	15	55		164	149

Asia

Sum of Quantity	Column Labels					
Row Labels	April 1997	April 1998	August 1996	August 1997	December 1996	December 1997
Alice Mutton		12	30			108
Aniseed Syrup		25	30			20
Boston Crab Meat	30	55	100	60		1
Camembert Pierrot	106	174	76	30	111	15
Carnarvon Tigers	25	53	12	82	9	30
Chai	40	104	63	40	15	
Chang	12	201		100	25	85
Chartreuse verte	10			41	20	148
Chef Anton's Cajun Seasoning	50	25		35		
Chef Anton's Gumbo Mix		100	20	15	32	
Chocolade	15					
Côte de Blaye	15	25			40	15
Escargots de Bourgogne		40		6	95	
Filo Mix	6	32		5	20	23
Flotemysost		197	20	65	161	28
Geitost	16	55	44	20	60	
Genen Shouyu	12		20			
Gnocchi di nonna Alice	154	25		80	38	100
Gorgonzola Telino	20	15	55		164	149

EMEA

Figure 1-3. *Breaking down orders by geographic region*

Now we have a third *dimension*: geographic region. Consider arranging the data as shown in Figure 1-4.

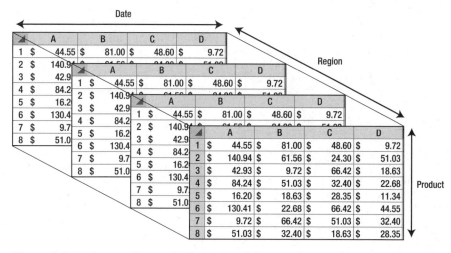

Figure 1-4. *Understanding multidimensional data*

With three dimensions to the data, we have a *cube*. We will be using that term a lot. Of course, we don't have to stop with three dimensions. We can break the orders down by customer, by discount level, by product type, and so on. As you start to consider the exponential impact of having multiple dimensions, with each dimension having dozens of members, you will start to appreciate that presenting a tool to a nontechnical user to deal with this type of analysis is a nontrivial problem.

Looking at Figure 1-4, what does it look like when we want to also break it down by warehouse location and shipping method? And yes, we still call it a *cube* when there are more than three dimensions.

Let's take a step back and look at another problem we can face when aggregating data. Let's take our order data and create a pivot table to show products ordered by customer (Figure 1-5).

Sum of Quantity	Column Labels								
Row Labels	Alice Mutton	Aniseed Syrup	Boston Crab Meat	Camembert Pierrot	Carnarvon Tigers	Chai	Chang	Chartreuse verte	Chef
Alfreds Futterkiste		6							21
Ana Trujillo Emparedados y helados				10					
Antonio Moreno Taquería	18		10				20		
Around the Horn				15			15		
Berglunds snabbköp	10	30	75	35		35	21		26
Blauer See Delikatessen				21	10				5
Blondesddsl père et fils	30					18	25		60
Bólido Comidas preparadas	40								
Bon app'	16		45			50			21
Bottom-Dollar Markets	56	20	50	75		60			
B's Beverages		30	10						
Cactus Comidas para llevar									
Centro comercial Moctezuma									
Chop-suey Chinese				10		15	30		
Comércio Mineiro									
Consolidated Holdings						10			
Die Wandernde Kuh			10			15	30		
Drachenblut Delikatessen									
Du monde entier	15					3			
Eastern Connection				50		25			
Ernst Handel	121	45	91	160	9	58			175
Familia Arquibaldo			4	12					50

Figure 1-5. Products ordered by customer

Obviously, few customers would order every product, so we see there is a lot of empty space. This is referred to as a *sparse data set*. The problem we have is that we are taking up space for every combination of customer and product, and these empty cells add up pretty quickly. OLAP servers are designed to optimize for storage of sparse data sets and aggregations.

Data Warehousing

A term that you'll frequently encounter in analytic processing—and that is horribly overused—is *data warehouse*. You may also hear *data mart*. (I will confess to also occasionally using *model* when I'm more interested in presenting the business problem we're trying to solve than getting into a debate about semantics.) Although these terms are not completely synonymous, they are often used interchangeably.

Data warehouse is easily the scariest word on the list. As this book will show you, working with OLAP technologies is relatively straightforward. I don't mean to trivialize the work necessary to build a robust dimensional solution, but the difficulty can be overhyped. A data warehouse is, most basically, a compilation of data fundamental to the business that has been integrated from numerous other sources. For example, you can pull salary data from an HR system, vendor data from an ERP system, customer billing information from a financial system, and create a financial data warehouse for the company (to start calculating profits and losses, profit margins, and so forth).

A data warehouse does not need to be dimensional. You could create a normalized relational database for querying and reporting. However, you will find it difficult to create many of the types of reports users expect from a very large storage of data (users and analysts will generally focus on aggregation and large-scale analysis as opposed to basic tabular reports).

Because *data warehouse* generally implies large-scale and cross-corporate information, people may perceive, as I mentioned, that data warehouses are heavy engineering efforts requiring a great deal of big up-front design. Where possible, I prefer to grow these types of resources organically: start with small projects, solve individual business problems, and grow the resulting *data marts* as necessary, finally bridging over to create an actual data warehouse if and when it is deemed necessary. I am very much about iterative design, and the work in this book supports that approach in design and maintenance of OLAP solutions.

Applications of OLAP

Multidimensional analysis of large quantities of data has applications in every line of business, as well as scientific analysis and research. Following are some examples:

- An executive might use a front-end analysis tool against corporate data to identify overall trends or causal relationships in corporate performance. In addition, aggregated data enables creation of corporate scorecards to provide "at a glance" views of how the business is doing (Figure 1-6).

Strategy Map Scorecard		
	Plan	Target
Financial Performance		
Increase Revenue		
Maintain Overall Margins		
Net Profit	18.00%	15.00%
Contribution Margin	64.44%	66.00%
YOY Revenue Growth	22.00%	15.00%
New Product Revenue	$2,463,887	$2,000,000
Control Spend		
Expense as % of Revenue	12.00%	10.00%
Expense Variance %	3.00%	1.00%
Customer Satisfaction		
Count of Complaints	127	200
Total Backorders	5,000	1,000
Avg Customer Survey Rating	7	3
Unique Repeat Customer Count	785	1,000
Acquire New Customers	3,547	3,000
New Opportunity Count	446	300

Figure 1-6. A corporate scorecard built on OLAP data

- A sales manager would find sales reports combining the transactional data we've been discussing (Figure 1-7) useful. These reports can show how products have been selling and can easily be configured to break down the data in a sensible way. Also, using a special OLAP query language, reports can be built to "roll" forward as time goes on—for example, a report may show just the last three quarters. Instead of having to write a complex query or constantly update the report, a simple statement such as [Dimensions].[Time].[FiscalQuarter].PrevMember would always show the last full quarter of data.

Sales Report

		FY 2002	FY 2003	FY 2004
Accessories		$36,815	$124,433	$410,050
Bikes	Mountain Bikes	$8,568,958	$9,184,859	$8,738,868
	Road Bikes	$6,449,576	$13,232,561	$9,676,070
	Touring Bikes			$10,451,490
Clothing	Bib-Shorts		$166,740	
	Caps	$5,081	$11,699	$14,762
	Gloves		$155,267	$52,508
	Jerseys	$54,673	$133,279	$391,357
	Shorts		$81,899	$260,304
	Socks	$6,573		$18,065
	Tights		$201,833	
	Vests			$223,801
Components	Bottom Brackets			$51,826
	Brakes			$66,019
	Chains			$9,378
	Cranksets			$203,943
	Derailleurs			$70,209
	Forks		$77,932	
	Handlebars		$77,104	$93,487
	Headsets		$60,942	
	Mountain Frames	$720,081	$1,728,122	$2,265,470
	Pedals			$147,484
	Road Frames	$446,685	$2,005,931	$1,397,237
	Saddles			$55,829
	Touring Frames			$1,642,328
	Wheels		$679,070	

Figure 1-7. A report produced from OLAP data

- A business user will prefer using a desktop tool such as Excel to perform ad hoc analysis of data (Figure 1-8). Excel, ProClarity, and other tools like these provide the ability to dig through data, produce reports, and perform analysis on large amounts of information from numerous data sources that have been aggregated, scrubbed, and normalized.

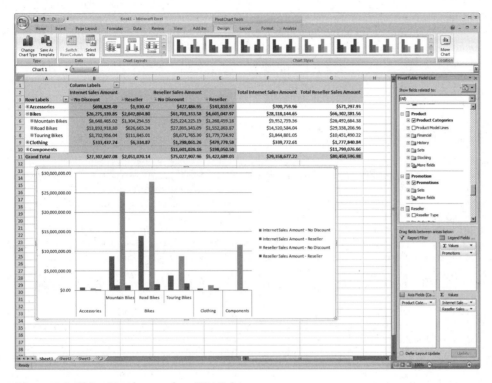

Figure 1-8. *Using Excel to analyze OLAP data*

Now that you have a rough feel of what OLAP is and why we're interested in it, let's take a look at where OLAP came from.

History of OLAP

Multidimensional analysis dates back to the 1960s and IBM's implementation of APL for managing large-scale arrays. However, although APL was very important, the requirement of a mainframe to run it on and its arcane language (including Greek letters) kept it firmly in a small niche. APL continues to run applications to this day, but when is the last time you encountered anyone discussing it?

In the 1970s, the next step in OLAP evolution came forth in an actual application: Express. Express was designed to provide data-aggregation services and eventually became Express Server. Express was a command-line-based application, and data was managed with scripts. Express Server was acquired by Oracle in 1996, and the OLAP technologies were folded into Oracle9i.

E.F. Codd (father of the relational database) coined the term OLAP in 1993. Although other products performed large-scale data aggregation before that, Codd's creation of the term formed a foundation for technologies to grow on. Around the same time, Arbor Software (which merged with Hyperion Software in 1998) released their OLAP product, Essbase (originally eSSbase). Essbase was a client-server product that presented a (proprietary) spreadsheet as the front end, and ran a PC-based server application on the back end.

A number of OLAP vendors followed through the 1990s, but by the close of the 20th century, Microsoft would change the game twice. First, although other vendors were the first to consider using a spreadsheet application as a front end for OLAP, Microsoft's Excel was the first platform that truly became a ubiquitous analysis front end and had the extensibility hooks enabling other vendors to leverage it for their own use. By 2004, most OLAP vendors leveraged Excel as a user-friendly front end. Microsoft's second game-changer was to enter the OLAP server market themselves.

SQL Server Analysis Services

In 1996, Microsoft acquired OLAP technology from Panorama Software. It shipped with SQL Server 7 as OLAP Services (`www.microsoft.com/technet/prodtechnol/sql/70/maintain/olap.mspx`). OLAP Services offered a Cube Creation Wizard to design cube schemas, a Microsoft Management Console (MMC) snap-in to manage the services, and Pivot Table Services for client-side OLAP. In SQL Server 2000, the feature was renamed SQL Server Analysis Services (SSAS).

OLAP capabilities were underappreciated in SQL Server 7 and 2000. Between the perception that SQL Server wasn't ready for larger database tasks, and the lack of an easy way of viewing the output of the aggregations, it pretty much sailed under the radar. (SQL Server 2000 had a community solution that allowed viewing cube info, but having to download and compile a viewer for an enterprise data solution can make one wonder.) Another drawback to SSAS 2000 was that the underlying relational data had to already be in a star schema, which you mapped the cube onto. This made a staging database (intermediate data storage) almost mandatory.

SQL Server 2005 really brought Analysis Services into the spotlight. The cube designer was united with Integration Services and Reporting Services in the Business Intelligence Development Studio, or BIDS (Figure 1-9), and Analysis Services was accessible via the unified environment of SQL Server Management Studio (Figure 1-10).

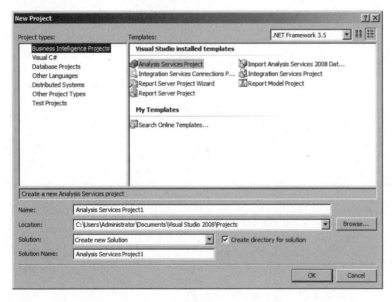

Figure 1-9. The New Project Wizard in BIDS

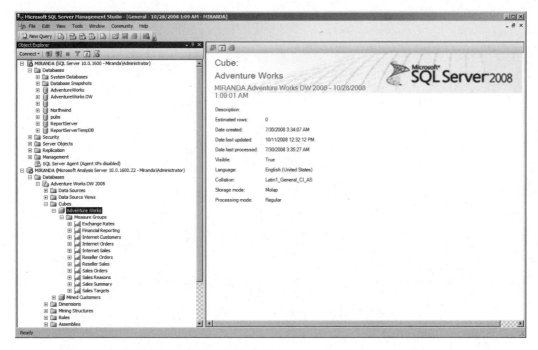

Figure 1-10. *Viewing the details of an SSAS cube in SQL Server Management Studio*

SSAS 2005 also made configuration management easier. In SSAS 2000, data connections to underlying databases were hard-coded and had to be manually changed when migrating a cube from development to test or production environments. SSAS 2005 provided tools to automate this process.

In 2005, Analysis Services also introduced the data source view (DSV), which allows a more robust management of how relational data sources are mapped within the cube solution. It also enables a developer to unify data from multiple sources (SQL Server, Oracle, DB2, and so forth). Figure 1-11 shows an example of a DSV from the AdventureWorks solution.

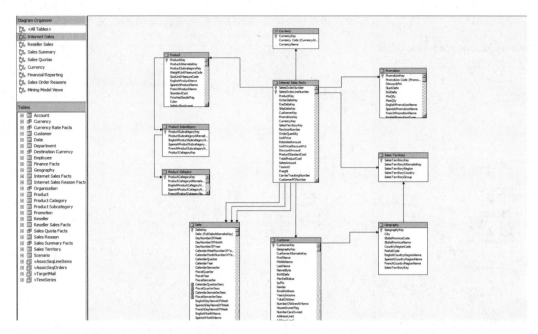

Figure 1-11. A data source view (DSV)

SQL Server Analysis Services 2005 also introduced the unified dimensional model (UDM). The UDM acts as a virtual model for data from a data warehouse or other RDBMS. With some planning and management, the UDM eliminates the need for various staging and normalized databases all containing the same data.

In 2005 we also saw the addition of a robust data-mining engine into SSAS. More important, Service Pack 2 added a data mining add-in for Excel 2007. This really made data mining accessible to the masses; instead of installing a database administrator (DBA) tool and having to understand all the data-mining models before really starting to work with them, Excel enables a more "point and play" approach (Figure 1-12).

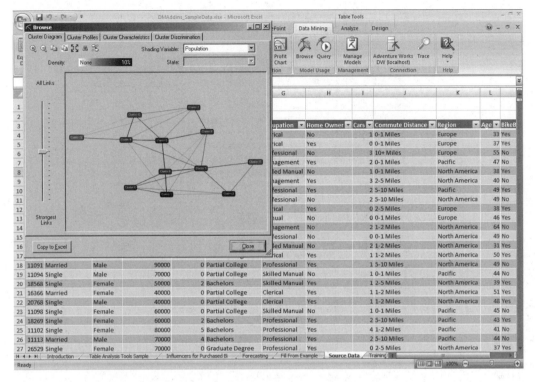

Figure 1-12. *Data mining in Excel leveraging SQL Server Analysis Services*

Microsoft established SQL Server as their business intelligence platform by adding key performance indicator (KPI) features to Analysis Services, as shown in Figure 1-13. *Key performance indicators* are structures for performance management or business intelligence. They allow defining a current value, a target value, and a trend indicator.

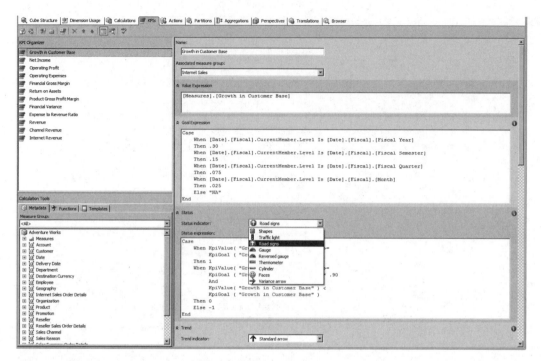

Figure 1-13. *The KPI designer in BIDS*

The real value in defining KPIs in a cube is that you will then have a uniform measurement to use for comparing various aspects of your business. Instead of every department defining *profit margin* with different business rules, they simply provide the underlying data, and the KPI does the work at any level of aggregation.

■ **Note** Chapter 11 covers KPIs to some degree, but for an in-depth examination of how SQL Server Analysis Services fits into business intelligence, check out Philo's book *Pro PerformancePoint Server 2007* (Apress, 2008).

Most notably, the Business Intelligence Development Studio included a browser (Figure 1-14), which made it easy to view the structure and content of the cubes you create. The drag-and-drop design is very intuitive for anyone who's worked with Excel pivot tables or most charting packages.

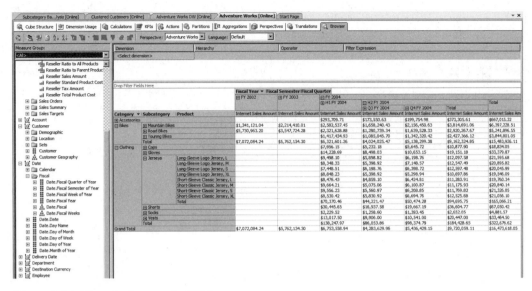

Figure 1-14. *The cube browser in BIDS*

SQL Server Analysis Services 2008 is somewhat more evolutionary than revolutionary. After the upheaval in the SQL Server 2000-to-2005 transition, Microsoft was intent not to stir the pot again so soon. As a result, SSAS 2008 focused on improving performance, implementing user feedback on the tools used, supportability, extensibility, and data-mining improvements. Of course, you'll be reviewing all these improvements in depth as you go through this book.

Data Mining

In its most basic form, *data mining* is the process of sifting through large quantities of information to gain insight into the underlying processes. A classic data-mining example is law enforcement, where officers may comb through reams of information (phone records, credit card receipts, noted meetings, and so forth) to identify the relationships in a crime syndicate.

■ **Note** This type of association, or combing through data, is generally referred to as *link analysis*.

Another form of data mining is running volumes of transactional data through a process to find patterns in the transactions. An example of this form of data mining is crunching through years of sales receipts for a grocery store to identify buying patterns of customers. This type of data mining is a perfect application of OLAP technologies, because it is dependent on aggregation of data.

An interesting aspect of this use of the OLAP engine is that you most likely won't be operating on a cube. Instead, you will create a data-mining model, train it on transactional data, and use it to process transactional data. To some degree, data-mining engines coexist in the same box as multidimensional cubes, but they are only tangentially related.

You will examine the data-mining capabilities of SQL Server 2008 Analysis Services in Chapter 13.

Summary

This chapter is simply meant to give you an idea of what this OLAP thing is all about, and where SQL Server fits in. You were introduced to data warehousing and data marts. You learned some OLAP history and were exposed to the various companies and individuals that contributed. You finished up the chapter with your first look at the SSAS environment.

In Chapter 2, you'll start to dig into the fundamental concepts behind how OLAP cubes work and what pieces make up an OLAP solution. You will learn the importance of cubes, dimensions, and measures. Finally, you'll see how these building blocks work together.

■ ■ ■

Cubes, Dimensions, and Measures

Cubes, dimensions, and measures are the fundamental parts of any OLAP solution. Let's consider the array of spreadsheets from Chapter 1, shown here again in Figure 2-1. At the most basic, the whole collection is a cube. In this chapter, you'll look at how this model relates to the concepts you need to grasp in order to understand OLAP and Analysis Services.

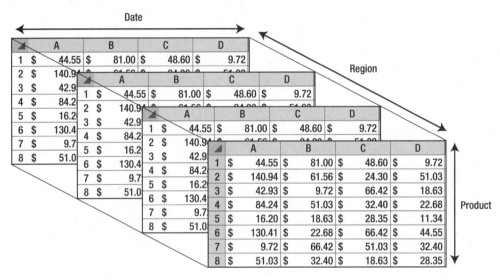

Figure 2-1. A cube

Cubes and Their Components

A *cube* is a collection of numeric data organized by arrays of discrete identifiers. Cubes were created to give a logical structure to the pivot table approach to analyzing large amounts of numerical data. Figure 2-1 shows how combining three variables creates a cube-like shape that we can work with. Although there are nuances to the design and optimization of cubes, all the magic is in that concept. A cube can (and generally will) have more than three variables defining it, but it will still always be referred to as a *cube*.

■ **Note** Shifting from a programming mindset to an OLAP mindset can be difficult. In programming, a variable can have many values, but represents only one distinct value at a time. In OLAP, we are interested in analyzing a set of the possible values, so we work with an array of discrete values. In other words, the *variable* Region may have a single *value* of Missouri or Florida or Washington, but in an OLAP analysis we want to examine the *set* or *array* of possible region values: {Florida, Missouri, Washington}.

Essentially, a cube is defined by its measures and dimensions (more on them in a bit). Where the borders are drawn to define a cube is somewhat arbitrary, but generally a cube focuses on a functional area. Consider our order details spreadsheet from Figure 1-1 in Chapter 1—we have information on customers, their orders, products, and so forth.

Perhaps we also want to calculate cost of sales (that is, the amount we pay in expenses for every dollar of revenue we bring in). To get that information, we need to add in company financial data: building expenses, advertising, manufacturing or wholesale costs, shipping, and personnel salary and benefits information. Do we just pile all this stuff into our sales cube? This quickly becomes an issue of "throwing all your tools in one box." The problem is that later changes to the cube may affect a myriad of applications, as we have to consider interoperability of such changes, and so on. In addition, personnel financial information has significant privacy considerations; we can report on pay data in aggregate, but nothing that could identify any one person.

Instead, we consider adding the data in two cubes: one for corporate financial data (without personnel pay data), and one for HR data. Then we can *link* the data and parameters from the cubes as necessary so they stay synchronized.

From the aspect of defining a cube, either solution would work—one big huge cube or three smaller, subject-specific cubes. We prefer the latter situation, as this is more in-line with the data mart, or business area approach mentioned in Chapter 1. Your cubes will be easier to enhance and maintain using this approach. In theory, we could have a sales cube and an HR cube and split corporate financial data between them. We would likely regret this decision later, but we would still have two *cubes*.

Thinking back, what makes each of these so-called cubes really a cube? Each is a collection of numeric data (for example, sales totals, expense costs, personnel pay) grouped by arrays of discrete identifiers. For example, the following is a list of possible identifiers in different categories:

Products: mutton, crabmeat, chai, filo mix

Countries: France, Italy, United Kingdom, United States

Dates: 1996, 1997, 1998

These categories are *dimensions*—the arrays of identifiers (*members*) that we use to analyze our data. By choosing different dimensions, we break up the data in different ways. Let's take another look at our "collection of spreadsheets" from Figure 2-1 with some more detail and a shift in perspective (Figure 2-2).

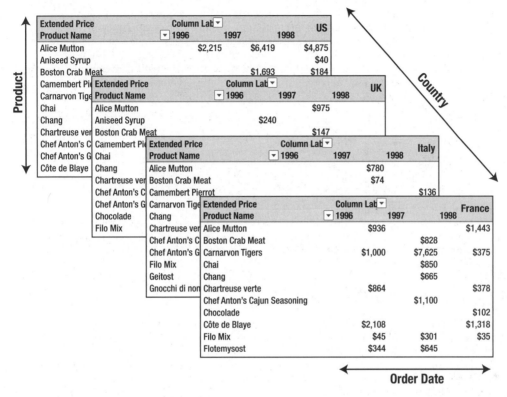

Figure 2-2. *Analyzing spreadsheets*

Again we can see that we have broken down our numbers by country, by year, and by product, and we see the cube shape that gives cubes their name. We can see that the amount of Alice mutton sold in France in 1996 is $936. So we have a member from each of three dimensions to get a single value. That value is a *measure*.

Measures are the numeric data "inside" a cube. They define the data we get by selecting members from dimensions. Generally, measures are added together to produce summary results. In the cube in Figure 2-2, for example, the measure of Alice mutton sold in France in 1996 is obtained by adding together that line item from every order that month during that year. If we wanted to know all the sales in France in 1996, we'd add in the amounts for all the products and get a total that way (that is, the value of the Sales measure for the France member of the Country dimension and the 1996 member of the Order Date dimension).

If you've done a significant amount of programming or database administrator (DBA) design work, you will recognize that the year values in Figure 2-2 are best represented as text, or strings. If you entered *1996* as a number in Excel, Excel may format it as 1,996. Why do I bring this up? To clarify the next point I'm going to make:

Dimensions are strings; measures are numbers.

The most confusing part about cube design is the question "What's a measure, and what's a dimension?" I've found that this single reminder—that dimensions are strings, and measures are numbers—always keeps me straight. (At times measures are text, but they are an edge case I'll cover later.) You might be tempted to ask, "What if I want to break down my data by age? Age is a number." This is true, but consider how we've defined a cube, and measures, and dimensions. When using age as a dimension, you don't add the numbers; you use them as discrete labels. You probably wouldn't even use the individual numbers. You would create brackets such as the following:

- 1–18 Years

- 19–40 Years

- 41–65 Years

- 65+ Years

If you look at it that way, each member is, indeed, a string.

Another approach I often use (and almost always use when designing a staging database for a cube) is that dimensions are often directly linked to lookup tables. Again, thinking in terms of lookup tables, you can see that what makes a dimension becomes pretty clear.

Defining Measures and Dimensions

Now that you have a basic understanding of cubes, measures, and dimensions, let's dig deeper into aspects of defining measures and dimensions in the OLAP world. In this section, you'll look at how dimensions relate to data, learning about *schemas* as the layer between a cube and a data source. Then you'll learn about dealing with hierarchical relationships (such as countries, states/regions, and cities) and time in OLAP. You'll also look at the question "If a member is a noun, where do we put the adjectives?" Finally, you'll dig more deeply into measures.

Schemas

Just as pivot tables derive from relational data tables, the underlying structure of a cube is a collection of tables in a relational database. In the OLAP field, we label tables as either fact tables or dimension tables. A *fact table* contains data that yields a measure. A *dimension table* contains the lookup data for our dimensions. If a table contains both fact and dimension information (for example, an order details table will probably contain both the dollar amounts of the order and the items that were ordered), we still just call it a fact table.

There are generally two types of *schemas*, or data structures, that cubes are built on: star and snowflake schemas. You'll see a lot of discussion devoted to these two structures, and might assume there is a huge amount of complexity to differentiating the two. Figure 2-3 shows a star schema with fact and dimension tables labeled. Figure 2-4 shows a snowflake schema.

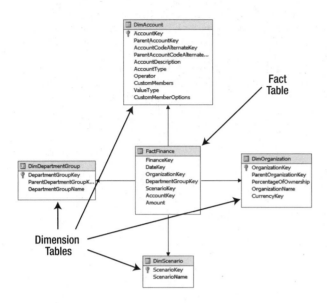

Figure 2-3. A star schema

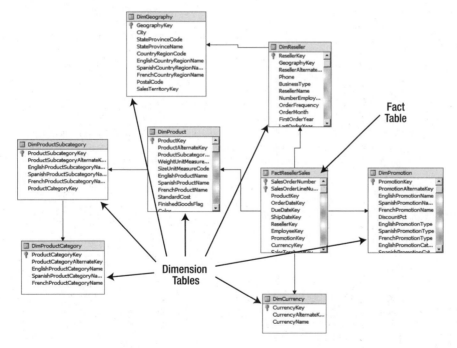

Figure 2-4. A snowflake schema

So what's the difference between a star schema and a snowflake schema? Yes, it's as easy as it appears. A *star schema* has one central fact table, with every dimension table linked directly to it. A *snowflake schema* is "everything else." In a typical snowflake schema, dimension tables are normalized, so hierarchies are defined via multiple related tables. For example, Figure 2-4 indicates a hierarchy for products, which is detailed in Figure 2-5. That's the entirety of the difference between star schemas and snowflake schemas—whether dimensions are represented by single tables or a collection of linked tables. (More-complex snowflake schemas can have multiple fact tables and interlinked dimensions. The AdventureWorks schema that we'll be using in Analysis Services is like this.)

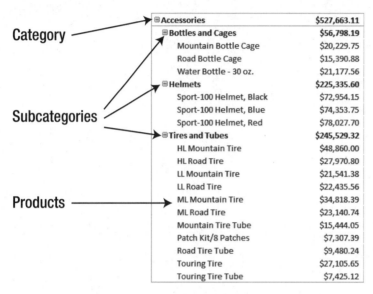

Figure 2-5. *The product hierarchy in AdventureWorks*

Basically, snowflake schemas can be easier to model and require less storage space. However, they are more complex and require more processing power to generate a cube. Snowflake schemas can also make data transformation and loading into your staging tables more cumbersome. In addition, working with your slowly changing dimensions may also become more complex. (You'll learn more about slowly changing dimensions later in this chapter.)

Star schemas are generally easier for end users to work with and visualize, because they can easily see the dimensions involved. Finally, some products designed to work with OLAP technologies may restrict which type of schema they will work with. (Microsoft's PerformancePoint Business Modeler will work only with star schema models, for example.)

Dimensions in Depth

As I mentioned previously, the best way to think of a dimension is as a lookup table. A string is the member of the dimension. The dimension is represented by the unique array of values in that string field, and they are linked to the fact table with a primary key–foreign key relationship.

The concept of what makes a *dimension*, however, isn't as restrictive as a lookup table. Dimensions can derive from columns within the fact table, from related tables that contain data, from lookup tables, or even from tables that are related through a third table (reference relationship). Remember that the focus is on relating the array of dimension members to the measures that derive from the fact table, and don't get hung up on the implementation under the covers. Conceptually, dimensions are fairly straightforward, but several possible twists are covered in the next few sections.

Attributes

Attributes are metadata that belong to a dimension, and are used to add definition to the dimension members. In our ongoing example of orders management, a product member may have attributes giving details about the product (color, weight, description—even an image of the product). A customer dimension may have attributes giving address information, the length of time that person has been a customer, and other related information. Figure 2-6 shows a dimensional report (using SQL Server Reporting Services) that has a tool tip showing the number of days it takes to manufacture a product, using an attribute from the product dimension.

		Touring Tire			$10,233	$16
		Touring Tire Tube			$2,949	$4
Bikes	Mountain Bikes	Mountain-100 Black, 38	$700,113	$639,884		
		Mountain-100 Black, 42	$660,893	$593,829		
		Mountain-100 Black, 44	$733,295	$6?? 4 Days to Manufacture		
		Mountain-100 Black, 48	$623,698	$6??,578		
		Mountain-100 Silver, 38	$688,158	$603,711		
		Mountain-100 Silver, 42	$592,814	$593,681		
		Mountain-100 Silver, 44	$629,140	$588,071		

Figure 2-6. *Creating a tool tip with a dimension attribute*

Attributes can also be used to calculate values. For example, consider a cube used to analyze consulting hours. Hours billed is the measure, and you may have dimensions for time, customer, consultant, and billing type. The consultant dimension could have an attribute of Hourly Rate that you can use to multiply by the hours to get the total amount billed, for analysis purposes.

Which brings us to another issue: hourly rates change over time. Other aspects of our business change over time (heck, what *doesn't* change over time?). To deal with that, we have slowly changing dimensions, often abbreviated as SCD.

Slowly Changing Dimensions (SCD)

*Slowly changing dimension*s deal with the reality that things change. Product definitions change; sales districts change; pricing changes; managers change; and customers change. Working with changing dimensions in a cube is fairly similar to working with changing lookup values in a relational database, with similar pros and cons.

ARE THERE FAST-CHANGING DIMENSIONS?

If we have slowly changing dimensions, you may be tempted to wonder whether there are fast-changing dimensions. Remember that with SCDs, the key word is *change*—a normal dimension can have members *added* on a regular basis without significantly impacting the cube design. SCDs come into play when a member or attribute *changes*. Consider the example of a salesperson and his relationship to a sales district. When the salesperson changes districts, we want to be able to look back at that person's performance over time, including in older districts.

On the other hand, if an employee gets married and changes his or her name, from a performance perspective we're not that concerned with keeping the old data. We just change the name in place (also known as a Type 0 SCD).

With that in mind, in terms of changes to dimensions we want to track, if you have something you would consider a fast-changing dimension, your first move should be to evaluate the dimension again and be absolutely sure that you've designed the dimensions properly. If you really do need to track the changes to a dimension on a rapid basis, you'll still need to follow the techniques for slowly changing dimensions but will need to ensure you're designing for scalability. You'll also most likely use a relational OLAP (ROLAP) architecture, which I describe later in the chapter. In other words, don't get hung up on the *slow* in slowly changing dimension.

Again, this is an edge case and I won't be covering fast-changing dimensions here.

Working well with SCDs requires methodologies for managing them—balancing manageability against the need to maintain (and report on) historical data. There are six methodologies for managing SCDs:

Type 0 isn't really a management methodology, per se. It's simply a way to indicate that there's been no attempt to deal with the fact that the dimension changes. (Most notably, when a dimension that was expected to remain static changes, it effectively becomes a Type 0 SCD.)

Type 1 simply replaces the old data with the new values. Thus any analysis of older data will show the new values. This is generally appropriate only when correcting dimension member data that is in error (misspelled, for example). This is obviously easy to maintain, but shows no historical data and may result in incorrect reports.

Type 2 uses To and From dates in the member record, with the current date having a null End Date field. OLAP engines can implement logic to bind measure values to the appropriate dimension value transparently. Obviously, the need to maintain beginning and end dates, with only one null end date, makes data transformations for this method more complex. Type 2 is the most common approach to dealing with slowly changing dimensions.

Type 3 is completely denormalized—it maintains the history of values and effective date in the record for the dimension, each in its own column. Given the need to track multiple columns of data, it's best to reserve this method for a small number of changes (for example, original value and current value without tracking interim changes). Because this is a simpler approach, it makes data transformations easier, but it's pretty restrictive.

Type 4 uses two tables for dimensions: the standard dimension table, and a separate table for historical values. When a dimension value is updated, the old value is copied to the history table with its begin and end dates. This structure makes for the fastest processing of OLAP cubes and queries, while allowing for unlimited history. But again, transforming data and maintenance are problematic.

Type 6 is a combination of Types 1, 2, and 3 (1 + 2 + 3 = 6; someone had a sense of humor). The dimension table has start and end dates (Type 2), adds attributes so that any historical data can be analyzed using the current "today" data (Type 1), or any mix of the two (Type 3).

Again, slowly changing dimensions are not that complex a topic to understand. The trick is in the analysis and implementation. You'll examine proper implementation of SCDs in Analysis Services later in the book. For our next challenge, what do we do when there is a hierarchy (parent-child relationship) in a dimension?

Hierarchies

In the section on schemas, you looked at stars and snowflakes, with the main difference being that snowflakes use multiple tables to define dimensions. As I mentioned, you use snowflake schemas to implement a hierarchy—for example, Categories have Subcategories, and Subcategories have Products, as shown in Figure 2-7.

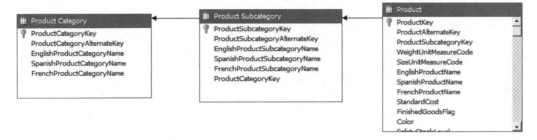

Figure 2-7. Hierarchical dimension tables

In a snowflake schema, the dimension could be built exactly like this: three tables, one for each level. In a star schema, we would denormalize the tables into a single dimension table similar to that shown in Figure 2-8.

ProductKey	Category	Subcategory	ProductName	Cost	Color	ListPrice	Weight	Style
448	Accessories	Pumps	Minipump	8.2459	NA	19.9900	NULL	NULL
449	Accessories	Pumps	Mountain Pump	10.3084	NA	24.9900	NULL	NULL
537	Accessories	Tires and Tubes	HL Mountain Tire	13.0900	NA	35.0000	NULL	NULL
540	Accessories	Tires and Tubes	HL Road Tire	12.1924	NA	32.6000	NULL	NULL
535	Accessories	Tires and Tubes	LL Mountain Tire	9.3463	NA	24.9900	NULL	NULL
538	Accessories	Tires and Tubes	LL Road Tire	8.0373	NA	21.4900	NULL	NULL
536	Accessories	Tires and Tubes	ML Mountain Tire	11.2163	NA	29.9900	NULL	NULL
539	Accessories	Tires and Tubes	ML Road Tire	9.3463	NA	24.9900	NULL	NULL
528	Accessories	Tires and Tubes	Mountain Tire Tube	1.8663	NA	4.9900	NULL	NULL
480	Accessories	Tires and Tubes	Patch Kit/8 Patches	0.8565	NA	2.2900	NULL	NULL
529	Accessories	Tires and Tubes	Road Tire Tube	1.4923	NA	3.9900	NULL	NULL
541	Accessories	Tires and Tubes	Touring Tire	10.8423	NA	28.9900	NULL	NULL
530	Accessories	Tires and Tubes	Touring Tire Tube	1.8663	NA	4.9900	NULL	NULL
348	Bikes	Mountain Bikes	Mountain-100 Black, 38	1898.0944	Black	3374.9900	20.35	U
349	Bikes	Mountain Bikes	Mountain-100 Black, 42	1898.0944	Black	3374.9900	20.77	U
350	Bikes	Mountain Bikes	Mountain-100 Black, 44	1898.0944	Black	3374.9900	21.13	U
351	Bikes	Mountain Bikes	Mountain-100 Black, 48	1898.0944	Black	3374.9900	21.42	U
344	Bikes	Mountain Bikes	Mountain-100 Silver, 38	1912.1544	Silver	3399.9900	20.35	U
345	Bikes	Mountain Bikes	Mountain-100 Silver, 42	1912.1544	Silver	3399.9900	20.77	U
346	Bikes	Mountain Bikes	Mountain-100 Silver, 44	1912.1544	Silver	3399.9900	21.13	U
347	Bikes	Mountain Bikes	Mountain-100 Silver, 48	1912.1544	Silver	3399.9900	21.42	U
358	Bikes	Mountain Bikes	Mountain-200 Black, 38	1105.8100	Black	2049.0982	23.35	U
359	Bikes	Mountain Bikes	Mountain-200 Black, 38	1251.9813	Black	2294.9900	23.35	U
360	Bikes	Mountain Bikes	Mountain-200 Black, 42	1105.8100	Black	2049.0982	23.77	U

Figure 2-8. Showing a hierarchy in a single table for a star schema

With the single-table approach, the Subcategory and Category fields will become attributes in the Product dimension. From there, you will create a hierarchy by using the attributes—using the Category attribute as the parent, the Subcategory as its child, and finally the Product as the child of the Subcategory attribute. Other attributes in the dimension will remain associated with the product. The end result is shown in Figure 2-9.

ProductKey	Category	Subcategory	ProductName	StandardC	Color	ListPrice	Weight	Style
212			Sport-100 Helmet, Red	12.0278	Red	33.6442		
213		Helmets	Sport-100 Helmet, Red	13.8782	Red	33.6442		
214			Sport-100 Helmet, Red	13.0863	Red	34.99		
487		Hydration Packs	Hydration Pack - 70 oz.	20.5663	Silver	54.99		
451			Headlights - Dual-Beam	14.4334	NA	34.99		
452		Lights	Headlights - Weatherproof	18.5584	NA	44.99		
450	Accessories		Taillights - Battery-Powered	5.7709	NA	13.99		
447		Locks	Cable Lock	10.3125	NA	25		
446		Panniers	Touring-Panniers, Large	51.5625	Grey	125		
448		Pumps	Minipump	8.2459	NA	19.99		
449			Mountain Pump	10.3084	NA	24.99		
537			HL Mountain Tire	13.09	NA	35		
541		Tires and Tubes	Touring Tire	10.8423	NA	28.99		
530			Touring Tire Tube	1.8663	NA	4.99		
348			Mountain-100 Black, 38	1898.094	Black	3374.99	20.35	U
349		Mountain Bikes	Mountain-100 Black, 42	1898.094	Black	3374.99	20.77	U
345			Mountain-100 Silver, 42	1912.154	Silver	3399.99	20.77	U
346	Bikes		Mountain-100 Silver, 44	1912.154	Silver	3399.99	21.13	U
311			Road-150 Red, 44	2171.294	Red	3578.27	13.77	U
312			Road-150 Red, 48	2171.294	Red	3578.27	14.13	U
313		Road Bikes	Road-150 Red, 52	2171.294	Red	3578.27	14.42	U
314			Road-150 Red, 56	2171.294	Red	3578.27	14.68	U
310			Road-150 Red, 62	2171.294	Red	3578.27	15	U
373			Road-250 Black, 44	1320.684	Black	2181.563	14.77	U

Figure 2-9. The same data showing a hierarchy from categories to subcategories to products

OLAP engines separate the creation of hierarchies from the underlying dimension data. Hierarchies are created by arranging relationships between attributes in the dimension. For example, in the table in Figure 2-8, you could create a hierarchy by linking Category to Subcategory, then to ProductName. Then either the designer or the engine would create a hierarchy by inferring the one-to-many relationships between categories, subcategories, and products. This approach means that fields can be reused in various hierarchies, so long as there is always a one-to-many relationship from top to bottom.

■ **Tip** When you use hierarchies with cubes, some tools require what is referred to as the *leaf level* of the hierarchy. This simply means that if you use a hierarchy, you must select it down to the lowest level. For instance, in our category-subcategory-product hierarchy we've been working with, the product level is the leaf level.

Another consideration regarding hierarchies is whether a hierarchy will be balanced or unbalanced. In a *balanced hierarchy*, every branch descends to the same level. Our category-subcategory-product hierarchy is a balanced hierarchy, because every branch ends with one or more products. We don't have subcategories without products, or categories without subcategories.

An *unbalanced hierarchy* has branches that end at different levels. Organizational hierarchies are often unbalanced. Although some branches may descend from CEO to VP to Director to Manager to Employee, there will be directors with nobody working for them. That branch will just be CEO to VP to Director.

One other hierarchy that comes to mind is a calendar, in which you'll have years, quarters, months, weeks, and days. However, the calendar is a special case, as it impacts many areas of a cube. Let's take a look at the time dimension. (Cue theremin. Yes, go look it up.)

Time Dimensions

A lot of analysis of OLAP data is financial in nature and performed by the green-eyeshade people. A lot of financial analysis turns out to be time based. They want calculations such as rolling averages, year over year change, quarter to date attainment, and so on. Rolling averages are actually fairly straightforward, as you just need to take the data from the last *x* number of days. But what about quarter to date? On February 15, we want to add the data from January 1 through today. On March 29, it's January 1 through March 29. But on April 1, we just use the data from April.

I have known groups that update their reports every quarter to deal with this situation. A better but still painful solution is to write a query that determines the current quarter, determines the first date of the current quarter, and then uses that to restrict the data in a subquery. However, you have to consider that this calculation has to be run for every value or set of values in the report.

OLAP engines provide a time-based analysis capability that makes these types of reports easy. First, as I mentioned in the previous section, is the hierarchical aspect of a time dimension: years, quarters, months, days. In addition, the ability to have multiple hierarchies on a single set of dimension members comes into use here; you can have a hierarchy of year-month-day coexist with year-week-day. More important, you can have multiple calendars: a standard calendar, in which the year runs from January 1 to December 31, and a fiscal calendar, which runs from the start of the fiscal year to the end (each with halves, quarters, weeks, and so forth). You might also have a manufacturing calendar, or an offset calendar for another reason. Figure 2-10 shows the attributes of a date dimension used in various hierarchies.

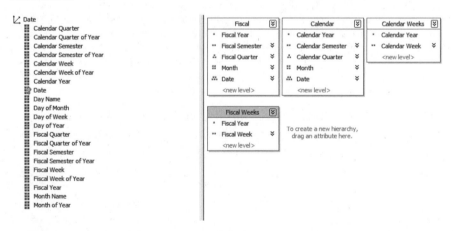

Figure 2-10. *Hierarchies built in a date dimension*

After we have a time dimension set up, we can leverage some powerful commands in the OLAP query language, MDX (more on MDX later). For now, it's sufficient to understand that MDX, when coupled with a time dimension, enables an analyst to create queries for aggregations of data such as year to date, a running average, or parallel periods (aggregate totaled over the same period in a previous year, for example). After the query is established, the OLAP engine automatically calculates the time periods as the calendar moves on.

You'll look at time analysis of data in Chapter 10. In the meantime, you need to understand what data, and what aggregations, are at the center of our attention. Let's take a look at the center of the cube: *measures*.

Measures

At the root of everything we're trying to do are measures, or the numbers we're trying to analyze. As I stated earlier, generally measures are numbers—dollars, days, counts, and so forth. We are looking at values that we want to combine in different ways (most commonly by adding):

- How many bananas do we sell on weekdays vs. weekends?

- What color car sells best in summer months?

- What is the trend in sales (dollars) through the year, broken down by sales district?

- How many tons of grain do our livestock eat per month in the winter?

In each of these cases, we'll have individual records of sales or consumption, which we can combine. We'll then create dimensions and select members of those dimensions to answer the question.

■ **Tip** OLAP engines will automatically provide a count measure for you based on the number of members of a dimension. It is effectively summing a 1 for every record in the collection of records returned. Extrapolating from this, you can see it's possible to create a subset count. For example, if you had a cube representing an automobile inventory, you could create a total of low-emission vehicles (LEVs) by having a column for LEVs and placing a 1 in the column for every LEV and a 0 for every other vehicle. By totaling the column in a query, you would have a count of the LEV vehicles.

Now that I've covered the fundamentals of a cube, let's visualize these aggregations. Figure 2-11 is our "collection of spreadsheets" cube. Note the dimensions: dates along the columns, products down the rows, and each page is a separate country.

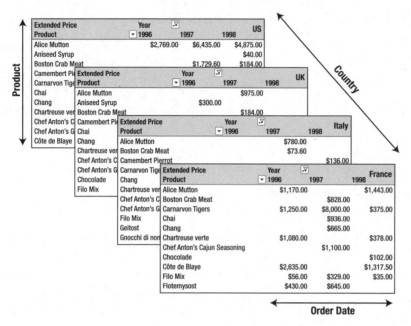

Figure 2-11. *A collection of spreadsheets representing a cube*

Now let's look at the more abstract representation in Figure 2-12. You should be able to see the analogue pretty clearly.

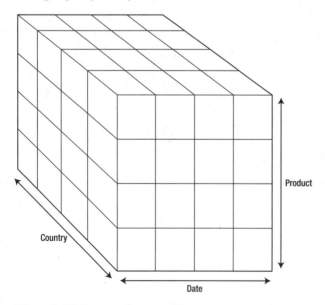

Figure 2-12. *A more abstract cube*

Imagine that each small cube has a single numeric value inside it. When we want to do analysis, we use the dimensions, which each have a single member in each cube unit, as shown in Figure 2-13.

Figure 2-13. *Members on the date dimension*

So, filling in the other dimensions, we have something similar to Figure 2-14.

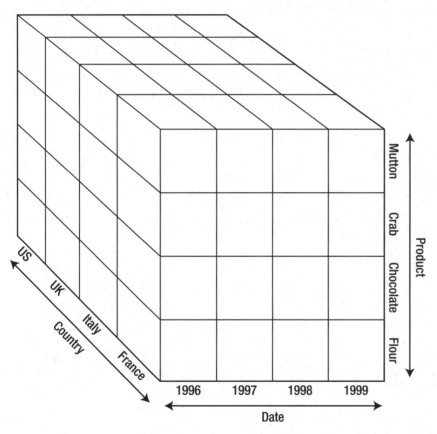

Figure 2-14. *A cube with dimensions and members. Each small cube contains a measure value.*

Given this structure, let's say we want to know the total sales of mutton in Italy from 1996–1998. Figure 2-15 should give you an idea of where we're going with this.

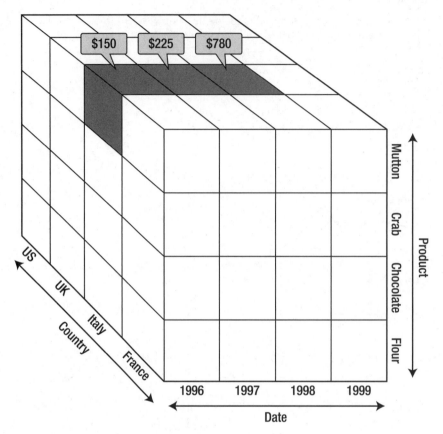

Figure 2-15. *Selecting members of the dimensions on the cube*

As you can see, by selecting the members of various dimensions, we select a subset of the values in the whole cube. The OLAP engine responds by returning a set of values corresponding to the members we've selected, that is, {$150, $225, $780}. Each of those values will be an aggregation. For example, *mutton sold in Italy in 1996* was returned as $150; that may consist of $75 bought by one customer, $35 by another, and $40 by a third. Now imagine that we want to see the amount of mutton sold in Italy and France in 1996 and 1997 (Figure 2-16). The OLAP engine will return an array we can use for reporting, calculations, analysis, and so forth. (You can already see that France outsold Italy in each year. Perhaps additional investigation is needed?)

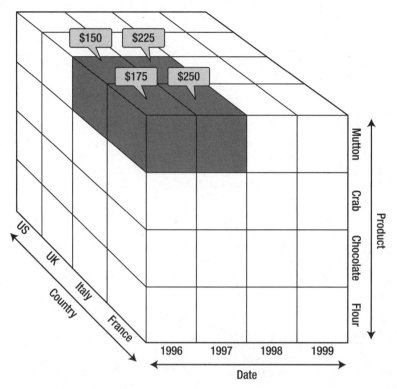

Figure 2-16. Selecting a different set of dimension members

To further analyze the results, we may drill into the date hierarchy to see how the numbers compare by quarter or month. We could also compare these sales results to the sales of other products or number of customers. Maybe we'd like to look at repeat customers in each area (is France outperforming Italy on attracting new customers, bringing back existing customers, or both?). All these questions can be answered by leveraging various aspects of this cube.

Incidentally, selection of various members is accomplished with a query language referred to as *Multidimensional Expressions*, or more commonly MDX. You'll be looking at MDX in depth in Chapter 9.

A question that may have come to mind by now: "Are measure values always added?" Although measures are generally added together as they are aggregated, that is not always the case. If you had a cube full of temperature data, you wouldn't add the temperatures as you grouped readings. You would want the minimum, maximum, average, or some other manner of aggregating the data. In a similar vein, data consisting of maximum values may not be appropriate to average together, because the averages would not be representative of the underlying data.

Types of Aggregation

OLAP offers several ways of aggregating the numerical measures in our cube. But first we want to designate *how* to aggregate the data—either additive, nonadditive, or semiadditive measures.

Additive

An *additive measure* can be aggregated along any dimension associated with the measure. When working with our sales measure, the sales figures are added together whether we use the date dimension, region, or product. Additive measures can be added or counted (and the counts can be added).

Semiadditive

A *semiadditive measure* can be aggregated along some dimensions but not others. The simplest example is an inventory, which can be added across warehouses, districts, and even products. However, you can't add inventory across time; if I have 1,100 widgets in stock in September, and then (after selling 200 widgets) I have 900 widgets in October, that doesn't mean I have 2,000 widgets (1,100 + 900).

Nonadditive

Finally, a *nonadditive measure* cannot be aggregated along any dimension. It must be calculated independently for every set of data. A distinct count is an example of a nonadditive measure.

■ **Note** SQL Server Analysis Services has a semiadditive measure calculation named `AverageOfChildren`. You might be confused about why this is considered semiadditive. It turns out that the way this aggregation operates is that it sums along every dimension except a time dimension; along the time dimension, it averages (covering the inventory example given earlier).

Writeback

Most of the time OLAP cubes are implemented, they are put in place as an analytic tool, so cubes are read-only. On some occasions, users may want to write data back to the cube. We don't want users changing inventory or sales numbers from an analysis tool, so why would they want to change the numbers?

A powerful analysis technique to offer your users is *what-if* or *scenario* analysis. Using this process, analysts can change numbers in the cube to evaluate the longer-term effects of those changes. For example, they might want to see what happens to year-end budget numbers if every department cuts its budget by 10 percent. What happens to salaries? Capital expenses? Recurring costs? Although these effects can be run with multiple spreadsheets, you could also create an additional dimension named *scenario*, which analysts can use to edit data and view the outcomes. The method of committing those edits is called *writeback*.

The biggest concern when implementing writeback on a cube is dealing with *spreading*. Consider our time dimension (Figure 2-17). An analyst who is working on a report that shows calendar quarters might want to change one value. When that value is changed, what do we do about the months? The days?

Figure 2-17. *A calendar dimension*

We have two choices. In our design, we can create a dimension that drills down to only the quarter level. Then the calendar quarters are the *leaf level* of the dimension, the bottom-most level, and the value for the quarter is just written into the cell for that quarter. Alternatively, some OLAP engines will allow the DBA to configure a dimension for spreading; when the engine writes back to the cube, it distributes the edited value to the child elements. The easiest (and usually default) option is to divide the new value by the number of children and divide it equally. An alternative that may be available if the analyst is editing a value is to distribute the new value proportionally to the old value.

Writeback in general, and spreading in particular, are both very processor- and memory-intensive processes, so be judicious about when you implement them. You'll look at writeback in Analysis Services in Chapter 11.

Calculated Measures

Often you'll need to calculate a value, either from values in the measure (for example, extended price calculated by multiplying the unit cost by the number of items), or from underlying data, such as an average.

Calculating averages is tricky; you can't simply average the averages together. Consider the data in Table 2-1, showing three classes and their average grades.

Table 2-1. *Averaging Averages*

Classroom	Number of Students	Average Score
Classroom A	20	100%
Classroom B	40	80%
Classroom C	80	75%

You can't simply add 100, 80, and 75 then divide by 3 to get an average of 85. You need to go back to the original scores, sum them all together, and divide by the 140 students, giving an answer of 80 percent. This is another area where OLAP really pays off, because the OLAP engine is designed to run these calculations as necessary, meaning that all the user has to worry about is selecting the analysis they want to do instead of how it's calculated.

Actions

Generally, an OLAP solution is the first-layer approach to analysis—it's where you start. After you find something of interest, you generally want additional information. One method of getting amplifying data for something you find in an analysis is to *drill through* to the underlying data. Some analysis tools provide a way of doing this directly, at least to the fact table; others don't.

A more general way of gaining contextual insight into the data that you are looking at is to create a structure called an *action*. This enables an end user to easily view amplifying data for a given dimension member or measure. You can provide reports, drill-through data sets, web pages (Figure 2-18), or even executable actions.

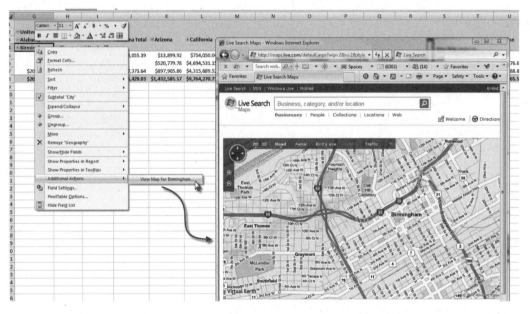

Figure 2-18. Using an action to open a map based on the member of the dimension

Actions are attached to objects in the cube—a specific dimension, hierarchy or hierarchy level, measure, or a member of any of those. If the object will have several potential targets (as a dimension has multiple members), you will have to set up a way to link the member to the target (parsing a URL, creating a SQL script, passing a parameter to a report). For example, Listing 2-1 shows code used to assemble a URL from the members selected in an action that opens a web-based map.

Listing 2-1. Creating a URL from Dimension Members

```
// URL for linking to MSN Maps
"http://maps.msn.com/home.aspx?plce1=" +

// Retreive the name of the current city
[Geography].[City].CurrentMember.Name + "," +

// Append state-province name
[Geography].[State-Province].CurrentMember.Name + "," +

// Append country name
[Geography].[Country].CurrentMember.Name +

// Append region parameter
"&regn1=" +

// Determine correct region parameter value
Case
    When [Geography].[Country].CurrentMember Is
        [Geography].[Country].&[Australia]
    Then "3"
    When [Geography].[Country].CurrentMember Is
        [Geography].[Country].&[Canada]
        Or
        [Geography].[Country].CurrentMember Is
        [Geography].[Country].&[United States]
    Then "0"
    Else "1"
End
```

This code will take the members of the hierarchy from the dimension member you select to assemble the URL (the syntax is MDX, which you'll take a quick look at in a few pages and dig into in depth in Chapter 9). This URL is passed to the client that requested it, and the client will launch the URL by using whatever mechanism is in place.

Other actions operate the same way: they assemble some kind of script or query based on the members selected and then send it to the client. Actions that provide a drill-through will create a data set of some form and pass that to the client.

All these connections are generally via XMLA.

XMLA

XML for Analysis (XMLA) was introduced by Microsoft in 2000 as a standard transport for querying OLAP engines. In 2001, Microsoft and Hyperion joined together to form the XMLA Council to maintain the standard. Today more than 25 companies follow the XMLA standard.

XMLA is a SOAP-based API (because it doesn't necessarily travel over HTTP, it's not a web service). Fundamentally, XMLA consists of just two methods: discover and execute. All results are returned in XML. Queries are sent via the execute method; the query language is not defined by the XMLA standard.

That's really all you need to know about XMLA. Just be aware of the transport mechanism and that it's nearly a universal standard. It's not necessary to dig deeper unless you discover a need to.

■ **Note** For more information about XMLA, see `http://msdn.microsoft.com/en-us/library/ms977626.aspx`.

Multidimensional Expressions (MDX)

XMLA is the transport, so how do we express queries from OLAP engines? There were a number of query syntaxes before Microsoft introduced MDX with OLAP Services in 1997. MDX is designed to work in terms of measures, dimensions, and cubes, and returns structured data sets representing the dimensional nature of the cube.

In working with OLAP solutions, you'll work with both MDX queries and MDX statements. An *MDX query* is a full query, designed to return a set of dimensional data. *MDX statements* are parts of an MDX query, used for defining a set of dimensional data (for use in client tools, defining aspects of cube design, and so forth).

A basic MDX query looks like this:

```
SELECT [measures] ON COLUMNS,
[dimension members] ON ROWS
FROM [cube]
WHERE [condition]
```

Listing 2-2 shows a more advanced query, and Figure 2-19 shows the results from a grid in Excel.

Listing 2-2. *A More Advanced MDX Query*

```
SELECT    {DrilldownLevel({[Date].[Calendar Year].[All Periods]})} ON COLUMNS,
          {DrilldownLevel({[Geography].[Geography].[All Geographies]})} ON ROWS
FROM
(
    SELECT
        {[Geography].[Geography].[Country].&[United States],
        [Geography].[Geography].[Country].&[Germany],
        [Geography].[Geography].[Country].&[France]} ON COLUMNS
    FROM [Adventure Works]
)
WHERE ([Product].[Product Categories].[Category].&[1],[Measures].[Reseller Sales Amount])
```

Product Categories	Bikes	⟨▼⟩				
Reseller Sales Amount	Column Labels ⟨▼⟩					
Row Labels ⟨▼⟩	CY 2001	CY 2002	CY 2003	CY 2004	Grand Total	
⊞ France		$654,238.20	$1,794,568.76	$1,111,858.69	$3,560,665.65	
⊞ Germany			$820,513.65	$722,502.00	$1,543,015.65	
⊞ United States	$6,024,627.35	$14,716,804.14	$16,139,984.68	$7,951,335.55	$44,832,751.73	
Grand Total	$6,024,627.35	$15,371,042.35	$18,755,067.08	$9,785,696.24	$49,936,433.02	

Figure 2-19. *The results of the MDX query in Listing 2-2*

When working with dimensional data, you can write MDX by hand or use a designer. There are several client tools that enable you to create MDX data sets by using drag-and-drop, and then view the resulting MDX. Just as with SQL queries, you will often find yourself using a client tool to get a query close to what you're looking for, then tweak it manually from the MDX.

Chapter 9 covers MDX in depth.

Data Warehouses

Data warehouse is a term that is loosely used to describe a unified repository of data for an organization. Different people may use it to refer to a relational database or an OLAP dimensional data store (or both). Conceptually, the idea is to have one large data "thing" that serves as a repository for all the organization's data for reporting and analytic needs.

The data warehouse may be a large relational data store that unifies data from various other systems throughout the business, making it possible to run enterprise financial reports, perform analysis on numbers across the company (perhaps payroll or absentee reports), and ensure that standardized business rules are being applied uniformly. For example, when calculating absenteeism or consultant utilization reports, are holidays counted as billable time? Do they count against the base number of hours? There is no *correct* answer, but it is important that everyone use the *same* answer when doing the calculations.

Many companies perform dimensional analysis against these large relational stores, just as you can create a pivot table against a table of data in Excel. However, this neglects a significant amount of research and investment that has been made into OLAP engines. It is not redundant to put a dimensional solution on top of the relational store. Significant reporting can still be performed on the relational store, leaving the cube for dimensional analysis. In addition, the data warehouse becomes a *staging database* (more on those in a bit) for the cube. There are two possible approaches to building a data warehouse: bottom-up or top-down.

Bottom-up design relies on departmental adoption of numerous small data marts to accomplish analysis of their data. The benefit to this design approach is that business value is recognized more quickly, because the data marts can be put into use as they come online. In addition, as more data marts are created, business groups can blend in lessons learned from previous cubes. The downside to this approach is the potential need for redesign in existing cubes as groups try to unite them later. The software design analogy to bottom-up design is the agile methodology.

Top-down design attacks the large enterprise repository up front. A working group will put together the necessary unifying design decisions to build the data warehouse in one fell swoop. On the plus side, there is minimal duplication of effort as one large repository is built. Unfortunately, because of the magnitude of the effort, there is significant risk of analysis paralysis and failure. Top-down design is similar to software projects with big up-front or waterfall approaches.

Data warehouses will always have to maintain a significant amount of data. So storage configuration becomes a high-level concern.

Storage

Occasionally, you'll have to deal with configuring storage for an OLAP solution. One issue that arises is the amount of space that calculating every possibility can take. Consider a sales cube: 365 days; 1,500 products; 100 sales people; 50 sales districts. For that one year, the engine would have to calculate $365 \times 1,500 \times 100 \times 50 = 2,737,500,000$ values. Each year. And we haven't figured in the hierarchies (product categories, months and quarters, and so forth).

Another issue here is that not every intersection is going to have a value; not every product is bought in every district every day. The result is that OLAP is generally considered a *sparse* storage problem (for every cell that could be calculated, most will be empty). This has implications both in designing storage for the cube as well as optimizing design and queries for response time.

Staging Databases

When designing an OLAP solution, you will generally be drawing data from multiple data sources. Although some engines have the capability to read directly from those data sources, you will often have issues unifying the data in those underlying systems. For example, one system may index product data by catalog number, another may do so by unique ID, and a third may use the nomenclature as a key. And of course every system will have different nomenclature for *red ten-speed bicycle*.

If you have to clean data, either to get everyone on the same page or perhaps to deal with human error in manually entered records (where is *Missisippi*?), you will generally start by aggregating the records in a *staging database*. This is simply a relational store designed as the location where you unify data from other systems before building a cube on top. The staging database generally will have a design that is more cube-friendly than your average relational system—tables arranged in a more fact/dimension manner instead of the normalized transactional mode of capturing individual records, for example.

■ **Note** Moving data from one transactional system into another is best accomplished with an extract-transform-load, or *ETL*, engine. SQL Server Integration Services is a great ETL engine that is part of SQL Server licensing.

Storage Modes

The next few sections cover storage of relational data; they are referring to caching data from the data source, not this staging database. It's possible to worry entirely too much about whether to use MOLAP, ROLAP, or HOLAP—don't. For 99 percent of your analysis solutions, your analysts will be using data from last month, last quarter, or last year. They won't be deeply concerned about keeping up with the data as it changes, because it's all committed and "put to bed." As a result, MOLAP will be just fine in all these cases.

ROLAP really becomes an issue only when you need continually updated data (for example, running analysis on assembly line equipment for the current month). Although it's important when it's needed, it's generally not an issue. Let's take a look at what each of these mean.

MOLAP

Multidimensional OLAP (MOLAP) is probably what you've been thinking of to this point—the underlying data is cached to the OLAP server, and the aggregations are precalculated and stored in the OLAP server as well. This approach optimizes response time for queries, but because of the precalculated aggregations, it does require a lot of storage space.

ROLAP

Relational OLAP (ROLAP) keeps the underlying data in the relational data system. In addition, the aggregations are calculated and stored in the relational data system. The benefit of ROLAP is that because it is linked directly to the underlying source data, there is no latency between changes in the source data and the analytic results. Some OLAP systems may take advantage of server caching to speed up response times, but in general the disadvantage of ROLAP aggregations is that because you're not leveraging the OLAP engine for precalculation and aggregation of results, analysis is much slower.

HOLAP

Hybrid OLAP (HOLAP) mixes MOLAP and ROLAP. Aggregations are stored in the OLAP storage, but the source data is kept in the relational data store. Queries that depend on the preaggregated data will be as responsive as MOLAP cubes, while queries that require reading the source data (aggregations that haven't been precalculated, or drilled down to the source data) will be slower, akin to the response times of ROLAP.

We'll review Analysis Services storage design in Chapter 12.

Summary

That's our whirlwind tour of OLAP in general. Now that you have a rough grasp of what cubes are and why we care about them, let's take a look at the platform we'll be using to build them—SQL Server Analysis Services.

CHAPTER 3

■ ■ ■

SQL Server Analysis Services

Now that you have a fundamental understanding of OLAP and multidimensional analysis, let's start to dig into the reason you bought this book: to find out how these OLAP technologies are implemented in SQL Server, specifically SQL Server Analysis Services (SSAS). SSAS really came into its own in SQL Server 2005, which was a massive overhaul of the entire data platform from SQL Server 2000. SQL Server 2008 Analysis Services is more evolutionary than revolutionary, but still has significant improvements and additions from the 2005 edition.

I wrote this chapter from the perspective of SSAS in the 2008 version (formerly code-named *Katmai*). If you're familiar with the 2005 version of SQL Server Analysis Services (formerly code-named *Yukon*), you may just want to skip to the last section, where I call out the specific improvements in SQL Server 2008 Analysis Services.

Requirements

Before I dive into the "all about SQL Server Analysis Services" stuff, you may want to install it. For a detailed overview and instructions regarding installation of SQL Server 2008 and SSAS, see the SQL Server 2008 Books Online topic "Initial Installation" at `http://msdn.microsoft.com/en-us/library/bb500469.aspx`. I'll cover some of the high points here.

Hardware

I get a lot of questions about what kind of hardware to use for an Analysis Services installation. The answer is, "It depends." It depends on these factors:

- How many users you plan to support (and how quickly)
- What types of users you need to support (lightweight read-only reporting, or heavy analysis?)
- How much data you have
- How fast you expect the data to grow

Generally, the hardware decision boils down to one of three scenarios:

New business intelligence initiative: Smallish amount of data, pilot group of users (fewer than ten).

Business intelligence initiative to satisfy large demand: For example, the current user base is using Excel spreadsheets or an outdated software package against a fairly large existing body of data. So although there's no current solution, you anticipate that when a solution is introduced, it will see rapid adoption.

Replacing an existing solution: In this case, there is generally a large body of existing data that sees heavy usage from a large number of users.

The first and third scenarios are the easiest to deal with. For the first scenario, you can start with a single server and either install all the software on the physical machine or set up a virtual environment reflecting a more mature architecture but on a single physical host (more on virtualization in a moment). In the third scenario, you'll have to do the hard-core analysis of the needs for data storage, data growth, and usage. In other words, you know the answers—you just have to translate them.

The second scenario is the scary one. Your options seem to be either spend a ton of money on a large-scale implementation, or run the possibility of setting up an architecture that your users outgrow very quickly. The best approach here is to plan a small pilot and measured growth to a full implementation, with provisions for scaling as necessary as usage and data storage needs grow.

Having said that, the minimum hardware requirements for SQL Server Analysis Services is a single-core, single-CPU 1GHz CPU with 512MB RAM. Obviously, this is fairly silly; it's almost impossible to buy a server that doesn't meet these specifications unless you're shopping on eBay. My personal recommendation for the hardware for a SQL Server Analysis Services implementation is as follows:

- Two dual-core CPUs. Multiple cores are great for multithreading, but the I/O and cache architecture around discrete physical CPUs provide better scalability. An alternative, should cost be an issue, would be to start with a single dual-core CPU and plan to add a second when necessary. (Also be sure to verify that your server will accept quad-core CPUs, and keep an eye on the coming advances in eight-core CPUs and higher.)

- 4GB RAM, with capability to grow to 64GB. SSAS is an extremely memory-hungry application.

- For the disk system, I'm partial to two drives in RAID 1 (mirrored) for the system drive, and then a RAID 5 array for data. Some consider this fairly complex for monitoring and management, so a single RAID 5 or RAID 10 array can also serve. Analysis Services reads more than it writes, so read speed is far more important that write speed.

ABOUT STORAGE-AREA NETWORKS

For large-scale storage, a lot of organizations immediately jump to storage-area networks, or SANs. A SAN is an abstraction that allows creation of a large network-attached storage array. The SAN is maintained and monitored by itself, and then various servers can attach to the SAN; they see it as a logical virtual drive (called a LUN).

The benefit of a SAN is that it's a single drive array that can be maintained with much closer attention than, say, arrays scattered across a large number of servers in a data center. The downside of a SAN is that it's expensive, complicated, and a single point of failure. In addition, for a lot of enterprise-class software, there needs to be a special infrastructure for supporting the software on a SAN.

Depending on your anticipated architecture, needs, and whether your organization already has a SAN, you might be better served by simply leveraging RAID arrays in servers and ensuring that you have capable monitoring software. Most important (whether you have a SAN or not) is, of course, that you have the processes in place to deal with hardware failures.

Virtualization

I mentioned virtualization earlier. Virtualization was made popular by VMware over the last ten years, and Microsoft now offers both Virtual Server for Windows Server 2003, and Hyper-V technologies on Windows Server 2008. I'm not sure that virtualization is a good idea with SSAS in the grand scheme of things. It's such a resource-intensive service that you'll generally lose more than you gain. The only time I would advocate it is if you're just starting out; you could set up a virtualized network on a single server, and then move virtual machines to physical machines as necessary (see Figure 3-1).

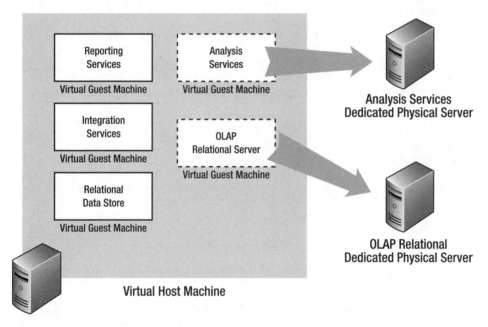

Figure 3-1. *Scaling up from virtual to physical*

In Figure 3-1, the solution was originally set up as five virtual servers on a single physical box. As the solution grew, the first place we started seeing limitations were on the SSAS box (RAM) and the OLAP relational store (hard drive space and I/O speed). So in a planned migration, we back up each server and restore it to a new physical server to give us the necessary growth and headroom.

■ **Note** The Microsoft support policy for virtualization can be found at `www.microsoft.com/sqlserver/` `2008/en/us/virtualization.aspx`. The basic support policy is that SQL Server 2008 is supported on Hyper-V guests. However, for other virtualization solutions (for example, VMware), Microsoft's support is *best effort* (if something can be resolved in the virtual environment, Microsoft will do its best to assist). However, if at any time it becomes possible that the problem is related to the virtualization environment, you'll be required to reproduce the problem directly on hardware.

Software

To answer the first question in the realm of the 2008 Servers: No, you can't install SQL Server 2008 on Windows Server 2008 Core. You *can* install it on Windows Server 2003 SP2 or later, or Windows Server 2008. SQL Server Standard Edition can also be run on Windows XP SP2 or Vista. SQL Server x86 (32-bit) can be run on either x86 or x64 platforms, while x64 (64-bit) can run only on x64 platforms.

■ **Tip** Although SQL Server 2008 is supported on a domain controller, installing it on one is not recommended.

SQL Server setup requires Microsoft Windows Installer 4.5 or later (you can be sure that the latest installer is installed by running Windows Update). The SQL Server installer will install the software requirements if they're not present, including the .NET Framework 3.5 SP1, the SQL Server Native Client, and the setup support files. Internet Explorer 6 SP1 or later is required if you're going to install the Microsoft Management Console, Management Studio, Business Intelligence Development Studio, or HTML Help.

■ **Note** Installation of Windows Installer and the .NET Framework each require rebooting the server, so plan accordingly.

Upgrading

Upgrading from SQL Server 2000 to 2005 was a fairly traumatic experience, because of the massive architecture changes in the engine, storage, and features. Although some features have been deprecated or removed in SQL Server 2008 as compared to 2005, the migration is far smoother.

The bottom line with respect to upgrading: If you have SQL Server 2005 installations that you have upgraded from SQL Server 7 or 2000, the migration to 2008 should be much easier. More important, if you have current SQL Server 2000 installations and you are evaluating migration to SQL Server 2005, you should move directly to SQL Server 2008.

Consider one more point when evaluating upgrading from SQL Server 2005 to 2008. A number of my customers have only recently finished upgrading to SQL Server 2005 and are understandably concerned about another migration effort so soon. There is no reason your server farm has to be homogeneous—for example, you could upgrade your Analysis Services server to 2008 while leaving the relational store at 2005. Evaluate each server role independently for upgrade, because each role offers different benefits to weigh against the costs.

Resources for upgrading to SQL Server 2008 can be found at http://msdn.microsoft.com/en-us/library/cc936623.aspx, including a link to the Upgrade Technical Reference Guide.

Standard or Enterprise Edition?

When you decide to go with SQL Server 2008 Analysis Services, a big decision to make is whether to go with Standard or Enterprise Edition. In general, Standard Edition is for smaller responsibilities, and Enterprise Edition is for larger, more mission-critical jobs. One easy way to differentiate is to ask yourself, "Can I afford for this server to go down?" If not, you probably want to look at Enterprise Edition.

■ **Note** The full comparison chart for SQL Server's Standard and Enterprise Editions is at www.microsoft.com/sqlserver/2008/en/us/editions.aspx.

With SQL Server 2000, the primary differentiator was that you could cluster Enterprise Edition while you couldn't cluster Standard Edition. That alone was pretty much the deal-maker for most people. In SQL Server 2005, you could cluster Standard Edition to two nodes, which seemed to remove a lot of the value of Enterprise Edition (not quite true—there are still a lot of reasons to choose Enterprise Edition).

SQL Server 2008 adds a lot of features, and a majority of them are only in the Enterprise Edition. From an Analysis Services perspective, features that are available only in Enterprise Edition include the following:

Scalable shared databases: In SQL Server 2005, you could detach a read-only database and park it on a shared cluster for use as a reporting database. In SQL Server 2008, you can do this with a cube after the cube is calculated. You detach the cube and move it to central storage for a farm of front-end database servers. Users can then access this farm by using tools such as Excel, ProClarity, or PerformancePoint for analysis and reporting.

Account intelligence: This feature enables you to add financial information to a dimension that specifies account data and then sets the dimension properties to be appropriate to that account type. For example a "statistical" account type would have no aggregation, whereas an "asset" account type would be set to aggregate the last nonempty member (similar to an inventory calculation).

Linked measures and dimensions: I've explained that instead of having one large cube, you often want to create several smaller cubes. However, you may have shared business logic or dimensions (who wants to create and maintain the corporate structure dimension over and over?). Instead, you can create a *linked measure* or *linked dimension*, which can be used in multiple cubes but maintained in one location.

Semiadditive measures: As I mentioned in Chapter 2, you won't always want to aggregate measures across every dimension. For example, inventory levels shouldn't be added across a time dimension. Semiadditive measures provide the ability to have a measure aggregate values normally in several directions, but then perform a different action along the time dimension.

Perspectives: When considering cubes for large organizations, the number of dimensions, measures, and members can get pretty significant. The AdventureWorks demo cube has 21 dimensions and 51 measures, and it's focused on sales. Browsing through dozens or hundreds of members can get old if you have to do it frequently. Perspectives offer a way of creating "views" on a cube so that users in specific roles get shorter, focused lists of dimensions, measures, and members suiting their role.

Writeback dimensions: In addition to being able to write back to measures, it's possible to enable your users to edit dimensions from a client application (as opposed to working with the dimension in BIDS). Note that dimension writeback is possible only on star schemas.

Partitioned cubes: Also mentioned in Chapter 2, the ability to partition cubes makes maintenance and scaling of large cubes much, much easier. When you can shear off the last 12 years of sales data into a cube that has to be recompiled on only rare occasions, you do a *lot* for the ability to rebuild the current cube more often.

Architecture

SQL Server Analysis Services runs as a single service (`msmdsrv.exe`) on the server. The service has several components, including storage management, a query engine, XMLA listener, and security processes. All communication with the service is via either TCP (port 2383) or HTTP.

The Unified Dimensional Model

A major underlying concept in Analysis Services is the *unified dimensional model*, or UDM. If you examine more-formal business intelligence, data modeling, or OLAP literature, you will often find something similar to Figure 3-2. Note the requirement for a staging database (for scrubbing the data), a data warehouse (for aggregating the normalized data), data marts (for severing the data into more-manageable chunks), and finally our OLAP store. I have seen architecture with even more data redundancy!

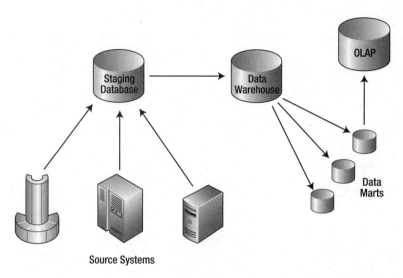

Figure 3-2. A traditional BI architecture

Apart from the duplication of data (requiring large amounts of disk space and processing power to move the data around), we also have the increased opportunity for mistakes to surface in each data translation. But the real problem we face is that systems like these often seem to end up like Figure 3-3. Various emergent and exigent circumstances will create pockets and pools of data, and cross connections, and it will all be a mess.

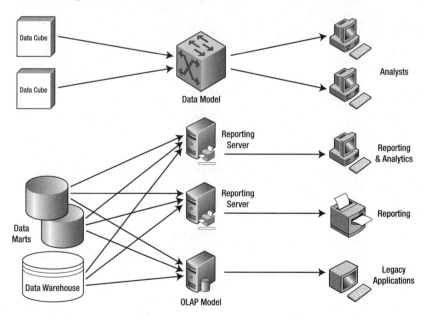

Figure 3-3. Does this look familiar?

SSAS is designed to conceptually unify as much of Figure 3-2 as possible into a single-dimensional model, and as a result make an OLAP solution easier to create and maintain. Part of what makes this possible is the data source view (DSV), which is covered in Chapter 5. The DSV makes it possible to create a "virtual view," collating tables from numerous data sources. Using a DSV, a developer can create multiple cubes to address the various business scenarios necessary in a business intelligence solution. The net result is Figure 3-4—less data redundancy and a more organized architecture.

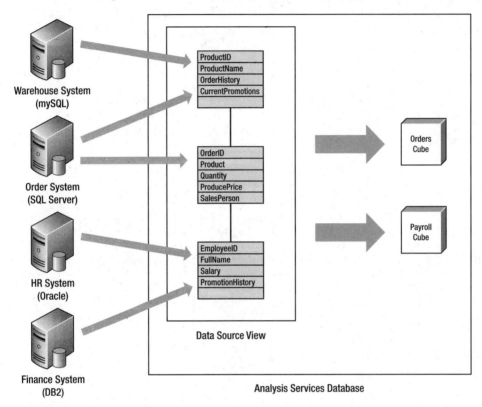

Figure 3-4. *How Analysis Services enables the unified dimensional model*

As I've mentioned previously, in many cases the data in the source systems isn't clean enough for direct consumption by a business intelligence (BI) solution. In that case, you will need a staging database, which is designed to be an intermediary between the SSAS data source view(s) and the source systems. This is similar to Figure 3-5, which also shows various clients consuming the cube data.

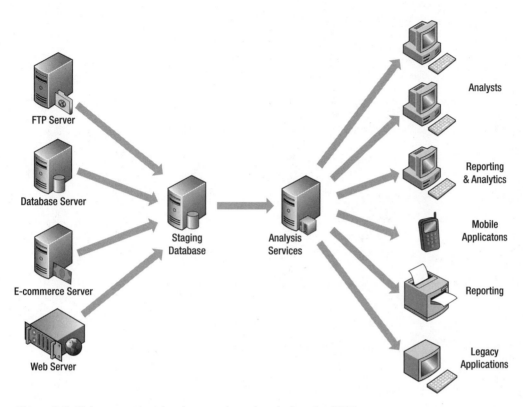

Figure 3-5. *Using a staging database to clean data before the SSAS server*

There is still a lot of potential for complexity. But I hope you see that by using one or more data source views to act as a virtual materialized view system, combined with the power of cubes (and perspectives, as you'll learn later), you can "clean up" a business intelligence architecture to make design and maintenance much easier in the long run.

Logical Architecture

Figure 3-6 shows the logical architecture of Analysis Services. A single server can run multiple instances of Analysis Services, just as it can run several instances of the SQL Server relational engine. (You connect to an Analysis Services instance by using the same syntax: [`server name`] \ [`instance name`].) Within each instance is a server object that acts as the container for the objects within.

Each server object can have multiple database objects. A database object consists of all the objects you see in an Analysis Services solution in BIDS (more on that later). The minimum set of objects you need in a database object is a dimension, a measure group, and a partition (forming a cube).

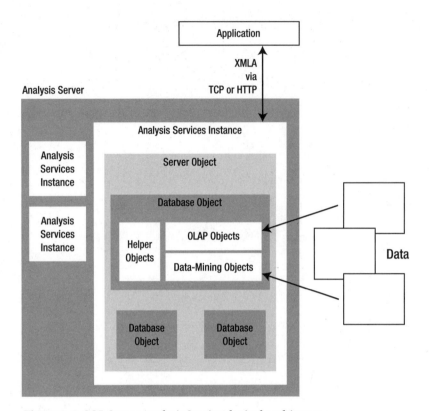

Figure 3-6. SQL Server Analysis Services logical architecture

I've grouped the objects in a database into three rough groups:

OLAP objects: Consisting of cubes, data sources, data source views, and dimensions, these are the fundamental objects that we use to build an OLAP solution. This is an interesting place to consider the object model as it relates to our OLAP world (Figure 3-7).

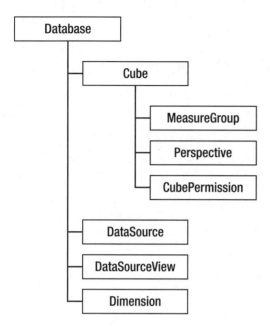

Figure 3-7. *The database object model*

Note that the Dimension collection is not a member of the Cube class, but an independent collection under the Database class. This speaks to the way that dimensions are created and can be shared among different cubes in the same database. The Cube class then has a collection of CubeDimension objects, which are references to the corresponding Dimension objects.

The Cube class does own its MeasureGroup, which is a collection of Measure objects. The same applies for the Perspectives collection and CubePermissions collection.

Data-mining objects: This is pretty much the MiningStructure collection and the subordinate object hierarchy. A mining structure contains one or more MiningModel objects, as well as the columns and bindings necessary to map a mining model to the data source view. Chapter 11 covers data mining in depth.

Helper objects: Something of an "everything else" catchall. The helper objects consist of a collection of Assembly objects, DatabasePermission objects, and Role objects for managing security. An Assembly object represents a .NET assembly installed in the database.

You may ask, "What do these object models do for me?" In SQL Server Analysis Services, you can have stored procedures to provide functions that implement business rules or requirements more complex than perhaps SSAS can easily accomplish. Perhaps you need to run a query that calls to a web service and then retrieves a data set from a relational database based on the results of that query. You could create a stored procedure that accepts parameters and returns a data set, and then call that procedure from MDX in a KPI-bound or a calculated measure.

Physical Architecture

As I've mentioned previously, SQL Server Analysis Services runs as a single Windows service. The service executable is `msmdsrv.exe`, the display name (*instance name*) is SQL Server Analysis Services, and the service name is `MSSQLServerOLAPService`. The default path to the executable is as follows:

```
$Program Files\Microsoft SQL Server\MSAS10.MSSQLSERVER\OLAP\bin
```

That service has an XMLA listener that handles all communications between the SSAS service and external applications. The XMLA listener defaults to port 2383, and can be changed either during setup or from SQL Server Management Studio (SSMS). The location of database data files can also be changed in SSMS; Chapter 4 covers that in more detail.

If you've ever had to root around the SQL Server file system, there's some great news with SQL Server 2008. With previous versions of SQL Server, folders for additional services (Analysis Services, Reporting Services, Integration Services) were simply added to the Microsoft SQL Server folder with incrementing suffixes (see Figure 3-8). You would have to open each folder to find the one you were looking for.

Figure 3-8. Folder hierarchy in SQL Server 2005

In SQL Server 2008, the folder-naming conventions are far more intuitive (see Figure 3-9). You will have folders for MSSQL10, MSAS10, and MSRS10. In addition, you can see that the service has the instance name in the folder, such as `MSAS10.MSSQLSERVER` (*MSSQLSERVER* being the tag for the default instance).

Figure 3-9. Folder naming in SQL Server 2008

The startup parameters for the SSAS service are stored here:

`MSAS10.<instance>\OLAP\Config\msmdsrv.ini`

This is an XML file. Most notable here are the `DataDir`, `LogDir`, and `AllowedBrowsingFolders` tags. In case of gremlins, it's good to verify that these entries are what you think they are. You should also verify *which* INI file the service is loading by checking the properties for the SQL Server Analysis Services service. You'll see Path to Executable, as shown in Figure 3-10.

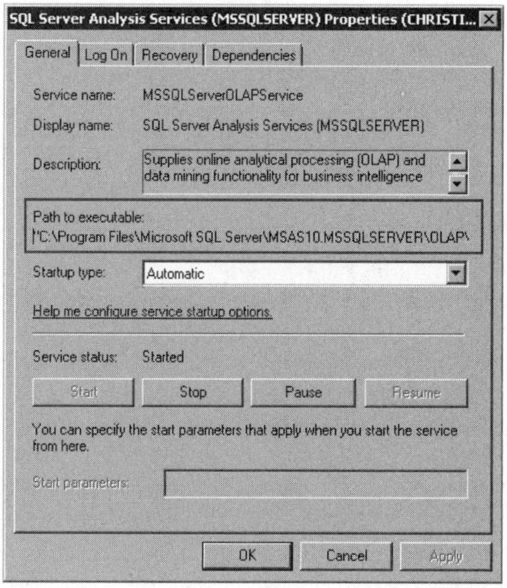

Figure 3-10. SQL Server Analysis Services Windows Service properties

You'll probably have to highlight and scroll to see the whole path. You should have something like "C:\[path]\msmdsrv.exe" -s "C:\[path]\Config", where the Config file is the Config.ini file location. If you need to change this file location, you can use msmdsrv.exe on the command line to unregister the service, and then re-register it with the new INI file location. (Use msmdsrv /? to see the command-line options.)

■ **Caution** Do not change the INI file location unless you absolutely need to to address a problem. You could easily put the SSAS service in an unusable state.

So now let's take a look at where Analysis Services stores all its data.

Storage

When considering storage of SSAS solutions, you have the actual data, the aggregation values, and the metadata of the solution. Each of these are handled separately by Analysis Services. How they're handled depends on the storage mode you choose—ROLAP, MOLAP, or HOLAP.

The default storage option in SSAS is MOLAP. The *M* is for *multidimensional*. In MOLAP storage, Analysis Services keeps everything in its data stores: the metadata defining the cube solution, a copy of the data, and the precalculated aggregations from the data.

In ROLAP (*relational*), the metadata defining the object is stored in the SSAS data store, but the data source isn't cached. The live data from the relational source is used, and the aggregations are calculated on-the-fly.

HOLAP is a mix of the two (*H* is for *hybrid*). The aggregations are stored in multidimensional format, but the source data is retained in the original data store. SSAS offers additional options in which the measure group data is stored in SSAS storage, but the source data is monitored for changes, and the cube is reprocessed dynamically based on the amount of data changed.

With the exception of ROLAP and the data for HOLAP, SQL Server Analysis Services stores its data in the file system. The administrative and developer access to all SSAS structures and data is through the SQL Server Management Studio and Business Intelligence Development Studio. As we've discussed, all these interfaces operate by using XMLA via the SSAS service. Although you may be used to SQL Server storing databases in a single data file (or a small number of files if you're using file groups), SSAS starts its optimization by storing its data in a structured system within Windows file folders.

The root for SSAS storage is going to be the location indicated in the StorageLocation setting for the structure selected. The default value is set at the server level in the DataDir property (Figure 3-11). You can access the property dialog box by right-clicking on the server in SQL Server Management Studio and selecting Properties.

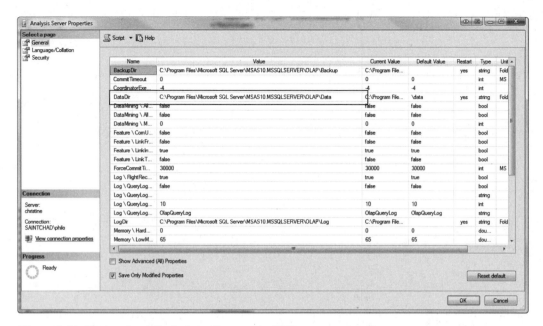

Figure 3-11. *Setting the default data directory in SSAS server properties*

The cube and measure group metadata locations can be set in the StorageLocation properties for each. This will open a dialog box that lists the folders available for locating files (Figure 3-12).

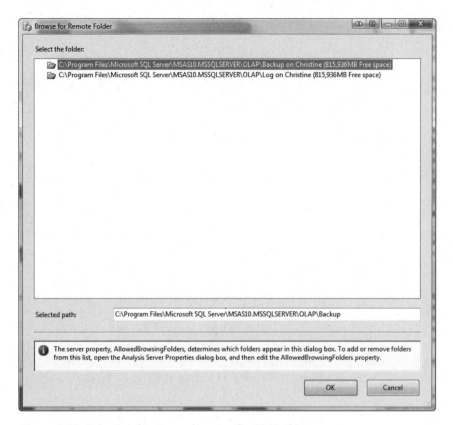

Figure 3-12. *Selecting the storage location for SSAS objects*

You can set the folders that are listed by changing `AllowedBrowsingFolders` in the advanced properties for the SSAS server in SQL Server Management Studio. (There is a check box for Show Advanced Properties near the bottom of the properties dialog.)

Under the `DataDir`, SSAS will create a folder for the database, or catalog (the logical structure equivalent to the SSAS project you'll discover in Chapter 4.) This folder will have the name of the catalog ending in an index (a version number) and a `.db` extension. Under this folder will be XML files representing each object in the solution. These are effectively header files, containing the properties and description of each object. In this folder, you'll see files ending in `.cub.xml` for cubes, `.ds.xml` for data sources, `.dms.xml` for data-mining structure, and so on.

Of more interest are the subfolders in our main catalog folder (Figure 3-13). There's a subfolder for each object in the catalog. We're primarily interested in two types of folders: cubes (`.cub`) and dimensions (`.dim`).

Figure 3-13. *File storage for SSAS catalogs*

There will be a `.cub` folder for every cube in the catalog. Inside the `.cub` folder you'll find folders for each measure group ending in `.det`. You'll also find XML files—one for each measure group again (`*.det.xml`), one for each perspective on the cube (`*.persp.xml`), and an info file. The `info.[version].xml` files are effectively the header files for the given folder they're in. Each `.det` folder will have a subfolder and XML header file for each partition in the measure group.

Now before we dive into the partition folders, let's take a look at how the partitions are defined for the AdventureWorks cube we'll be working with in this book (see Table 3-1). Note the estimated number of rows in each partition—just over a thousand in 2001, and almost three thousand in 2002. But in 2003 and 2004, we have 32,265 rows. (These are the same because the aggregation design set an upper limit.) So how does this affect our storage?

Table 3-1. *Internet Sales Partitions*

Partition Name	Size	Rows	Storage Mode
Internet_Sales_2001	28.9KB	1,013	MOLAP
Internet_Sales_2002	77.1KB	2,677	MOLAP
Internet_Sales_2003	1.4MB	32,265	MOLAP
Internet_Sales_2004	1.5MB	32,265	MOLAP

Let's check the `prt` (partition) folder for these partitions (Figures 3-14 and 3-15).

Figure 3-14. *Folder contents for the 2002 partition*

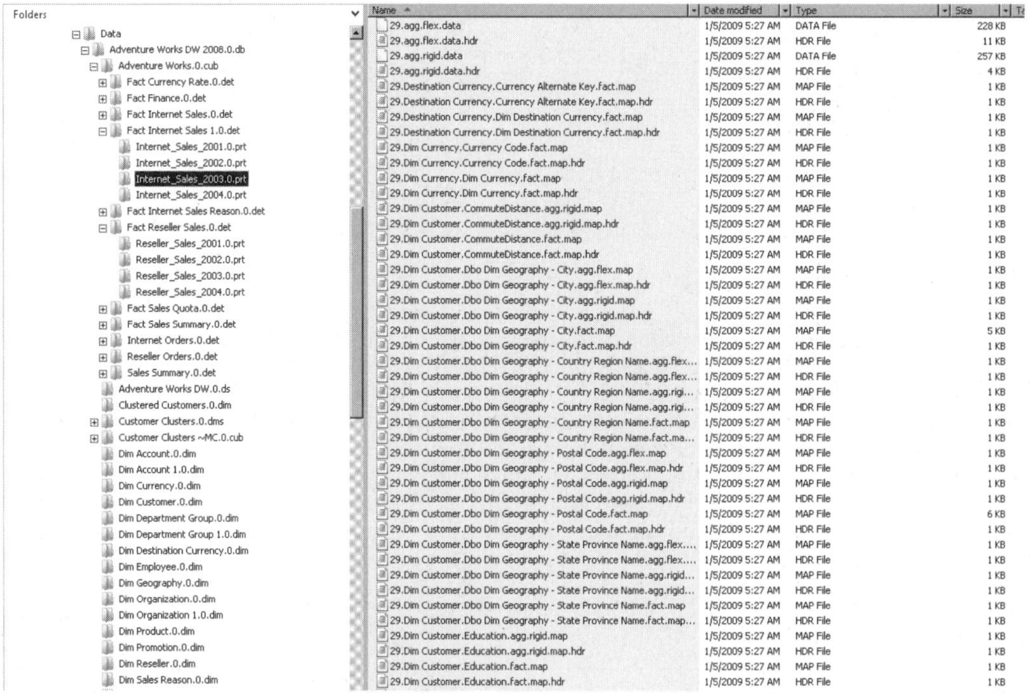

Figure 3-15. *Folder contents for the 2003 partition*

What happened? Remember the difference in row counts? For smaller data sets, SSAS may have only three files: the ever-present info.[version].xml, a data file, and a header (hdr) file. The header file is the index indicating where the data is stored in the .data file. However, when we get into larger data sets, scanning files looking for data becomes very inefficient. So Analysis Services uses bitmap indexes for each dimension. Instead of scanning through gigabytes of data files, SSAS parses the dimensions and members needed from the query, and then reads the dimension map file for each dimension, determining the data page number for each member.

SSAS then joins the lists of page numbers, and has to scan only the data pages that resolve from all dimensions queried. For example, if the queried members of the products dimension needs values from pages 12, 34, and 46, but the queried members from the geography dimension need values only from 34, 46, and 57, then the engine needs to retrieve only the values from pages 46 and 57, where the sets intersect.

Dimensions work in a similar fashion. Each `.dim` folder will have the ubiquitous `info.[version].xml` file as well as a number of files for each level of each attribute or hierarchy of the given dimension. The files (which are encoded binary) all have extensions ending in *store* for storage. The extensions and their meanings are in Table 3-2.

Table 3-2. *Dimension Storage File Extensions*

Extension	Meaning
.kstore	Key store
.ksstore	Key string store
.khstore	Key hash store
.astore	Property store
.asstore	Property string store
.ahstore	Name hash table
.hstore	Hole store
.sstore	Set store
.lstore	Structure store
.ostore	Order store
.dstore	Decoding store

The most interesting thing to understand is that Analysis Services consists of a windows service and all these files for indexing and storage. To run SQL Server Analysis Services, *you do not need a SQL Server RDBMS instance running on the server.* (You'll still need one somewhere if you plan to keep your source or staging data there.)

Now that you have a solid understanding of what SSAS looks like from the underside, let's look at an OLAP solution from the perspective we'll usually see it.

Cube Structures in SSAS

When you're creating an Analysis Services cube, you'll pretty much always be working through the Business Intelligence Development Studio, or BIDS. I'll be using BIDS here for a basic walk-through of the developer's eye view of an SSAS solution, but I'll cover BIDS in depth in Chapter 4.

Figure 3-16 shows the AdventureWorks cube open in BIDS. The central area shows the data source view (more on that in a moment.) To the left are panes that show measure groups and measures, as well as dimensions and hierarchies. On the right is the Solution Explorer, which details all the objects in our solution, including data sources, data source views, cubes, dimensions, data-mining structures, security roles, and assemblies.

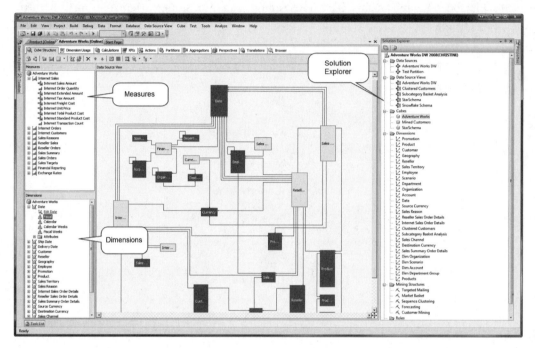

Figure 3-16. *An Analysis Services cube in the Business Intelligence Development Studio*

Figure 3-17 shows the Solution Explorer rolled up.

Figure 3-17. The items in the Solution Explorer

What I really like about the Solution Explorer is that it's like a checklist for creating a cube; you create data sources, and then a data source view by using the tables from the data sources. You can then create a cube directly and use the cube wizard to generate dimensions. Finally, you can create mining structures on your cube, assign security roles, and add assemblies for advanced capabilities in the cube.

Data Sources

Data sources are the reason we're going through this exercise in the first place! You create data sources to connect to the places you'll be pulling data from. You can essentially connect to anything you can create an OLE DB connection to. The connections are used when the data is read. For MOLAP, that's generally only when the cube is processed; for HOLAP or ROLAP, it may be any time a user performs analysis on the cube.

After you have one or more data sources, you need a way to weave the data together. Similar to a relational database where you can combine multiple flat tables into a view, in Analysis Services we have the data source view.

Data Source View

The easiest way to think of a data source view is to picture the database diagram tool in SQL Server Management Studio. A data source view (DSV) lets you add multiple tables and views from data sources to a canvas (Figure 3-18). You can either maintain existing relationships from their original data source, or you can create relationships in the DSV.

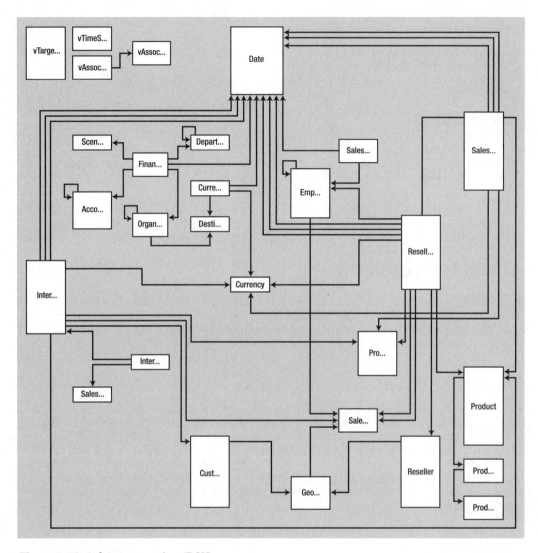

Figure 3-18. *A data source view (DSV)*

You can even create new views (known as *named queries* here) directly in the DSV from existing tables in data sources. Creating a new named query will give you the old faithful query designer and allow you to add existing tables and join them, and then select the fields for your DSV.

The nice thing about the DSV is that you don't have to aggregate all your data into a staging database just to build a cube. If your data is in a state to do so, you can effectively create a "virtual" staging database in the DSV. All the data will be read in accordance with the schema in the DSV, but you didn't have to create a second copy just to stage the data. You may also hear this capability referred to as the unified dimensional model, or UDM.

The Cube Structure Itself

After we have a data schema, we can build one or more cubes. BIDS has a very nice cube design wizard that can walk you through creating a workable cube. The wizard will evaluate the data in your DSV, recommend measures and dimensions, and create the supporting structures. As you gain more experience, of course you'll want to fine-tune things better.

Just a reminder that while the cube "owns" the measures and measure groups, it does not own the dimensions. Dimensions are created as equivalent objects and then associated with the cube. After a dimension is associated with a cube, you will need to indicate the relationship between the dimension and each of the measures in the cube (see Figure 3-19 for how this works).

Dimensions ▾	Internet Sales	Internet Orders	Internet Customers	Sales Reasons
Date	Date	Date	Date	
Date (Ship Date)	Date	Date	Date	
Date (Delivery Date)	Date	Date	Date	
Customer	Customer	Customer	Customer	
Reseller				
Geography				
Employee				
Promotion	Promotion	Promotion	Promotion	
Product	Product	Product	Product	
Sales Territory	Sales Territory Region	Sales Territory Region	Sales Territory Region	
Sales Reason	Sales Reasons	Sales Reasons	Sales Reasons	Sales Reason
Internet Sales Order Details	Internet Sales Order	Internet Sales Order	Internet Sales Order	Internet Sales Order

Figure 3-19. Associating dimensions with measures

When you associate a dimension with a measure, you have a choice of ways to relate the two:

No relationship: There is no relationship between the dimension and the measure. Attempting to slice the measure with the dimension will have no effect (all the values will be the same).

Regular: The most common, this is a standard relational relationship between two tables.

Fact: This means that the measure and the dimension are based on the same table. For example, in a purchase order scenario, the table with line items may have both cost (measure) and product (dimension) information.

Referenced: In this case, there is an intermediate table between the measure table and the dimension table. For example, if we want to break down sales by geography, we will have to relate the Orders table to the Geography table via the Customer table (orders.customerID → customer.customerID; customer.StateID → states.StateID).

Many-to-many: Beware, this kind of arrangement can cause problems with local cubes! In a many-to-many join, the fact table is joined to an intermediate dimension table, which is joined to an intermediate fact table, which is joined to the final dimension table.

Data mining: This connection leverages a mining model to make the connection. I'll cover the implications of this in Chapter 11.

Cubes can also have calculated measures, key performance indicators (KPIs), and actions. A cube can have multiple perspectives, as I've mentioned, to simplify the user interface for an especially complex cube. You can have translations to offer multilingual cube solutions. These are all features of cubes and found in the cube designer in BIDS.

Dimensions

Although the cube wizard may automatically create some dimensions, dimensions are generally designed separately from cubes in SSAS. Figure 3-20 shows a dimension in the dimension designer. You can see the data source view for the dimension on the right (dimensions are data-driven and so will need a reference to the data sources). The dimension itself, with attributes, is on the left. In the middle are the hierarchies for the dimension.

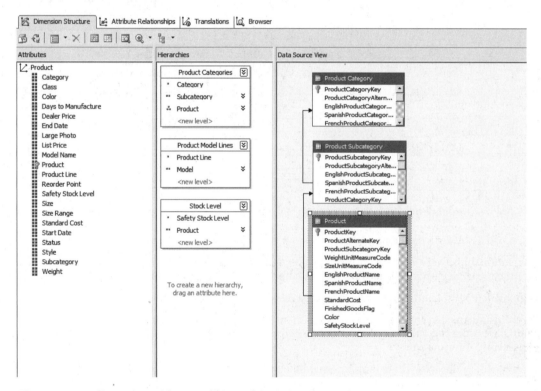

Figure 3-20. *A dimension with several hierarchies in BIDS*

Mining Structures

Mining structures contain mining models—data structures that implement one of the many data-mining algorithms in SSAS to identify patterns in data or predict values in new data. Chapter 11 covers mining structures.

What's New in SQL Server 2008

Most of this chapter has been fairly agnostic between SQL Server 2005 and 2008. The basics and the architecture didn't change that much. So let's talk about some of the things that *have* changed.

■ **Note** Most of the changes in SQL Server 2008 are incremental and additive, so think in terms of "adding a few features" instead of "massive migration pain."

Performance

Performance was one of the two areas the SSAS team focused heavily on (developer and administrator experience being the other). The goal was both to improve and optimize the engine, but also to provide more tools for cube developers and DBAs to get the maximum use out of their hardware. Following are just some of the performance enhancements to SQL Server 2008 Analysis Services.

Management Data Warehouse

The *Management Data Warehouse (MDW)* is a table inside SQL Server that provides for the collection of performance-related statistics. You can extend the statistics collection to include any metric. After you have the data in a table, you can build reports on it, or even build a cube for analysis.

To set up the MDW, you need access to an instance of SQL Server 2008. Under the Management folder, find the Data Collection node. Right-click on it and select Configure Management Data Warehouse (see Figure 3-21).

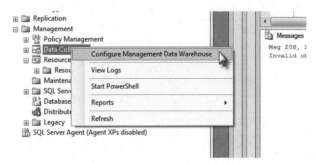

Figure 3-21. Configuring the MDW

This starts a wizard through which you can either create an MDW, or start a collection to an existing MDW (so you can centralize statistics from several servers). After you create a database and assign user permissions, you'll need to run the configuration wizard again, but select the Configure Data Collection option.

■ **Note** The SQL Server Agent must be running on the SQL Server that is hosting the MDW, or the data collection configuration will fail.

Now you'll have data collectors running on disk usage, query statistics, and server activity. You could prompt a collection and upload manually, you could script it, or you could use SSIS or a job to run the collection.

After some data has been collected, you can right-click on the Data Collection node, choose Reports → Management Data Warehouse, and then select a report. The Server Activity History report is shown in Figure 3-22. I've found the best source of information about Performance Studio is the hands-on lab at http://go.microsoft.com/?linkid=8316556.

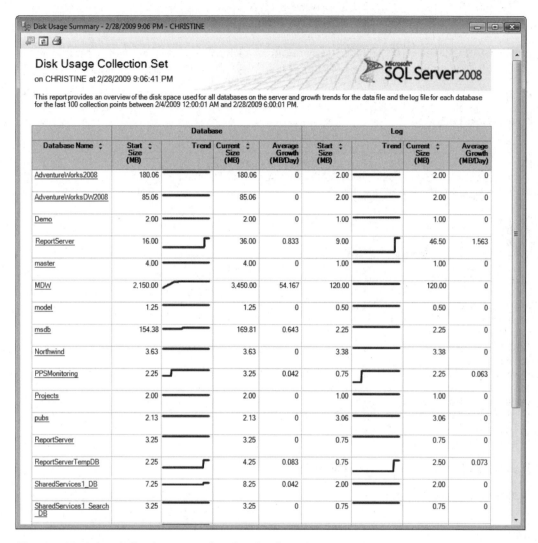

Figure 3-22. *A data collection report showing database sizes*

Reliable Query Cancel

In SSAS 2005, trying to get a query to stop often took minutes (and ended up with the admin just stopping the service and risking data corruption). In SQL Server 2008, canceling a query or connection should stop the query immediately.

Dynamic Management Views

Dynamic management views (DMVs) have been around for a while, but SSAS 2008 is the first time we've been able to use them in Analysis Services. Simply put, these are dynamic views that return management data. For example, open an MDX browser on an Analysis Services cube and query SELECT * FROM $SYSTEM.MDSCHEMA_DIMENSIONS. You should get a result of all the dimensions in the current server. This is simply easier than having to write a query to extract data from the system tables. You can find a list of DMVs in the Books Online.

■ **Tip** If you use DMVs a lot, you should check out the Analysis Services Stored Procedure Project on CodePlex at www.codeplex.com/ASStoredProcedures.

Block Computing—Optimization

Once again, let's consider our purchase order scenario. We have thousands of products and customers, and 365 days a year, and we want to calculate the sales tax for all purchases in 2008. Do we go cell by cell to calculate each value and then add them all together? Think in terms of customers shopping. Did most of our customers come by the store every day and buy most of our products? No—on average, say somewhere between 1 percent and 10 percent of our customers shop in the store, and they each buy a few products.

Most of our cube is empty.

Do we really want to go cell by cell? "On January 1, Mr. Smith did not buy any products—zero. On January 1, Mrs. Green did not buy any products—zero…" No, it makes more sense to calculate only the non-default values (we need to calculate the default value only once, not every time). SSAS 2008 now performs aggregations this way, optimizing the queries so that there is minimal grind through a sparse matrix of empty cells.

Writeback Performance Improvements

In SSAS 2005, writing back to a cube required updating the ROLAP (writeback) partition and then querying the MOLAP and ROLAP partitions together to effectively rebuild the picture. In SSAS 2008, the change is made to the ROLAP and MOLAP partitions, and the query is made to the MOLAP partition only afterward.

Change Data Capture

Change Data Capture (CDC) is a new feature in SQL Server 2008 that provides a method of tracking changed data in relational tables. The benefit is that you can have a job running that slowly updates any cubes without having to reprocess the whole thing.

■ **Tip** Don't confuse Change Data Capture with Change Tracking. The latter captures the row that's changed, but not the data that is explicitly changed.

You will have to enable CDC on the server, database, and tables you want to track by writing stored procedures. See Books Online for details on enabling CDC and also how to leverage Change Data Tracking in an Integration Services package.

Tools

SQL Server Management Studio (SSMS) and the Business Intelligence Development Studio (BIDS) truly benefited from being in the field since 2005. The SQL Server team took all lessons learned to heart and used them to improve the tools. Again, because this version is incremental instead of a full overhaul, we get the benefit of the improvements and added features without the learning curve of adopting another new user interface.

Here are some of the significant improvements in BIDS for Analysis Services.

Dimension Design

Designing dimensions was a little tricky in BIDS. It was one of those "once you get it, you get it, but getting there isn't easy" things. The wizard has been cleaned up in 2008, and dimension configuration is much, much easier. Compare the first page of the new dimension wizards from BIDS 2005 (Figure 3-23) and BIDS 2008 (Figure 3-24).

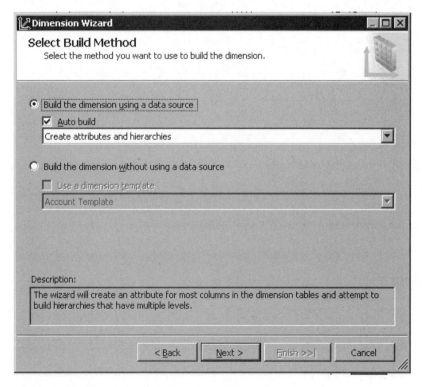

Figure 3-23. *The Dimension Wizard in BIDS 2005*

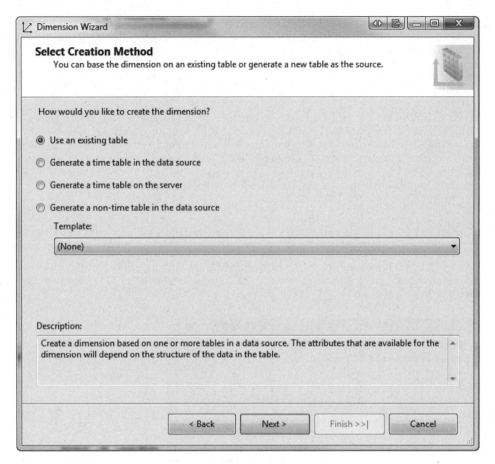

Figure 3-24. *The Dimension Wizard in BIDS 2008*

You can see that from the first page, the wizard is cleaner and easier to follow. In addition to the wizard, there is a full-featured attribute relationship designer in the Dimension Editor (Figure 3-25). Creating attribute hierarchies used to be a serious pain, but with this designer, it's much more straightforward.

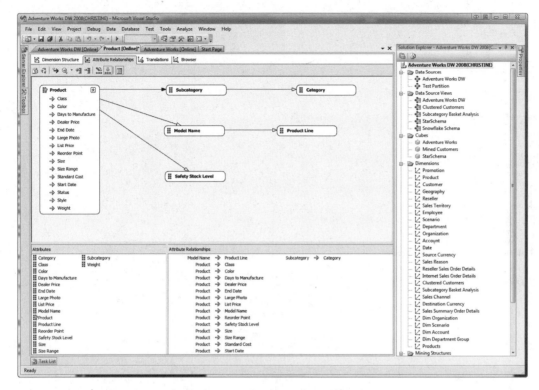

Figure 3-25. Attribute hierarchy designer in the Dimension Editor

Chapter 6 covers dimension design.

Aggregation / UBO Designer

In BIDS 2005, we didn't have a lot of control over the storage of measures and measure groups. In 2008, there is a dedicated designer for aggregations (Figure 3-26), which also provides a wizard for usage-based optimization, or UBO (Figure 3-27).

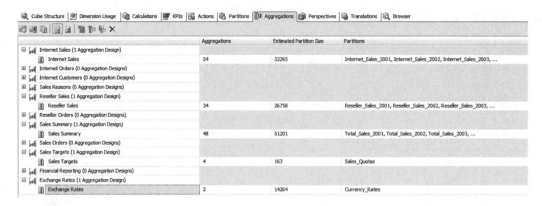

Figure 3-26. *Aggregation Designer in BIDS 2008*

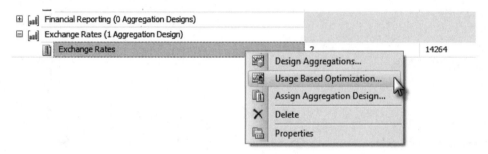

Figure 3-27. *The Usage-Based Optimization Wizard*

The Aggregation Designer lets you easily combine partitions into single measures, as well as controlling the storage for the measures. The UBO Wizard helps you adjust storage requirements based on past query history to optimize your cubes based on your users' usage patterns. Of course, because the SQL Team also invested significantly in improving initial aggregation designs, maybe you won't need these optimizations as much as you used to.

AMO Design Warnings

With SSAS 2005, I was often either stuck with a cube that didn't work as expected, or suspicious that I was doing something wrong. So I'd have to research the subject area I was having problems with and try to understand the whole field so I could figure out what I was doing wrong.

In BIDS 2008, there more than 40 "best practice" design warnings that let you know when you're doing it wrong. You can access the warnings by choosing Database ➤ Edit Database, and then selecting the Warnings tab. Here you can either choose to disable particular warnings, or enable them if you've disabled them from the UI.

The warnings show up as a blue squiggle—pretty recognizable to anyone who's spent any time in Visual Studio (or Office!). See Figure 3-28 for an example.

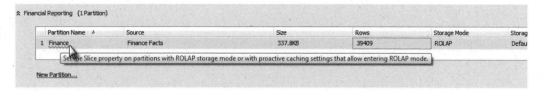

Figure 3-28. An AMO "best practice" warning

Summary

That's a quick tour of some of the many improvements in SQL Server Analysis Services 2008. I'll review what's new in data mining in Chapter 11. Of course, a cube designer can also reap the benefits of the improvements in Integration Services, Reporting Services, and the relational engine, so don't neglect those!

In the next chapter, you will discover the SSAS developer and administration interfaces. I will introduce you to the Business Intelligence Development Studio, the SQL Server Management Studio, and PowerShell.

CHAPTER 4

■ ■ ■

SSAS Developer and Admin Interfaces

Now that you have a feeling for how Analysis Services works, you need to figure out how to work with it. In this chapter, I'll review the ways of interacting with SSAS as a developer or administrator. (I'll cover user interfaces in Chapter 14.) Three interfaces are available: Business Intelligence Development Studio (BIDS), Management Studio, and PowerShell. Of these three, BIDS is perhaps the one you'll use most often.

Business Intelligence Development Studio

BIDS was introduced as the development platform for SSAS with SQL Server 2005 and Visual Studio 2005. For SQL Server 2008, BIDS is based on Visual Studio 2008. As the *business intelligence* part indicates, BIDS is the development front end for all the BI services in SQL Server: Integration Services, Analysis Services, and Reporting Services.

Using Visual Studio as the platform for SQL Server projects means that cube developers have a familiar environment that has a common support base and broad community. It also reduces the footprint on the desktop when cube developers need to write code with Visual Studio. Finally, this also means that the BI development platform is easily integrated with Team Foundation Server for project management and source control.

BIDS Is Visual Studio?

A common question seems to be "Since Visual Studio is the front end for SQL Server BI development, do I need an MSDN license for every DBA?" The answer is no.

When you install the SQL Server client tools, one of the options is to install Business Intelligence Development Studio. If you don't have Visual Studio installed, then installing BIDS will install the shell of Visual Studio with the BI project templates. Figure 4-1 shows BIDS 2008 as installed without a prior installation of Visual Studio. If you have Visual Studio 2008 already installed, the BIDS installer will just add the BI projects to Visual Studio.

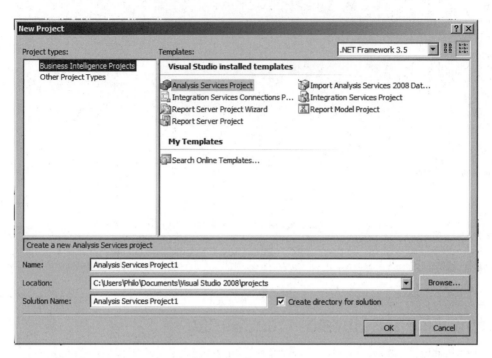

Figure 4-1. *BIDS 2008 as a stand-alone install*

In any event, the bottom line is that SQL Server licensing is covered at the server level—whether you license per processor or per server/client access license (CAL). If the server is licensed correctly, you can install the client tools anywhere you need to.

Panes

If you haven't worked with Visual Studio before, let's take a quick tour. Figure 4-2 shows BIDS 2008 with a cube open. The panes on the left (for measures and dimensions) are fixed in the cube designer, while the panes on the right (Solution Explorer and Properties) are Visual Studio standard panes and can be detached and moved around.

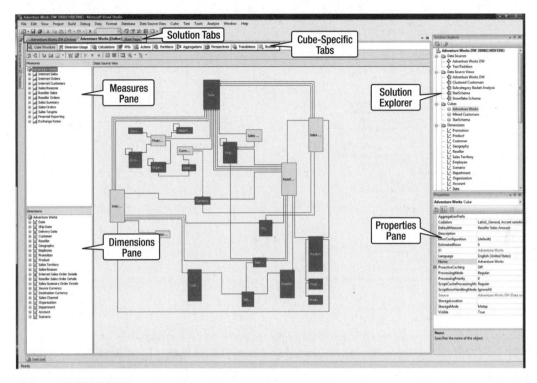

Figure 4-2. *BIDS 2008*

The solution tabs at the front are where you'll be able to select different objects as you open them (data source views, cubes, dimensions, and so forth). Underneath those tabs are a row of tabs specific to the object you have open. You can see the tabs for the cube in Figure 4-2; for a dimension, you'd have just four tabs: Dimension Structure, Attribute Relationships, Translations, and the Browser.

You'll be diving into the designers in depth in future chapters. My goal in this chapter is just some basic familiarization with Visual Studio itself.

You can arrange the Visual Studio panes to suit your needs, or tuck them out of the way. Clicking the little thumbtack in the pane's title bar makes the pane autohide to a tab along the side of the studio. Clicking the tab slides the pane back in. You can click the thumbtack again to pin the pane in place. You can also "tear off" a pane so it floats, or dock it somewhere else. Just grab the title bar and drag the pane, and it will detach (Figure 4-3).

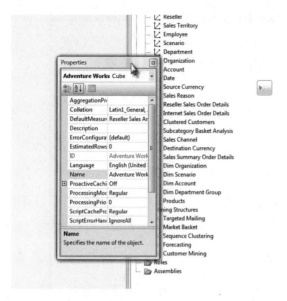

Figure 4-3. Tearing off a pane

In the past, after you tore a pane off, getting it back into place could be frustrating. It seemed like it was a matter of finding the right single pixel to drag the pane over for it to glue back into place. In Visual Studio 2008, there's a new UI for docking panes; when you drag a pane close to an edge, a docking icon will show up (Figure 4-4). Drag the pane to one of the tabs to dock it where you want it.

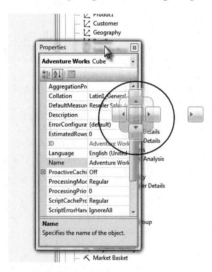

Figure 4-4. Icon for docking panes

You can also close panes by clicking the X in the top-right corner of the pane. To open the Solution Explorer, press Ctrl+Alt+L. To open the Properties window, press F4 (or right-click any object and select Properties). Most important to remember is that you can always get to any pane via the View menu.

Solution Explorer

The Solution Explorer (Figure 4-5) is the organizer for the objects that make up your OLAP solution. The objects generally map to files in the solution. Double-clicking an object opens it in the designer (except for Data Sources—double-clicking it just opens the Data Source Designer dialog box). Right-clicking the solution name and selecting Properties opens the solution property pages, which is where you can set the deployment location.

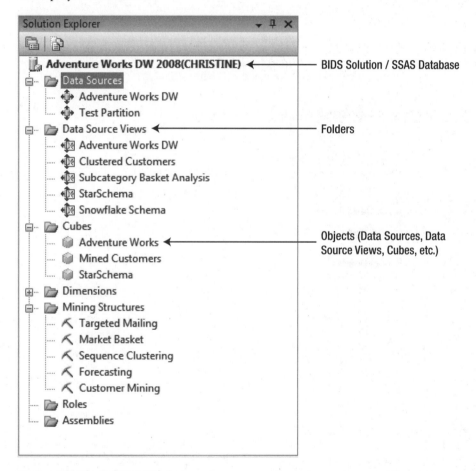

Figure 4-5. The Solution Explorer pane

The top item, in bold, is referred to as the *solution* in BIDS, and represents a database in Analysis Services. Note in this case our database is named Adventure Works DW 2008. The name in parentheses after the database is the name of the server this database was opened from. You won't see the server label if you are working on an SSAS project. It's present only when you open a database from a server (more on this later in the chapter). The database, or solution, is the logical container for all the other objects that make up our OLAP solution. It's somewhat easy to think of a cube as the primary object in SSAS, but remember that the database is the primary container we'll see on the server.

If you're used to working in Visual Studio, the simplicity of an Analysis Services solution may startle you. There are only seven types of objects, as you can see in Figure 4-5. With the exception of assemblies, each can be created in BIDS by right-clicking the folder and selecting New. Double-clicking an item opens it in BIDS. You can right-click an item and select Properties to see its properties. However, the properties of an object will generally be the file name, path, object name, and ID. Properties of the things we're interested in are more fully accessible after you have an item open in the designer.

Properties Pane

The Properties pane may be in the lower right, or parked on the right side of the environment (Figure 4-6). If it's parked, you can open it by clicking the Properties tab. The pane will stay open while it's active and then slide back to the side. If you want to keep it open, click the thumbtack icon in the upper right of the pane.

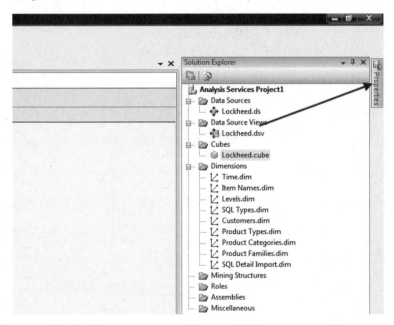

Figure 4-6. The Properties pane parked in Visual Studio

To show the properties for an object, you can either right-click the object and select Properties, or select the object if the Properties pane is already open. You can also drop down the selector in the top of the pane (Figure 4-7).

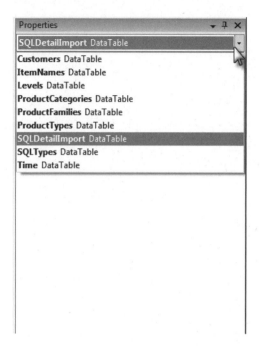

Figure 4-7. Selecting an object in the Properties pane

The toolbar on the Properties pane has buttons to sort the properties alphabetically or to group them by category. When the properties are grouped by category, some will be rolled up. Click the [+] icon to open the group (Figure 4-8).

⊟ **Source**	
CustomRollupColumn	(none)
CustomRollupPropertiesColum	(none)
⊟ KeyColumns	Time.Fiscal_Day (Date)
⊞ Time.Fiscal_Day (Date)	Time.Fiscal_Day (Date)
NameColumn	(none)
ValueColumn	(none)

Figure 4-8. Grouped properties

One final note about working with properties: There is a Dimensions pane in the Cube Editor; however, clicking a dimension or attribute opens only a handful of properties. To edit the full collection of properties for a dimension or attribute, you'll have to edit the dimension in the Dimension Editor (more on this in Chapter 6). There's a link to open the dimension in the editor immediately under the dimension in the Cube Editor (Figure 4-9).

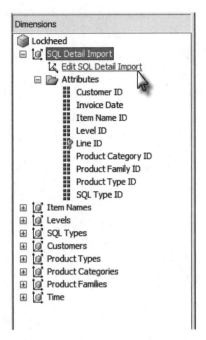

Figure 4-9. Opening the Dimension Editor from the Cube Editor

With all this talk about cubes, let's take a look at how we create an Analysis Services project in Visual Studio/BIDS.

Creating or Editing a Database Solution

Before you can work with an OLAP solution in BIDS, you need to either create a new database solution or open an existing one. In this section, I'll walk through how to create a new SSAS solution in BIDS and two ways of opening an existing Analysis Services database.

Create a New Analysis Services Project

Creating a new project is how you basically start from scratch. When you open BIDS or Visual Studio, you'll see the Start Page. You can either click the Project link next to Create or choose File → New → Project. Figure 4-10 shows the Start Page, highlighting the link to use in creating a project.

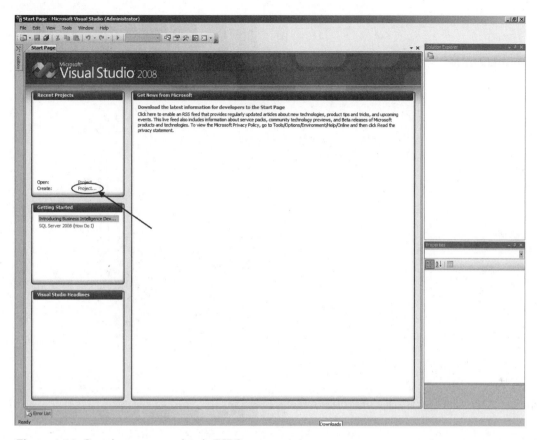

Figure 4-10. *Creating a new project in BIDS*

Next the New Project dialog box will open (Figure 4-11). If you're using BIDS (installed without Visual Studio), you'll have only Business Intelligence Projects and Other Project Types in the left pane. If you're running Visual Studio, you'll have several other project types. In either case, select Business Intelligence Projects.

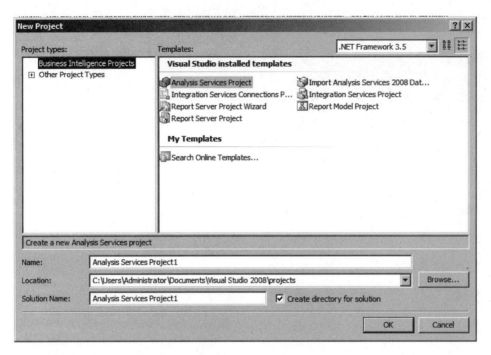

Figure 4-11. *The New Project dialog box*

Select Analysis Services Project to create a new OLAP solution. Give the project a name, select the location in the file system, click OK, and you're all set.

Open an Existing SSAS Database

You may have a server with an existing database that you wish to work on, but you don't have the solution files. If you just need to look at the structures in that database, or make some minor changes, you can open the database from the server by choosing File → Open → Analysis Services Database. You can then view and manipulate that database from BIDS without having to create a project.

Asking to open a database will get you the Connect to Database dialog box (Figure 4-12). Enter the SSAS server name and instance (just the server name if it's the default instance, or [server]\[instance]). If you can connect to the server and have the appropriate permissions, the database list will be populated with the databases on the server.

■ **Note** SSAS can connect only by using integrated authentication, so you must be on either the same domain or a domain with trust with the Analysis Services server.

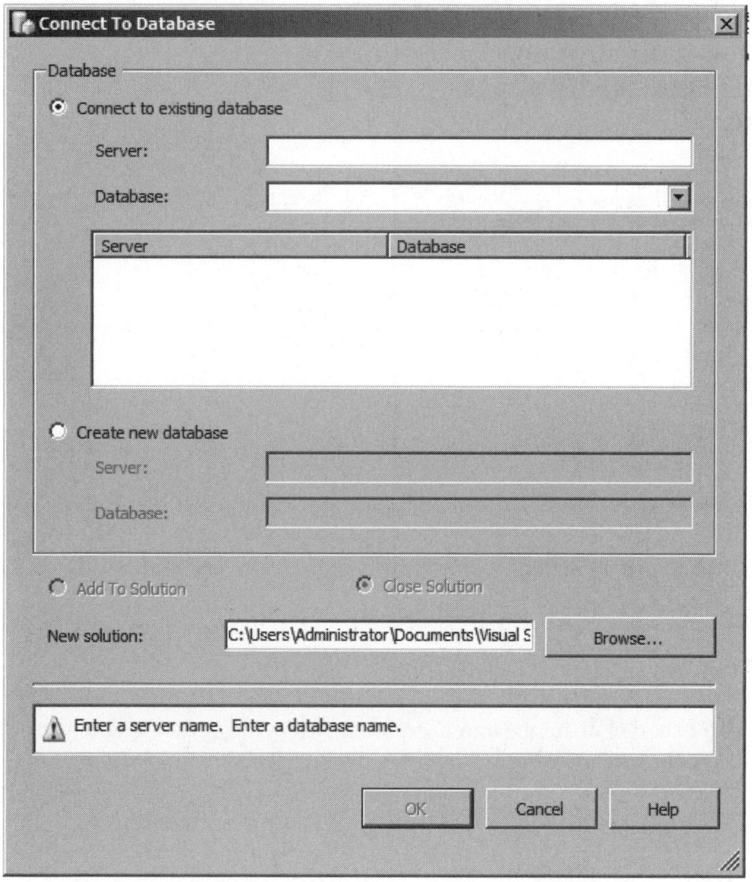

Figure 4-12. Connecting to an Analysis Services database

After you've selected the database, click OK, and BIDS will open the database. Note that if you make changes, when you save them, they will be committed directly back to the server. There's no way to do a "save as" after you've opened the database this way. If you need to create a project from a database on the server, see the next section.

Open an Existing SSAS Database as Part of a Project

Your final option is to open an existing database and simultaneously create a new Analysis Services project that includes that database. This is something of a stealth option. If you need to create a project from an Analysis Services database, open BIDS and create a new project. In the New Project dialog, select Import Analysis Services Database. Give the solution a name and select the location. When you click OK, you'll get the Import Analysis Services Database Wizard (Figure 4-13).

Figure 4-13. Importing an Analysis Services database into a new project

On the next page, enter the server or server\instance name, and then select the database you want to import. When you click the Next button, the wizard will import all the objects in the database and close, leaving BIDS open with the solution.

SQL Server Management Studio

Management Studio is the primary tool for DBAs working with Analysis Services. If the last time you looked at SQL Server was the 2000 version, SQL Server Management Studio (SSMS) replaces Enterprise Manager and Query Analyzer. It is the administration side of working with SQL Server. From SSMS, an administrator has access to Analysis Services databases and their subordinate objects—data sources, data source views, cubes, dimensions, mining structures, roles, and assemblies.

■ **Note** If you *have* used Management Studio in SQL Server 2005 or 2008, be advised that a lot of the features you may be used to using with the relational engine aren't available with Analysis Services. For example, the resource governor, system data collection, and custom reports either won't show up or will be disabled when connected to an Analysis Services server, as those features are not available for Analysis Services yet.

Another benefit of SSMS is that you can have connections open to SQL Server relational servers, Analysis Services, and Reporting Services at the same time (Figure 4-14). This can help when working with Analysis Services solutions that interact with a relational database (either as a data source or repository for ROLAP storage).

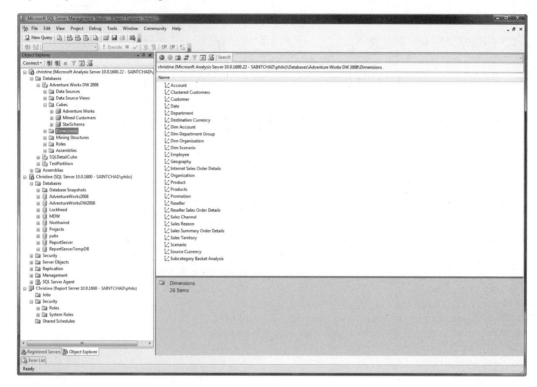

Figure 4-14. SQL Server Management Studio

Managing Analysis Services

The view of an Analysis Services server in SSMS will start with a folder for databases and a folder for assemblies. The Assemblies folder is a collection of .NET assemblies that provide serverwide functions. You can add assemblies here by right-clicking the folder and selecting New Assembly. (I'll cover the use of assemblies in Analysis Services in Chapter 11.)

If you open the Databases folder, you'll have a list of databases installed on the server. A *database* in Analysis Services is the equivalent of a project in BIDS, and the database object you looked at in Chapter 3. Each database can have multiple data sources, data source views, cubes, dimensions, and so forth.

Data Sources

From SSMS, you can edit a data source's credentials. You can also edit the connection string (Figure 4-15), and so manage which servers a database connects to.

■ **Note** If you edit a database in SSMS and someone subsequently attempts to deploy an edited project over it, they will get a warning that the database has changed since they last deployed it.

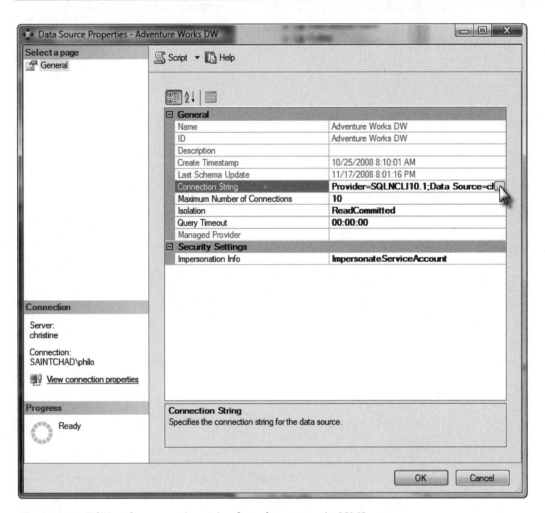

Figure 4-15. Editing the connection string for a data source in SSMS

Data Source Views

Data source views (DSVs) are relational maps for Analysis Services to use as the data structure for its cubes and dimensions. Because of their complexity, management of data source views via SSMS is through XMLA (XML for Analysis). XMLA is an XML-based structure for interacting with OLAP and data-mining providers.

If you right-click on a data source view in SSMS, you'll have the option to generate an XMLA script of the DSV to a query editor window, a file, or the clipboard. You'll have the traditional options to script it as a CREATE script, an ALTER script, or a DELETE script. You can execute these scripts from an SSMS query editor window.

■ **Tip** This is really of use only if you need to store creation scripts for an SSAS database. I wouldn't ever suggest actually trying to manage data source views via XMLA.

Cubes

Cubes are generally the primary reason we're interested in Analysis Services, and here is where most of our capabilities are in SSMS. In the properties for each cube, you can change the location of the storage files, the processing mode, and proactive caching (ROLAP, MOLAP, HOLAP), among other things.

Under the cube object is a folder of measure groups (Figure 4-16)—each measure group will be here. Although you can't work with individual measures, for each measure group you can process the group, work with writeback options, and manage partitions and aggregation designs.

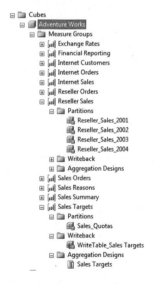

Figure 4-16. Cubes and measure groups in SSMS

When you're managing several servers with multiple databases and possibly dozens of cubes, this management capability is a great feature. You can manage the storage locations, partitions, and aggregations from an administrative console in order to balance response time against storage requirements and hardware restrictions. Of course, the ability to process OLAP data, at the database, cube, or measure group level is also a great administrative tool, especially for troubleshooting or performance management.

If you're troubleshooting cubes, you're going to want to look at the data to evaluate it. SQL Server Management Studio includes a cube browser to, well, browse cubes (Figure 4-17). Right-click a cube and select Browse to open the browser. The same browser is available in the cube designer in BIDS. You'll take a closer look when you build a cube later.

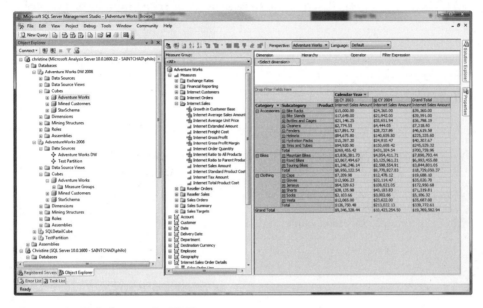

Figure 4-17. *Cube browser in SSMS*

Dimensions

You can also browse dimensions in SSMS, which will let you examine all the members in the dimension or hierarchy. You can also view member attributes. From an action perspective, you can process the dimension, or set the storage or proactive caching for the dimension. There's not much more here, so let's look at mining structures.

Mining Structures

SSMS provides great access to data-mining structures and mining models. Using Data Mining Extensions (DMX) queries, you can script, browse, run predictions, and process your mining models. You can also use test data to evaluate the accuracy of your mining models. (If this doesn't make a lot of sense now, I'll be covering data mining in depth in Chapter 13.)

Roles

The Roles folder in SSMS gives you full control over user roles and membership. Roles can be controlled in BIDS, but more properly here, where an administrator can manage roles and access. Chapter 10 covers roles.

Executing MDX Queries

Chapter 9 covers MDX in depth, but I just want to quickly point out here that you can run MDX queries in SSMS, which is the most convenient query editor for MDX queries. To open a query window, you can either right-click on the database (not the cube!) and select New Query ➤ MDX. You also have options to open windows for DMX or XMLA queries. Note that XMLA queries have to be in the full XML format.

The MDX query editor (Figure 4-18) is a free text query editor that color-codes the queries and highlights syntax errors. (You don't get IntelliSense, however.) You'll also see an object browser; you can select cube objects here and drag them to the query window to get the proper MDX syntax for the object. The Functions tab will give you a list of standard MDX functions.

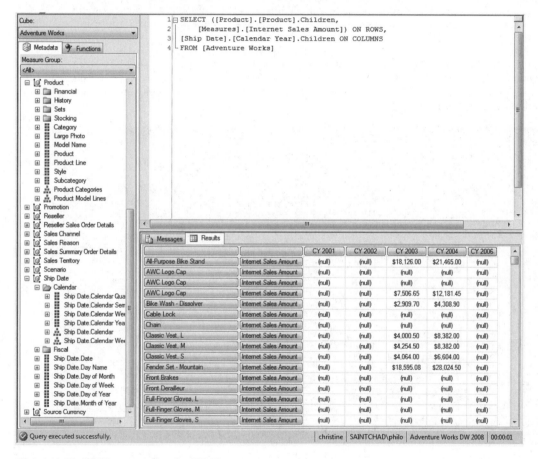

Figure 4-18. MDX query editor in SSMS

When you execute a query, the results will be displayed in the query results window. You won't get any drill-through or actions, and hierarchies won't fold up, but it's good enough to verify the query.

Hopefully, this brief overview will help you appreciate SSMS as an administrator's tool for working with Analysis Services. Often folks believe that every DBA needs BIDS on their desktop. Although BIDS has its purpose, a large amount of DBA admin work can be done from SSMS, which is far more convenient when dealing with multiple servers and multiple databases.

PowerShell

Now let's take a look at our final administrative tool, which should be very comfortable for DBAs who come from a command-line-oriented world: PowerShell. Microsoft introduced PowerShell to enable a standard scripting environment for Windows and application management. Every server Microsoft ships in the future will be PowerShell enabled, giving administrators a unified interface for scriptable management. For Windows XP, Vista, and Server 2003, you can download PowerShell from the PowerShell site. Installation is painless. Windows Server 2008 and Windows 7 have PowerShell installed by default.

PowerShell is, essentially, a command-line interface, as shown in Figure 4-19. However, instead of just being a DOS prompt, PowerShell is built on top of the .NET Framework. Instead of running and returning text, PowerShell operates in the world of .NET objects. When you execute a command, an object is returned. You can then operate on that object, manipulating properties, iterating through collections, and calling methods.

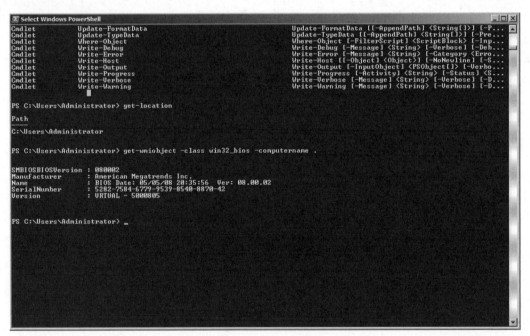

Figure 4-19. Windows PowerShell

One great thing about PowerShell is that the product team established a very structured design pattern for *cmdlets* (PowerShell commands) of verb-noun. So, for example, some cmdlets are `Get-Help`, `Start-Service`, or `Add-Member`. If you're familiar with .NET, after you understand the basics of PowerShell, you have a pretty good chance of guessing the cmdlets you need for a given task.

■ **Note** For an in-depth introduction to Windows PowerShell, be sure to start at the PowerShell site at www.microsoft.com/powershell. You can also check out Hristo Deshev's book, *Pro Windows PowerShell* (Apress, 2008).

A Convincing Example

If you want an idea of why you should care about PowerShell, one quick example may convince you of its value. Open a PowerShell window and type this command:

```
Get-WmiObject –class Win32_BIOS –computername .
```

Then press the Enter key. You should see something like the following:

```
SMBIOSBIOSVersion    : 080002
Manufacturer         : American Megatrends Inc.
Name                 : BIOS Date: 05/05/08 20:35:56  Ver: 08.00.02
SerialNumber         : 5282-7584-6779-9539
Version              : VRTUAL - 5000805
```

Yep—BIOS data right there, courtesy of the Windows Management Instrumentation (WMI) objects. I'm not sure about you, but I often struggle with figuring out how to get BIOS data without having to reboot the PC. There's your answer.

PowerShell for SQL Server

By now, either you're asking why there's a section on PowerShell in a SQL Server Analysis Services book, or you've figured out it has to do with managing SQL Server. But it's more than that: SQL Server 2008 installs PowerShell by default and has its own collection of SQL Server snap-ins. To run PowerShell with the SQL snap-ins, you can either run sqlps from a command prompt, or follow the instructions at http://msdn.microsoft.com/en-us/library/cc281962.aspx to load the snap-ins into your PowerShell installation.

After the snap-ins are installed, you'll have access to the SQL objects and providers on the host machine. For example, simply navigating to sql\localhost\default\databases and running Get-ChildItem will show you the listing in Figure 4-20.

```
Name                      Status    CompatibilityLevel
----                      ------    ------------------
AdventureWorks            Normal    Version100
AdventureWorks2008        Normal    Version100
AdventureWorksDW          Normal    Version100
AdventureWorksDW2008      Normal    Version100
AdventureWorksLT          Normal    Version100
AdventureWorksLT2008      Normal    Version100
```

Figure 4-20. Listing databases by using PowerShell

Now you can actually change *directory* into a database, and then tables, and so forth. As you learn about PowerShell, you'll see that you can pipe output of these listings into a text file or XML. So you could run a survey of all installed SQL servers, pull the statistics you need, and dump an XML file to a file share to be processed into a report. You can also run management tasks with PowerShell, so verifying jobs, backing up databases, rebuilding indexes can all be scripted.

PowerShell with SSAS

So how can we use PowerShell with Analysis Services? Do we get the nice easy syntax we saw with SQL Server? Sadly, no. While the SQL PowerShell snap-ins include SQL Server Management Objects (SMO), they don't include Analysis Management Objects (AMO), so we have to map them in on our own. Luckily, this is not difficult.

Type the following into PowerShell (you can actually do this on your client if you have the SQL Server 2008 client tools installed):

```
[Reflection.Assembly]::LoadWithPartialName("Microsoft.AnalysisServices")
```

This loads the AMO objects. We can now connect to an SSAS instance with the following code:

```
PS C:\> $ ServerName = New-Object Microsoft.AnalysisServices.Server
PS C:\> $ ServerName.connect("[Server Name]")
```

Now if you type **$ServerName** and press Enter, you'll see the server properties as shown here:

```
ConnectionString      : Christine
ConnectionInfo        : Microsoft.AnalysisServices.ConnectionInfo
SessionID             : 43013366-3256-48B6-B7E0-28529DA97C1E
CaptureXml            : False
CaptureLog            : {}
Connected             : True
SessionTrace          : Microsoft.AnalysisServices.SessionTrace
Version               : 10.0.1600.22
Edition               : Enterprise64
EditionID             : 1804890536
ProductLevel          : RTM
Databases             : {Adventure Works DW 2008, ~P
SQLDetailCube, TestPartition, AdventureWorks 2008}
Assemblies            : {System, VBAMDXINTERNAL, VBAMDX, ExcelMDX}
Traces                : {FlightRecorder, MicrosoftProfilerTrace1232304284}
Roles                 : {Administrators}
ServerProperties      : {DataDir, TempDir, LogDir, BackupDir...}
ProductName           : Microsoft SQL Server code name "Katmai" Analysis Services
IsLoaded              : True
CreatedTimestamp      : 1/4/2009 3:00:57 PM
LastSchemaUpdate      : 1/4/2009 3:00:57 PM
Description           :
Annotations           : {}
ID                        : CHRISTINE
Name                      : CHRISTINE
Site                      :
```

```
SiteID                    :
OwningCollection          :
Parent                    :
Container                 :
```

Again, you can see that this is an efficient way to pull a lot of data about a server remotely. And the true beauty is the ability to script tasks, making it easy to run logs, reports, or poll servers. Consider the following command (typed on a single line) and listing:

```
PS C:\> foreach($database in $ServerName.databases)
{foreach($cube in $database.Cubes)
{$cube | Select Name, LastProcessed}}
```

```
Name                          LastProcessed
----                          -------------
Adventure Works               3/15/2009 9:28:48 PM
Mined Customers               3/15/2009 8:53:33 PM
StarSchema                    12/30/1699 7:00:00 PM
Adventure Works               1/18/2009 4:45:19 PM
StarSchema                    1/18/2009 4:45:08 PM
Mined Customers               1/18/2009 4:45:29 PM
```

Look at that—a quick report of all the cubes on the server, and when they were last processed. And we can poll all our servers! Now using the PowerShell parsing syntax, we can filter out any database processed in the last week. We can then use the database names output from that to drive a loop to process databases. The net result: a script that processes any database that hasn't been processed in the last week. (For additional complexity, you could add an exception list.)

Summary

Hopefully, this chapter has given you a solid understanding of the tools we have available to use to manage SQL Server Analysis Services. You'll be spending most of your time in BIDS from here on out—for instance, when you start creating data source views in the next chapter.

CHAPTER 5

■ ■ ■

Creating a Data Source View

Step 1: Get data. That's what we have to do if we're going to build a cube.

To get data into an Analysis Services cube, we need to build a *data source view* (DSV). The data source view is how we represent complex relational data models for our dimension and cube design. Before we create a DSV, we'll need to create the data sources we will use to populate it. And before we do that, we'll need to create a project in BIDS.

We'll do everything I've just mentioned in this chapter, after we cover some introductory material about the data for our cube.

Cubes Need Data

A *data source view* (Figure 5-1) is the layer of abstraction between a cube and the underlying data source. Although it looks like a simple ERD from a database, the important thing to note is that we're able to map tables from different data sources together into a single unified schema. Traditionally, an OLAP solution would require an OLAP-specific data store, providing the views and data structures necessary to build the cube and dimension structure. SSAS does this virtually—by building the data structure in a data source view, we can skip the step of building it physically.

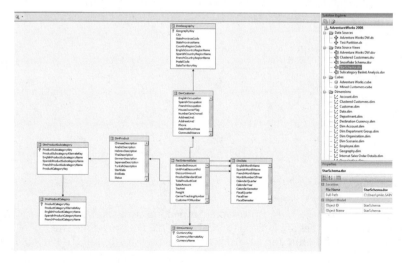

Figure 5-1. A data source view

In addition, by having multiple data source views, we can keep the alignment between cubes and data sources clearer. As I pointed out in Chapter 3, SSAS uses the concept of a universal data model to get away from the need for multiple data marts; using one or more data source views is part of that architecture.

You should notice something familiar about Figure 5-1: it looks like a diagram for a relational database. That's effectively what it is. We use the data source view to create a virtual schema representing relational data we can use to build our cube from. But again, because we're doing this in the Analysis Services server, we can map to tables from various servers. If we have the keys necessary to link the tables in our source systems, we can build a cube directly from those data sources—no staging database, no data warehouse, no data marts!

Mind you, in all likelihood you'll still need a staging database. The data needs to be cleaned and normalized (the key in one database is very unlikely to match to the corresponding key in another). So when you're aggregating the data, set up a staging database to act as the data source for our data source view.

Data Sources

Before we can start building a data source view, we need data sources. An SSAS database can have numerous data connections (Figure 5-2), so we can have multiple data source views, and individual views can draw from multiple databases.

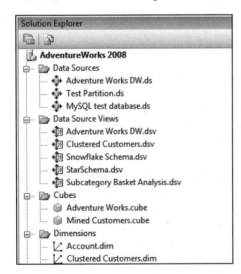

Figure 5-2. Data sources in the Solution Explorer in BIDS

Data sources are pretty much as you'd expect; they capture the connection string and authentication info for a server. SSAS data sources are limited to .NET and OLE DB providers—no ODBC. When you install the provider, you'll find that the wizard just picks it up directly and it's available in the selector in the wizard (Figure 5-3).

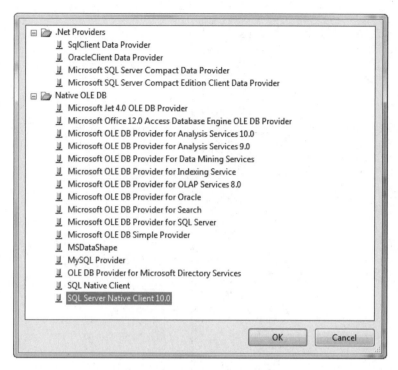

Figure 5-3. *Selecting a provider for a new data source*

You can see in Figure 5-3 that in addition to the SQL .NET providers, there's also an Oracle .NET provider. Under OLE DB we have providers for Jet (Access), SSAS, Data Mining, DataShape (for hierarchical record sets), Oracle, MySQL, and SQL Server. Selecting a provider will load the appropriate UI for the connection information.

Creating a data source is pretty straightforward—the designer is just two panels. The first page (Figure 5-4) enables you to build the connection string with the standard Windows connection manager. The Data Source References section in the center of the page can maintain a reference to another data connection in the same project or even another SSAS project. So, for example, if you want to have two data connections with the same connection information but using different impersonation properties (to use different user accounts on a database), you could link them this way and just make changes in one location.

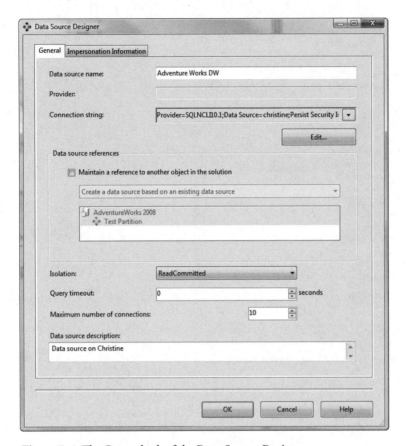

***Figure 5-4.** The General tab of the Data Source Designer*

Remember that the next set of options are about an OLE DB/.NET connection, not OLAP. This is a simple data connection to a database. The Isolation level lets you enable snapshot isolation on the SQL connection, reducing row locking and contention. You can also set the timeout on the query (by default, 0 seconds means the query won't time out). Finally, you can set the maximum number of connections and a logical description.

The second tab of the designer is dedicated to the identity the connection will use when connecting. Let's look at the four options here:

Use a specific Windows username and password: SSAS will use the provided credentials for connecting to the data source.

Use the service account: The default selection, this connection will use the account information set for the Analysis Services service account. To change that account, choose Start → Administrative Tools → Services. Alternatively, press the Windows key with the R key to open the Run dialog box, and then type **services.msc** and press Enter. Select the SQL Server Analysis Services service, right-click and select Properties, and then select the Log On tab.

Use the credentials of the current user: Analysis Services will use the current user's credentials for Data Mining Extensions (DMX) queries, local cubes, and mining models. However, this option isn't valid for any action that interacts with actions that may have to cross machine boundaries—processing, ROLAP queries, linked objects, remote partitions, and synchronization from target to source.

Inherit: This option will pass through the credentials of the user accessing the cube or data-mining model. (See the following sidebar, "That Old Double-Hop Problem.")

THAT OLD DOUBLE-HOP PROBLEM

If you work with SSAS for any length of time, you'll run into the error "Invalid login for user (null)." This will often happen when you deploy to production after a long, successful development process where the architecture was all installed on a single machine. The problem is that Windows has a security restriction. When you authenticate to a process, that process creates a security token with your credentials. By default, Windows will not allow that process to forward that token to another server.

The danger is that an intruder (person or virus) could manage to find a hole and authenticate to one server (say, a public-facing web front end). Then the intruder could pass that authentication to another server and could actually "hop" their way across the network with potentially powerful credentials.

When you work with multiple processes on a single server, you can pass the auth token back and forth with no problem. It's when you split to multiple servers that you run into problems. If you have a SharePoint site that's using SSAS cubes as a data source, the user will authenticate against SharePoint, and then the SharePoint process will negotiate a connection with SSAS. However, because the authentication token is restricted to the SharePoint server, the new connection will simply have no credentials—thus, user (null).

You can, however, enable servers to delegate credentials. You can explicitly authorize a server to delegate credentials, which is something of a quick fix, because it's managed on a server-by-server basis. A more scalable approach is to use Kerberos (an alternate spelling for the three-headed hound that guards the gates of Hades). Kerberos is an authentication protocol that allows the secure authentication of a user within a network. The downside is that to implement Kerberos, you do need to have a strong identity management program in place, and the Kerberos architecture is nontrivial. (Microsoft has a large amount of documentation on MSDN and TechNet regarding implementation of Kerberos.)

Our first exercise will be a short one—let's set up a data source for the AdventureWorks DW database (see Appendix A to set up AdventureWorks). Follow the steps in Exercise 5-1.

Exercise 5-1. Create a Data Source

In this exercise, you'll create a data source that you'll eventually use to build a cube. Here are the steps to follow:

1. Open BIDS (choose Start → All Programs → Microsoft SQL Server 2008 → SQL Server Business Intelligence Development Studio). Alternatively, you can open Visual Studio 2008 if you have that installed—you'll end up in the same place.

2. Create a new project (File → New → Project). This opens the New Project dialog box, shown in Figure 5-5.

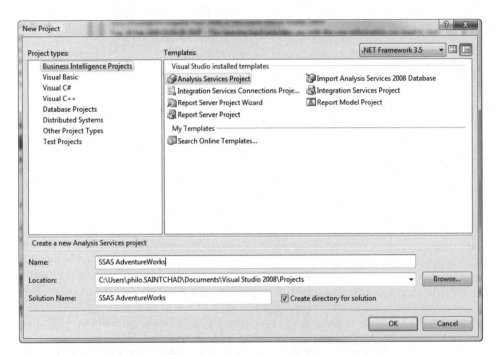

Figure 5-5. *Creating a new project*

3. Select Analysis Services Project, name it **SSAS AdventureWorks**, and click the OK button.

4. In the Solution Explorer (which is at the top right by default, or press Ctrl+Alt+L, or under the View menu if you don't see it), right-click the Data Sources folder. Click New Data Source to open the Data Source Wizard (Figure 5-6).

Figure 5-6. *Creating a new data source*

5.   Click the Next button.

6.   On the Select How to Define the Connection page, click the New button to create a new connection.

7.   In the Connection Manager (Figure 5-7), ensure that the Native OLE DB\SQL Server Native Client 10.0 option is selected as the provider.

8.   For the server name, enter the name of the server where the AdventureWorks DW database is located. Note that you can use **localhost** here if the database is on the same machine, but be wary—you will be deploying the cube to SSAS, and the server name has to resolve there as well.

Figure 5-7. *The Connection Manager*

9. After you enter the server name, the Connect to a Database option will be enabled. Select AdventureWorksDW2008.

10. Click the OK button to close the Connection Manager.

11. Ensure that your new connection is selected in the Data Source Wizard, and then click the Next button.

12. Select Use the Service Account for the Impersonation Information page. (Note that the SSAS Service Account will have to have permissions to access the AdventureWorks database in SQL Server—see the sidebar "That Old Double-Hop Problem" earlier.)

13. Click the Next button.

14. On the final page, note the data source name and then click the Finish button.

Data Source Views

Now that we have a data source, we'll need a data source view on which to build our dimensions and cubes. As I've mentioned before, you can build a data source view from multiple data sources. Simply create the additional data sources in the same manner.

The data source view defines the relational schema on which you build your cubes (consisting of measures and dimensions). As an abstraction point between an underlying database schema and the cube, the DSV allows the joining of tables that do not have relationships defined in the database, creation of primary keys on tables, calculated fields, and named queries to act as a view on data.

For the most part, the DSV is about building this relational model. If you've used a relational modeling tool in the past, it should be pretty familiar. We'll take a quick tour and then create a data source view for our cube.

Designer Tour

As I mentioned, the DSV designer is similar to any relational model designer—add tables and link them. If you add tables that already have relationships in the underlying data source, those relationships will be shown in the DSV. You can also add relationships in the DSV, but this is to inform links; it won't create any actual constraints on the tables.

The DSV designer is shown in Figure 5-8. The central pane is the actual design surface, as you can see from the tables and relationships shown there. To the top left is the Diagram Organizer. When you get large, complex relational structures in your data source view, you can create additional diagrams to focus on subsets of the tables. The AdventureWorks SSAS solution has several functional subviews.

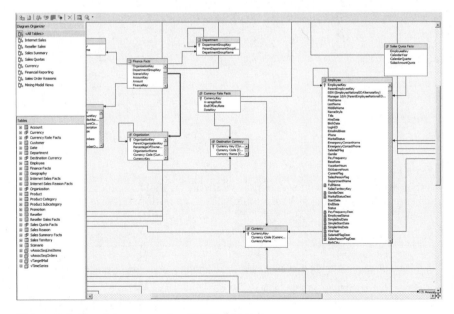

Figure 5-8. *The data source view (DSV) designer*

To the lower left of the designer is a table browser. You can use this for reference or drag tables from the browser to the design surface. You can also check the properties of fields by right-clicking on a field and selecting Properties. BIDS also enables viewing the data in the table: right-click on the table and select Browse Data.

Finding Tables

If you do have a large, complex diagram and frequently need to move around looking for various tables, there are a few easy navigation methods. First and easiest—if you click on a table name in the tables pane, the diagram will immediately show that table if it's in the view. You can also click the Find Table

button on the toolbar. This will open the Find Table dialog box, which lists all the tables in the current view. You can select a table and be taken to it.

The other (really cool) way to navigate a large diagram when you're zoomed in is via the arrow in the lower-right corner of the DSV designer. When you click on the arrow, you'll see an overview of the data source view with an outline of the current view you can move to change the view (Figure 5-9).

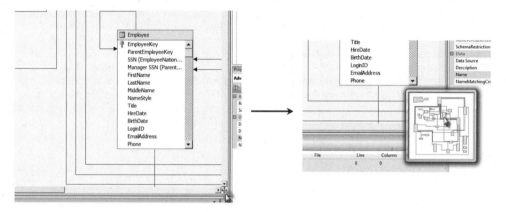

Figure 5-9. *Using the DSV table navigator*

Replacing a Table

You may need to replace a table in a data source view—perhaps with a view that has other information, or a table from another data source. If you right-click on a table, there's an option to replace the table with another table or a named query.

■ **Caution** If you replace the table with another table that is significantly different, you stand a good chance of corrupting any dimensions or cubes that are based on that table.

Now let's take a look at creating a data source view, in Exercise 5-2.

Exercise 5-2. Create a Data Source View

In this exercise, you'll build the data source view we're going to use for the rest of the book. Follow the steps in the Business Intelligence Development Studio.

1. Open the project from Exercise 5-1.

2. In the Solution Explorer in the top right, right-click on Data Source Views and select New Data Source View to open the Data Source View Wizard (Figure 5-10).

Figure 5-10. Starting the Data Source View Wizard

3. Click the Next button.

4. Select the data source you created, and then click the Next button to proceed to the Select Tables and Views page (Figure 5-11).

Figure 5-11. *Selecting tables for the data source view*

5. Add the tables listed in Table 5-1 by selecting them and clicking the single right-arrow (>) button.

Table 5-1. *Tables for the Data Source View*

Table Name	Schema	Type
FactResellerSales	dbo	Table
DimEmployee	dbo	Table
DimProduct	dbo	Table
DimProductSubcategory	dbo	Table
DimProductCategory	dbo	Table
DimPromotions	dbo	Table

6. Click the Next button.

7. The final page enables you to give the data source view a name. Leave the default name and click Finish.

8. The DSV designer should open—you should see the tables with their relationships shown. Figure 5-12 shows an arrangement; yours may be somewhat different.

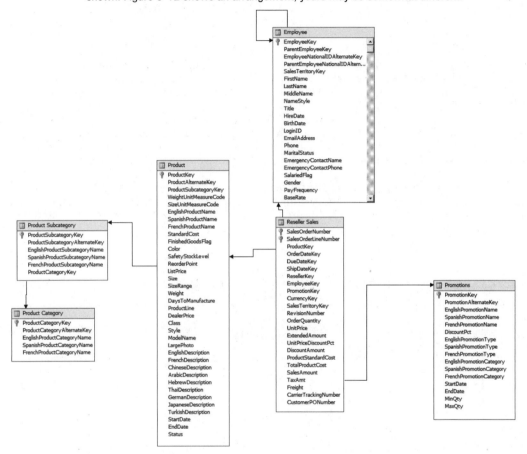

Figure 5-12. *The completed data source view*

9. You can rearrange the tables to a more logical layout if you choose.

10. Let's give the tables more user-friendly names. Right-click on the FactResellerSales table and click Properties.

11. In the Friendly Name field, type **Reseller Sales**.

12. Add friendly names to each of the other tables in the same way.

13. Save the project. Now let's move on to named calculations and queries.

Named Calculations and Queries

Occasionally you may need a column that can be derived from existing values, or a single table composed of columns from other tables. You can accomplish these in the data source view with named calculations and queries.

Named Calculations

For example, invoice line item data often doesn't include the subtotal; business systems calculate it on the fly. We could calculate it in the cube, as you'll see in Chapter 7, but for something that's relatively fixed (such as a subtotal value), why not calculate it at the data source and not waste processing power recalculating it over and over?

Analysis Services lets you calculate a value in the data source view by creating a *named calculation*. We effectively define a dynamic field in the table in the data source view that does the math when the data is read. Using this, we can create concatenated fields (full name consisting of first name plus last name), adjusted values (age from date of birth, aligning date/time values to join tables), or various calculated values.

Named calculations are defined in the data source view. You can create a named calculation by right-clicking on a table in the designer and selecting New Named Calculation. This will open the Named Calculation dialog box (Figure 5-13). It's pretty straightforward—name and description, and the expression for the calculation.

■ **Note** The calculation must be a standard SQL expression that returns a scalar value.

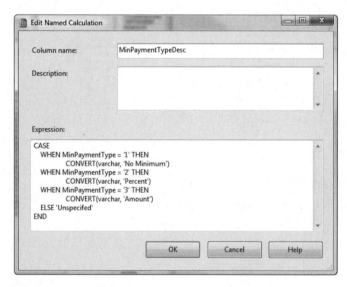

Figure 5-13. Creating a named calculation

In our data source view, we have a table for Employees. The table has a start date and end date, but perhaps we want to know the number of years the employees have been on the job. For employees still in the position, the end date is null, so we'll have to use today's date. If we're looking for the number of years, we can create a named calculation, which will recalculate whenever we process the dimension. So let's look at how we create the named calculation in Exercise 5-3.

Exercise 5-3. Create a Named Calculation

In this exercise, you're going to create a calculated field to show the number of years an employee has been in a role, using the start date and end date fields. If the end date field is null, you'll use today's date.

1. Open the SSAS AdventureWorks project again if you don't have it open.

2. Open the Adventure Works DW2008.dsv data source view.

3. Right-click on the Employee table and select New Named Calculation.

4. For column name, enter **YearsInRole**.

5. In the Expression field, enter the following:

```
CASE
    WHEN EndDate IS NULL
        THEN DATEDIFF(year, StartDate, GETDATE())
    ELSE
        DATEDIFF(year, StartDate, EndDate)
END
```

6. Click the OK button. You'll see the new field in the Employee table.

7. To see the results, right-click on the table and click Explore Data. The new column will be at the end. See Figure 5-14 to see some of the data.

StartDate	EndDate	Status	BirthCity	YearsInRole
1996-07-31 00:00:00Z		Current		13
1997-02-26 00:00:00Z		Current		12
1997-12-12 00:00:00Z		Current		12
1998-01-05 00:00:00Z	2000-06-30 00:00:00Z			2
2000-06-30 00:00:00Z		Current		9
1998-01-11 00:00:00Z		Current		11
1998-01-20 00:00:00Z	1999-08-15 00:00:00Z			1
1999-08-16 00:00:00Z		Current		10
1998-01-26 00:00:00Z		Current		11
1998-02-06 00:00:00Z		Current		11
1998-02-06 00:00:00Z		Current		11
1998-02-07 00:00:00Z		Current		11
1998-02-24 00:00:00Z		Current		11
1998-03-03 00:00:00Z		Current		11
1998-03-05 00:00:00Z		Current		11
1998-03-11 00:00:00Z		Current		11
1998-03-23 00:00:00Z		Current		11
1998-03-30 00:00:00Z		Current		11
1998-04-11 00:00:00Z		Current		11

Figure 5-14. A calculated field

8. Save all open files.

Named Queries

The counterpart of a named calculation is a *named query*. This is effectively a SQL view as part of the data source view. The named query designer (Figure 5-15) will look very familiar to anyone who has designed a query in Access, SQL Server Management Studio, Reporting Services, and so on.

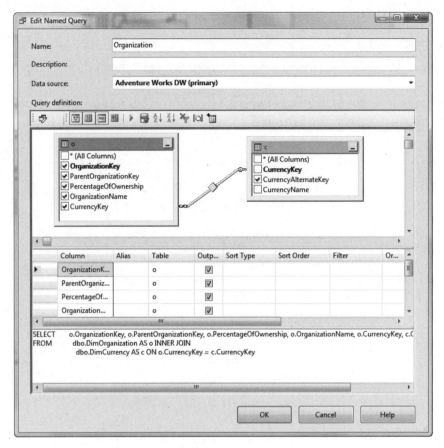

Figure 5-15. The named query designer

One of the top uses of a named query is to make data more readable. As shown in Figure 5-15, we want to show a label in the table instead of a numeric foreign key. Another use would be to merge tables for use in a star schema. Although a normalized database would have separate tables for county, state, and country, in our star schema we want a single table listing counties and indicating the state and country value for each. (You'll learn how to break this back down in Chapter 7.) After you create a named query, it will appear to subsequent wizards just as another table.

Let's add the Reseller table to our data source view, and include the geographic information about the reseller. Follow along in Exercise 5-4.

Exercise 5-4. Create a Named Query

In this exercise, you'll create a named query to represent our resellers, pulling some of their geographic information into the table (instead of bringing in the whole geography set of tables).

1. Right-click on the design surface and select New Named Query.

2. Name the query **Reseller**.

3. Right-click in the table area under Query Definition and then click Add Table.

4. In the Add Table dialog box, select DimReseller and DimGeography and then click Add.

5. Note that the tables are already connected. Select the fields listed in Table 5-2 by selecting the check box next to each field name.

Table 5-2. *Adding Fields to the Named Query*

Table	Field
DimReseller	ResellerKey
DimReseller	BusinessType
DimReseller	NumberEmployees
DimReseller	OrderFrequency
DimReseller	AnnualSales
DimReseller	MinPaymentType
DimReseller	MinPaymentAmount
DimReseller	YearOpened
DimGeography	City
DimGeography	StateProvinceName
DimGeography	EnglishCountryRegionName
DimGeography	PostalCode

6. You can click the Run button to see the query results and ensure that the query is okay. The result should look like Figure 5-16.

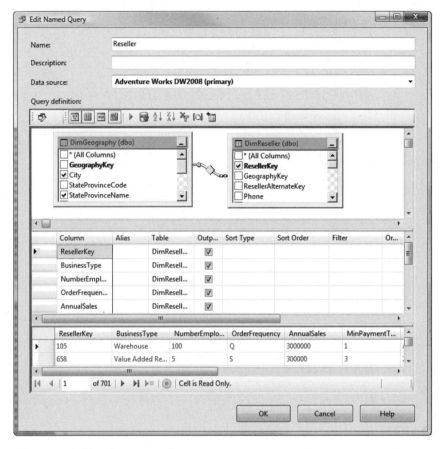

Figure 5-16. Creating a named query

7. Click the OK button.

8. Note that in this case our Reseller view still needs to be connected to the Reseller Sales table.

9. First create a logical primary key on the view: right-click on ResellerKey and select Set Logical Primary Key. This marks this field as the primary key for the view.

10. Click and drag the field ResellerKey from Reseller Sales to the ResellerKey field in the Reseller table.

■ **Note** If you try to drag the field the other way, you will get a warning that you are setting a relationship in an unusual direction, and you will have the option to switch the relationship.

When you're finished, the two tables should look like Figure 5-17.

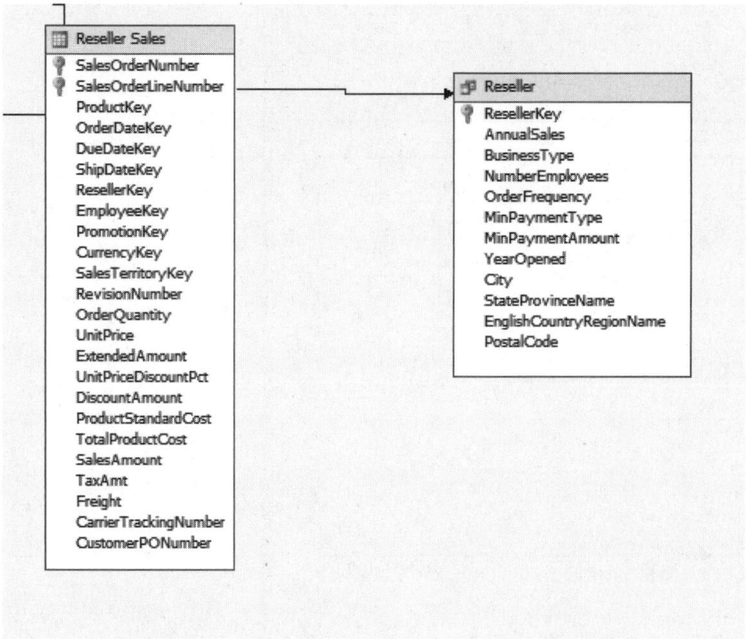

Figure 5-17. *The final named query*

11. Save all files—we're finished with our data source view!

Summary

Now that you understand data source views, you can see how they can be used to aggregate data from multiple data sources. Also, by using multiple data source views, you can provide different ways of looking at data without actually creating different databases and duplicating the data itself. The data source view is a key concept in the universal data model, by providing a way of severing data without duplicating it. The "model" part is formed by the dimensions and measures of the cube itself. Let's move on to understanding dimensions by creating some.

CHAPTER 6

■ ■ ■

Creating Dimensions

This is possibly the most important chapter in the book. Although *measures* are what we're after, *dimensions* are the framework on which we'll build our cubes. Dimension names give us the nomenclature we'll use to break down our data, attributes will give us additional information about the members, and the members themselves provide the semantic points against which our measure data will be arrayed. Finally, well-designed dimension and attribute relationships will ensure the scalability of our cube solutions.

Dimensional Analysis

Before we can start designing dimensions, we have to figure out what dimensions we need. What problem are we trying to solve? In my opinion, this is a driving foundation of a successful business intelligence effort—identifying the problem domain.

Many BI initiatives start as "we need a data warehouse" without real direction. This is akin to someone standing in Madison, Wisconsin and declaring one day that she wants to visit every country. A wise traveler will then buy a bunch of maps, identify locations, and start working on solving the traveling salesman problem. It should soon become apparent that a huge amount of analysis is necessary—even more so if our traveler has never actually been on a train, airplane, or boat. In this research, the traveler will find that she needs a passport, luggage, and various supplies depending on the country she visits. She also needs visas. But wait. What visa she needs for a country depends on which border she's using to enter the country (and some borders can't be crossed!).

This is certainly a huge problem, and will take a very long time to plan out properly. However, all the time she spends planning is time she's not traveling. Meanwhile, she has bills to pay and a life to live, and because she hasn't actually gone anywhere, over time she'll just lose interest in the whole project.

Now what if, instead, our traveler decides that what she really wants to do is visit Europe, and specifically the Eiffel Tower? It's much easier to plan a short trip to Paris, find a cheap round-trip flight, book a hotel, figure out the visa situation, and so on. Soon she's on her way to Paris, enjoys her trip, and she's home again. Now she decides to visit Japan, and so does some more research—but this time the research doesn't take nearly as long, because she can build on what she learned on the last trip, and add in her personal experiences with what worked and what didn't.

My analogy is probably too thin, but hopefully will make the point. Don't try to sit down and design the be-all, end-all ultra data warehouse that will answer any question anyone could possibly ask. Instead, identify smaller business problems that can be solved with smaller cubes. Build a collection of cubes, and grow the solution iteratively. In this way, your users get value from what you've already built, and each project draws from the experience the team gained on the previous problem. With that in mind, let's look at some concepts behind dimensions that will serve our analysis, and then how we'll build them in Analysis Services.

Review of the Dimension Concept

Let's look at our notional "cube" again, represented by spreadsheets. In Figure 6-1, Order Date, Country, and Product are each dimensions. Alice Mutton, Aniseed Syrup, Boston Crab Meat, and so on are members of the Product dimension. We use dimensions to organize collections of attributes into logical groupings (just as we have products, country, and time). You can see that the collection of dimensions effectively defines the collection of measures that make up our cube. In SSAS, the concept of a dimension actually exists independent of a cube. Dimensions can be the result of a single cube design, but they can also be in a database solution with no cubes. A single dimension can be used in multiple cubes, or even linked between multiple databases.

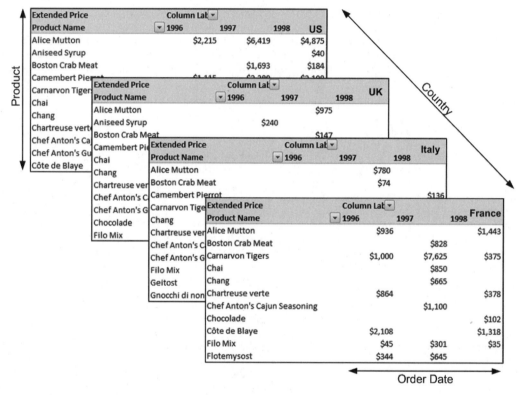

Figure 6-1. *Notional cube structure*

In our dimensional analysis, after identifying a problem domain, we'll have to figure out what types of questions users are likely to ask and how they will seek answers. I won't go into requirements analysis here. If you're interested in methods for gathering requirements in the data warehouse world, I recommend *The Data Warehouse Lifecycle Toolkit*, by Ralph Kimball, Margy Ross, Warren Thornthwaite, Joy Mundy, and Bob Becker (Wiley, 2008) or *The Microsoft Data Warehouse Toolkit: With SQL Server 2005 and the Microsoft Business Intelligence Toolset* by Joy Mundy and Warren Thornthwaite (Wiley, 2006). There is one big technical decision we need to make, however: do we want a star schema or a snowflake?

Star or Snowflake?

We generally have two options when creating a cube: a star or a snowflake schema. To review, a *star schema* has a single fact table with every dimension table linked directly to it (Figure 6-2).

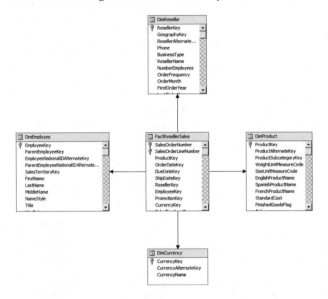

Figure 6-2. *A star schema*

Compare this to a *snowflake schema*, which keeps the tables more normalized, similar to Figure 6-3.

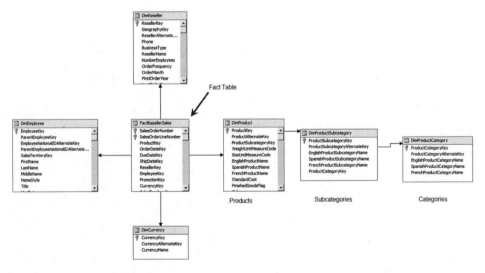

Figure 6-3. *A snowflake schema*

Λ snowflake schema will likely give way to what is often termed a *multisnowflake* schema, which has multiple fact tables. From here we can see a creeping problem that adds fact tables, dimension tables, supporting tables, and so on. You can see a similar result in the AdventureWorks schema in Figure 6-4, which I've annotated with arrows pointing to each fact table.

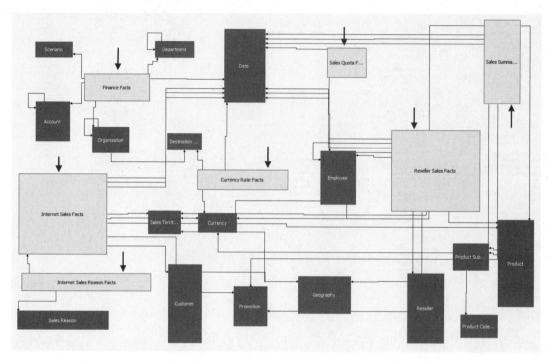

Figure 6-4. *The AdventureWorks schema*

The primary advantage of a star schema is that it keeps development focused on a simple approach—one fact table, and every dimension linked to the fact table. It's important when creating a cube to keep focused on end-user usability, especially because ease of use for our analyst end user is a major reason we're building these cubes in the first place.

The Advantage of Simplicity

As an example of how a complex schema can defeat end-user usability, consider the pick list for measures and dimensions we see if we open the AdventureWorks cube in Excel (Figure 6-5).

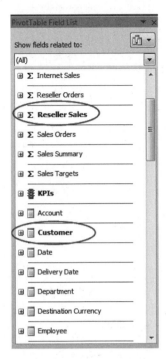

Figure 6-5. Measures and dimensions from the AdventureWorks cube in Excel

Perhaps our analyst wants to examine the effectiveness of an ad campaign on in-store buyers based on their geographic region. So it seems logical to examine reseller sales by customer location, right? Let's continue down that path. My argument will be easier to follow if you build the report yourself. So let's build a quick Excel report in Exercise 6-1.

■ **Note** You'll need the AdventureWorks 2008 demo cube installed and processed for this exercise. See Appendix A.

Exercise 6-1. Build an Excel Report

The following steps walk you through building a report to examine reseller sales by customer location, based on the schema in Figure 6-4, and on the measures and dimensions highlighted in Figure 6-5. As you develop the report, you'll see how the complexity of the schema works against you.

1. Open Excel 2007.

2. Click the Data tab. On the Data tab, click the Get External Data button, then From Other Sources, and finally From Analysis Services (Figure 6-6).

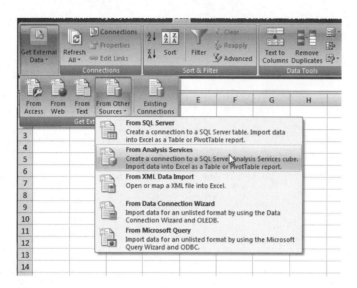

Figure 6-6. Inserting an Analysis Services data table in Excel 2007

3. On the first page of the Data Connection Wizard, enter the server name (**localhost** will do if you have a single dev machine). Then click the Next button.

4. Select Adventure Works DW 2008 from the database drop-down list, and Adventure Works in the specific cube selector (Figure 6-7).

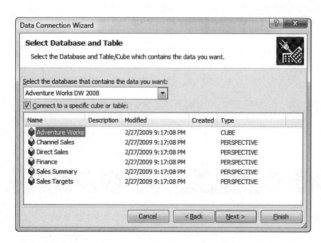

Figure 6-7. Connecting to a specific cube

5. Click Next and then click Finish.

6. In the Import Data dialog box, select PivotTable Report and then click the OK button. You'll get a pivot table area and the PivotTable Field List (Figure 6-8).

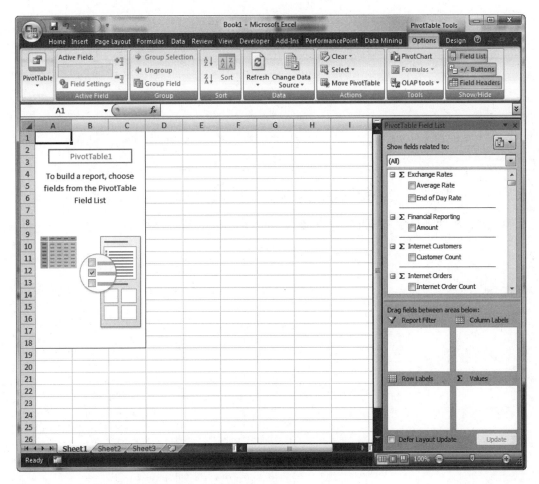

Figure 6-8. *A pivot table in Excel*

7. In the field list on the right, scroll down to Reseller Sales and select the Reseller Sales Amount check box (Figure 6-9).

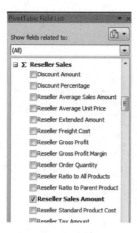

Figure 6-9. *Selecting Reseller Sales*

8. Now scroll down to Customer and drag Customer Geography to Row Labels
    (Figure 6-10).

Figure 6-10. *Dragging Customer Geography to the row selector*

When you're finished, you should have an area in the Excel spreadsheet that looks like Figure 6-11.

	A	B
1	Row Labels ▼	Reseller Sales Amount
2	⊞ Australia	$80,450,596.98
3	⊞ Canada	$80,450,596.98
4	⊞ France	$80,450,596.98
5	⊞ Germany	$80,450,596.98
6	⊞ United Kingdom	$80,450,596.98
7	⊞ United States	$80,450,596.98
8	Grand Total	$80,450,596.98
9		

Figure 6-11. *The finished report in Excel*

9. Click the [+] icons next to the countries to look at the region values. What happened? Read on.

Well, it's a safe bet that those six countries didn't sell exactly the same amount of product (down to the penny), so it looks like our Customer Geography dimension isn't splitting the measure data. Why is this? A little examination of the cube structure will show us why. First we note from the pivot table picker in Excel (Figure 6-10 in the exercise) that we used the Customer Geography hierarchy from the Customer dimension.

If we open up the AdventureWorks project in BIDS, we want to look at the AdventureWorks cube. Once in the cube, we're interested in the Dimension Usage tab (Figure 6-12). Note that we have measure groups across the top, and dimensions down the left side. If we track Reseller Sales and look where it intersects the Customer dimension, we see something interesting.

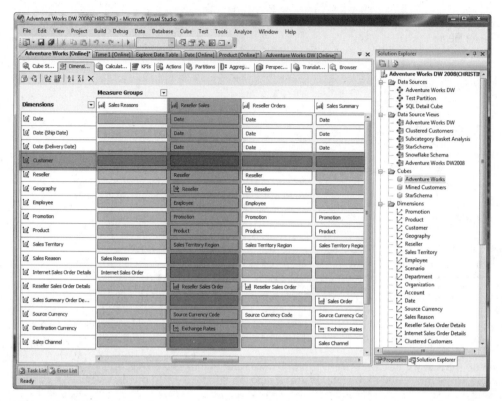

Figure 6-12. Relationship between the Reseller Sales measure group and the Customer dimension

They—the sales and customer dimensions—have no relationship defined. So any use of the Customer dimension against the Reseller Sales measure group won't slice the underlying data, which is the behavior that we saw in Exercise 6-1. At this stage, there most likely doesn't exist any way to identify customers from the reseller data we have. If that turns out to be a critical business need, an analyst will have to go back to the data warehouse group and have that data added to the cubes, if it even exists (the resellers may not report customer data back through the supply chain).

So we see that a complex cube structure can lead to confused and frustrated end users. That is one of the key reasons for preferring a star schema—to easily ensure that every dimension does in fact link to the fact table.

■ **Note** The Enterprise Edition of Analysis Services offers a powerful feature called *Perspectives* that allows us to address this potential complexity without creating numerous additional cubes. You'll look at this feature in Chapter 7.

Now that you have a basic understanding of where dimensions fit in the grand scheme of things, let's dive down and start looking at working with dimensions in Analysis Services, specifically focusing on building dimensions in BIDS.

Dimensions in SSAS

You can create a dimension at any time in BIDS just by right-clicking the Dimensions folder and selecting New Dimension, which will open the Dimension Wizard. Alternatively, you can link a dimension, which enables you to create a pointer to a dimension defined in another database. By using linked dimensions, you can maintain "standard" dimension definitions in a database, and then link to them from other cubes as necessary. Finally, dimensions can be implicitly created when you run the cube wizard (you'll look at this in Chapter 7).

Creating a Dimension

Dimensions can either be defined from the data in a data warehouse, or they can be designed in the Analysis Services database solution. In the latter case, the dimension and its attributes are laid out in BIDS, and then the schema generation wizard will create the underlying data tables to support the dimension. (The next step will be to design the ETL to move data from the data sources into the designed schema.) Follow along in Exercise 6-2 and create a dimension that you'll use later in this chapter.

Exercise 6-2. Create a Dimension

Follow the steps in this exercise to create a dimension on promotion. You'll use the dimension for later exercises in this chapter.

1. Open the SSAS AdventureWorks project you created in Chapter 5.

2. In BIDS, open the AdventureWorks DW2008 data source view.

3. Let's add the table we're going to use for our dimension: right-click on the design surface and select Add/Remove Tables. (Alternatively, open the Data Source View menu and select Add/Remove Tables.)

4. In the Add/Remove Tables Wizard, select the DimPromotion table and click the right-arrow (>) button to move that table to the Included Objects list.

5. Click the OK button. The DimPromotion table should appear in the data source view, with a relationship to the Reseller Sales table. (If its connector runs behind another table, you can move it to a more convenient location.)

6. Right-click the DimPromotion table and select Properties. Change the Friendly Name to Promotions.

7. Save the data source view.

8. In the Solution Explorer, right-click Dimensions and click New Dimension.

9. If you see the Welcome to the Dimension Wizard page, click the Next button.

10. In the Select Creation Method page (Figure 6-13), keep the default of Use an Existing Table and then click the Next button.

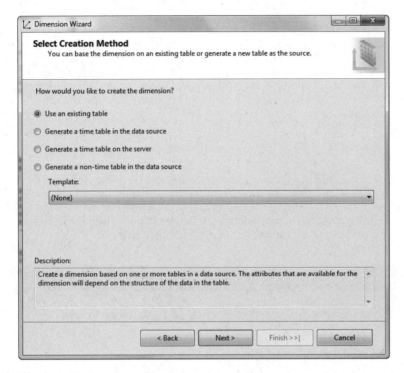

Figure 6-13. Creating a dimension

11. In the Specify Source Information page, the Adventure Works DW2008 data source view should be selected. Select Promotions for the Main table. Leave the default—PromotionKey—as the key column. Change the Name column to **EnglishPromotionName** (Figure 6-14).

Figure 6-14. Selecting the data table for the dimension

12. Click the Next button.

13. In the Select Dimension Attributes page, select attributes in accordance with Table 6-1.

Table 6-1. Setting Attributes

Attribute Name	Enable Browsing	Attribute Type
Promotion Key	Yes	Regular
Discount Pct	Yes	Regular
English Promotion Type	Yes	Regular
English Promotion Category	Yes	Regular
Start Date	No	Date Start
End Date	No	Date Ended
Min Qty	No	Quantity Range Low
Max Qty	No	Quantity Range High

14. Click the Next button.

15. Review the Completing the Wizard page and then click the Finish button. You should end up in the Dimension design surface, which should look like Figure 6-15.

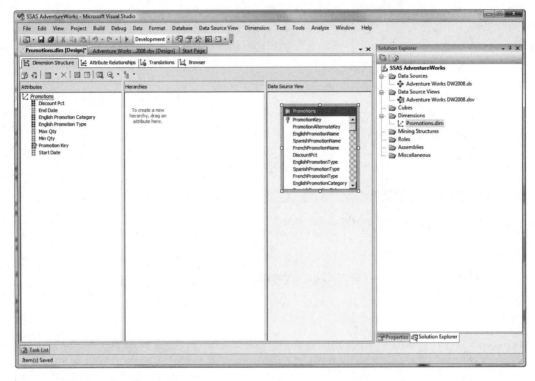

Figure 6-15. Our new dimension

Note that on the left side, the attributes that we indicated weren't browsable are grayed out.

16. Let's process this dimension. First you'll need to set your deployment server: right-click on SSAS AdventureWorks in the Solution Explorer and select Properties.

17. In the property pages (Figure 6-16), select Deployment and then enter the name of your server for the Server property (or leave it as localhost if you are working on your SSAS server).

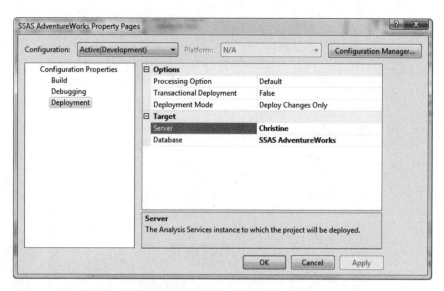

Figure 6-16. *Setting the deployment server*

18. Select the Build menu and then Deploy SSAS AdventureWorks to deploy the solution to the server.

19. Now either click the Process button ( ) or choose Dimension ➤ Process.

20. When the Process Dimension window comes up, click Run. (We'll dive into this window in depth in Chapter 8.)

21. Now we can browse our dimension. In the dimension designer, click the Browser tab.

22. You should see a single member labeled All. Expand it by clicking the [+] icon next to it. You should see something like Figure 6-17.

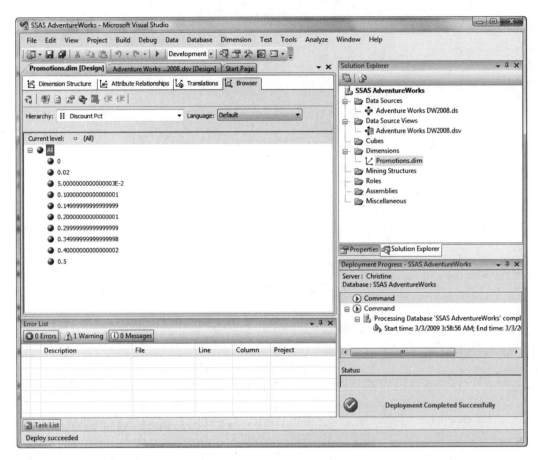

Figure 6-17. Our new dimension?

23. That's odd. Hold on—look at the Hierarchy selection. These are the available Discount Percentages. Change the selection to Promotion Key and open the tree again (Figure 6-18).

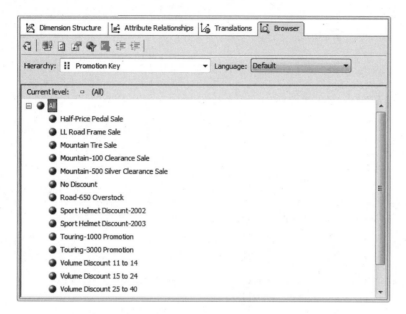

Figure 6-18. *The promotions in the dimension*

24. This looks more like a list of dimensions. So how do we make them come up first? We'll use a little trickery. Let's also clean up the dimension a bit. Click the Dimension Structure tab again.

25. Right-click English Promotion Category, select Rename, and rename it **Category**.

26. Similarly, rename English Promotion Type to **Type**, and Promotion Key to **Promotion**.

27. Now we're going to create a hierarchy. Drag Category to the Hierarchies section in the middle of the designer. This will create a Hierarchy container. Drag Type to the <new level> area under Category.

28. Finally, drag Promotion under Type.

29. Right-click Hierarchy and select Rename. Rename the hierarchy **Promotions**. It should look like Figure 6-19.

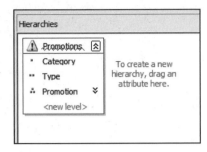

Figure 6-19. *The Promotions hierarchy*

30. Note that little blue squiggle under Promotions in the hierarchy. See the upcoming section, "Analysis Management Objects (AMO) Warnings," for more about that. For now don't worry about it.

31. Save the dimension and process it again.

32. Click the Browser tab. (You will likely have to click on the Reconnect link in the yellow bar at the bottom to refresh the connection.)

33. In the Hierarchy tab, select Promotions.

    Note that now the promotions are all structured in an easy-to-browse tree (see Figure 6-20).

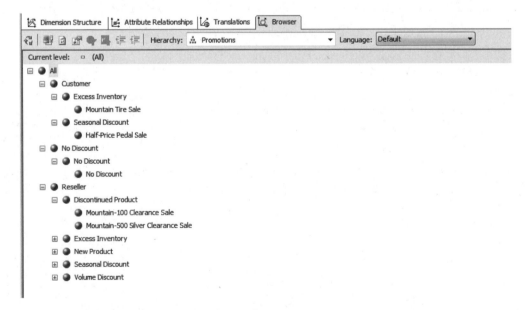

Figure 6-20. *The finished Promotions hierarchy*

In this exercise, we talked a lot about attributes and hierarchies. Let's take a deeper look at what they are and how they factor into cube design.

Analysis Management Objects (AMO) Warnings

In Exercise 6-2, we see a little blue squiggle in a hierarchy. When we mouse over it, we see a warning (Figure 6-21). This is an *Analysis Management Objects (AMO) warning*, a new feature in BIDS 2008. AMO warnings are best-practice recommendations to aid in design of Analysis Services cubes. In this case, the warning is pointing out that there is no relationship defined between the Type and Promotion attributes (more on this in the "Attributes" section later in this chapter).

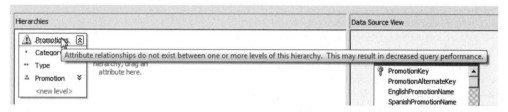

Figure 6-21. An AMO warning in BIDS

BIDS has dozens of design warning rules that it runs against OLAP designs as you work. The warnings will show up as you work. AMO warnings won't keep you from saving or processing cubes or dimensions. You can choose to ignore warnings, either on a case-by-case basis, or you can dismiss a warning globally.

To dismiss a warning, open the Error List pane (View menu ➤ Error List). Find the error and then right-click on it. Click Dismiss, and the Dismiss Warning dialog box will open (Figure 6-22). Note that you can enter comments. These comments will be stored with the database for future reference. For example, this same warning was dismissed in the AdventureWorks database. Let's take a look at why.

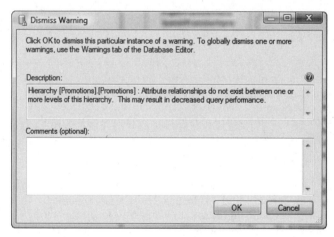

Figure 6-22. Dismissing an AMO warning

To view the status of warnings in a database, right-click the database name (the project in the Solution Explorer) and select Edit Database. In the designer that opens, click the Warnings tab (Figure 6-23).

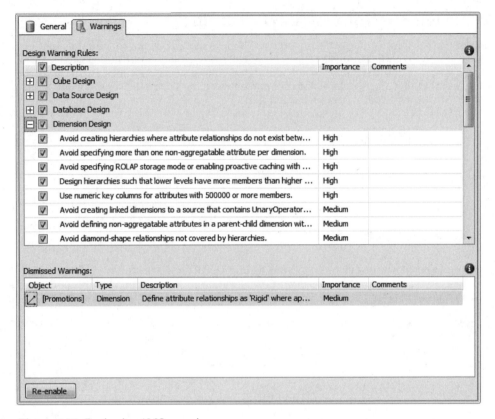

Figure 6-23. Reviewing AMO warnings

The top section lists the design warning rules available. They're grouped by type of object (Cube, Data Source, and so on), and ranked by importance. You can come in and deselect rules that you don't want enforced. In the lower window are the warnings you've dismissed for the current database (including comments you made when you dismissed them). You can highlight a dismissed warning and click the Re-enable button to put the warning back in place.

Dimension Properties

Now that you've built a dimension, let's dig into the features and properties. These are accessible by right-clicking on the dimension name in the attributes pane on the left, in Visual Studio, and selecting Properties. The following subsections describe each of the properties that you see.

Dimension Type

You can flag a dimension with a type—the default is Regular. For the most part, the dimension type is an indicator for client applications on how to treat a dimension. For example, if a dimension is the Customer type, a client application may provide a way to map the dimension to contact cards or a personnel selector-type control. Note that Account, Currency, and Time dimension types do have special handling built into SSAS—we'll look at them separately.

The dimension types are as follows:

Account: This dimension represents a chart of accounts (revenues, losses, profits, expenses, and so forth). An Account dimension type can be managed such that specific account types are positive or negative solely based on the account definition.

Bill of Materials: A Bill of Materials type will contain manufacturing or inventory information (unit of measure, cost, packaging).

Channel: Used in retail sales, a Channel dimension will indicate the sales channel used (wholesale, retail, partners).

Currency: The Currency dimension is designed to reflect the financial requirements of dealing with international transactions. SSAS has a currency-conversion feature to manage these transactions.

Customer: A Customer dimension will have standard attributes for customers and contact information.

Geography: If you want to designate attributes as addresses, cities, and zip codes, then set the dimension type as Geography. You can also have spatial data types in a Geography dimension type.

Organization: Employee names, positions, offices, and branches can all be defined in an Organization dimension.

Product: A Product dimension will have attribute types for brands, SKUs, and product categories. You can also add attributes for product start and end dates.

Promotion: If you need to group facts by the type of deal the buyer got, you can use a Promotion dimension type, including minimum and maximum quantities, start and end dates, percentage discounts, and so on.

Quantitative: Quantitative is more a general description than an actual dimension type, and more of a collection of measurement-oriented attributes (volume, weight, color, maximum, minimum, and so forth) than anything else.

Rates: Generally a dimension for measuring exchange rates.

Scenario: Scenarios are a fascinating capability of an OLAP solution. Because any dimension can "slice" the fact data into exclusive collections (sales in Europe, North America, or Asia), we can create a dimension solely for creating scenarios. For example, a scenario may have members for "5 percent growth," "No growth," and "5 percent decline." Selecting each member then groups the data for that scenario. You'll revisit the concept of scenarios when you look at writeback in Chapter 7.

Time: A Time dimension represents a calendar—days, weeks, months, quarters, years. The Time dimension is very special, as we've seen previously, and we'll cover time dimension types specifically later in the chapter.

Utility: A Utility dimension is basically a catchall for some aspect of the data that may not be reflected in the business rules.

Each of these dimension types has a host of matching attribute types (Figure 6-24). In general, you can "mix and match" attribute types with dimension types, though some of the special types can be used only in their matching dimension type. If you mismatch types, you'll get an AMO warning on the dimension.

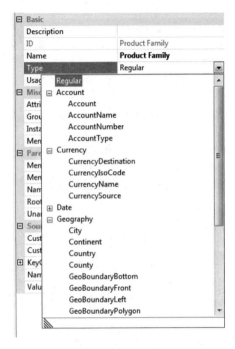

Figure 6-24. *Selecting attribute types*

After you've set a dimension type and assigned attribute types, OLAP client applications can leverage known dimension types in business logic. For example, a sales and marketing analysis tool could pull promotion and channel dimensions from your cubes and align them with the analysis system.

I mentioned that Account, Currency, and Time dimension types are special. In addition to being able to be leveraged by client applications, each of these dimensions can be specified within SSAS to work in special ways. Let's open AdventureWorks and look at the Account dimension (Figure 6-25). Note that the table in the data source view is self-referential. It has a field that is a foreign key for the table's primary key, to indicate the parent account. This creates the recursive hierarchy of accounts in the dimension.

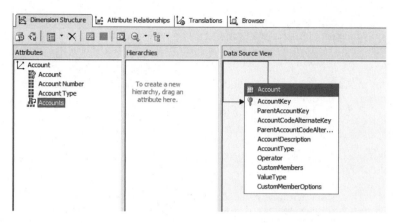

Figure 6-25. *The Account dimension in AdventureWorks*

Now let's look at the dimension browser (Figure 6-26). You might not notice at first, but various accounts have plus or minus signs on their icons in the browser. That's because the DBA set up *account business intelligence* on the dimension. When you run the Add Business Intelligence Wizard (Dimension menu), it walks you through mapping an accounts table to built-in logic for standard accounts. In addition, the UnaryOperatorColumn was set to the Operator column of the Account table. In a parent-child hierarchy dimension, setting the Unary Operator on the parent attribute indicates to Analysis Services to use that column to determine how to aggregate that member.

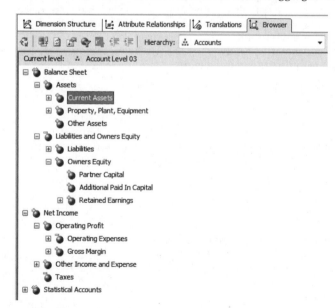

Figure 6-26. *Browsing the Account dimension*

Currency dimensions have similar logic that can be added via the business intelligence wizard. The wizard is pretty straightforward. You will want a table of exchange rates by time period to handle the automatic currency conversion. One aspect you will have to know is your preferred method of handling local currencies. SSAS will want one of the following three scenarios:

Many-to-many: All transactions are recorded in the local currency. Currency conversion is applied at the cube level and can target many currencies.

Many-to-one: All transactions are recorded in the local currency. Conversions are applied at the cube level to the corporate standard currency.

One-to-many: All transactions are recorded in a single currency. The conversion is applied at the cube level with multiple target currencies.

So if all your sales are reported in the local currency, you can set up a currency dimension and conversion rate table, apply the business intelligence wizard, and you can roll up aggregated sales figures with conversions automatically applied.

ErrorConfiguration

This section of properties refers to how Analysis Services should handle errors that crop up while processing the dimension. Think of a dimension as a lookup table—one thing that's necessary is a primary key value to link the dimension and subordinate attributes to records in the cube. So when processing the dimension (reading in the data, parsing it for attributes and relationships, and then caching the dimension data), if multiple records have the same key value, SSAS will throw an error.

You can indicate how to handle the error here. The default action is to ignore the error and continue processing. You can see how this may cause problems, so you might want to change this to either ReportAndContinue or ReportAndStop if you have a data source where you expect you may get duplicates. Both IgnoreError and ReportAndContinue will leverage the KeyErrorAction setting—either converting the key to an Unknown catchall value or discarding the record.

MdxMissingMemberMode

This setting indicates how Analysis Services should handle queries or calculations that reference a member that doesn't exist in the dimension. The default setting is Default, which means the error will be handled in accordance with the settings in Analysis Services and the cube. You can also specifically set this to error out or ignore the error.

Processing

These options govern how the dimension is processed. You should generally leave the setting for ProcessingGroup at the default value of ByAttribute. ByTable is used for a small subset of cases generally involving very large dimensions (millions of rows). Setting ProcessingMode to LazyAggregations will make the dimension available for use while it's still being processed, but as a result, processing will take longer.

ProcessingPriority allows you to prioritize the order in which dimensions are processed. You might want certain dimensions (such as Time) available sooner, for example, while other less-used dimensions are processed afterward.

Storage

There are two major sections here: StorageMode and ProactiveCaching. Recall the discussion about MOLAP and ROLAP; in Analysis Services you can specify the storage setting for each dimension and measure partition. The two settings here are MOLAP and ROLAP, reflecting the two major options for where dimensions are stored—either in the multidimensional repository or in the relational database.

Where is HOLAP? Well the "hybrid" part of the equation is where the proactive caching comes in. Remember that MOLAP stores the dimension data in the Analysis Services cube, while ROLAP stores the data in the relational store. However, with MOLAP we can also specify *proactive caching*. With proactive caching, SSAS rebuilds the dimension either when the source data changes, when signaled by a client application, or on scheduled intervals.

If you click the builder button [...] next to the ProactiveCaching setting, you'll open the Dimension Storage Settings dialog box (Figure 6-27). Here you can choose either one of the standard settings or a custom setting. The custom options give you some granular control over the cache, and then allow you to control how SSAS should be notified of changes in the underlying data.

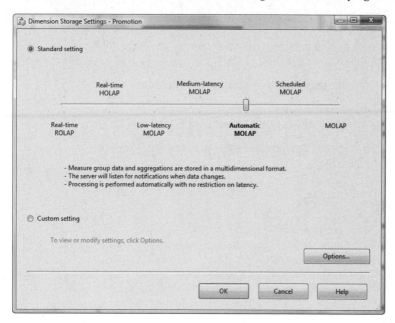

Figure 6-27. Setting dimension storage settings

The options for notification are as follows:

SQL Server: SSAS simply sets a trace on the necessary tables in SQL Server and tracks for data changes. You can select the Specify Tracking Tables check box and explicitly list the tables SSAS should track for changes (remember, all that will happen is that SSAS will rebuild the dimension if any changes to the data in the indicated tables are detected). If you don't specify the tables, Analysis Services will try to determine from the dimension structure which tables to track.

■ **Note** If you select the SQL Server option, the account that connects to the data sources tables must have ALTER TRACE permissions on the database.

Client Initiated: This is pretty close to regular MOLAP—a client calls for the processing to occur.

Scheduled Polling: Specify a polling interval, and SSAS will query the data source for changes; if the data has changed, SSAS will rebuild the dimension. You can also specify incremental updates if you need to process only part of a dimension based on data changes.

Let's take a look at how dimension storage works in Exercise 6-3.

Exercise 6-3. Specify Dimension Storage Modes

Follow the steps in this exercise to work through a detailed example showing how to specify dimension storage.

1. Open the SSAS AdventureWorks project. Double-click on the Promotions dimension to open that.

2. First let's check out the dimension as it currently works. Click the Browser tab.

3. Select the Promotion hierarchy, and then open the tree by clicking the [+] symbol next to the All member (Figure 6-28).

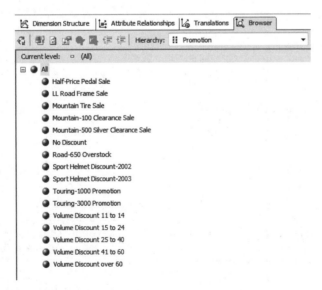

Figure 6-28. Viewing the promotion members in the dimension

4. We want to rename the Mountain Tire Sale to **Mountain Tire Sale 2009**, so let's go change the data in the underlying data table.

5. Open SQL Server Management Studio.

6. Connect to the server where the Adventure Works DW2008 database is located (Database Engine connection).

7. Open the AdventureWorks DW2008 database and then the Tables folder.

8. Right-click on the dbo.DimPromotion table and select Edit 250 Rows.

9. Find the promotion where the EnglishPromotionName is Mountain Tire Sale. Change the name to **Mountain Tire Sale 2009** (Figure 6-29).

Figure 6-29. Editing the Promotion

10. Click the down arrow to leave the row and write the record change to the database.

11. Now that we've changed the record, let's look at the dimension. Go back to BIDS. (Leave the edit table window open in SSMS—we'll come back to it.)

12. Refresh the dimension browser (Dimension menu ➤ Refresh). Note that the Mountain Tire Sale promotion hasn't changed.

13. Process the dimension (Dimension ➤ Process). Now refresh the browser view, and you should see the new promotion name.

14. Let's change the storage. Go back to the Dimension Structure tab.

15. In the left pane (Attributes), right-click on the Promotions at the top of the tree and select Properties (Figure 6-30).

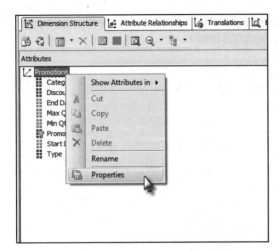

Figure 6-30. Opening the Properties pane for the dimension

16. In the Properties pane, under Storage, click the Off option next to ProactiveCaching.

17. Click the Builder button [...].

18. In the Dimension Storage Settings dialog box, select Custom Setting and then click the Options button.

19. At the top, under the Storage mode drop-down, select the check box for Enable Proactive Caching.

20. Select the check box labeled Update the Cache When Data Changes. Note the warning at the bottom (Figure 6-31).

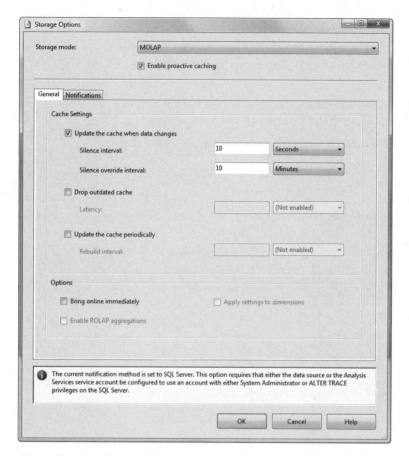

Figure 6-31. *Setting proactive caching*

21. Click the Notifications tab. Note that SQL Server is selected and Specify Tracking Tables is not selected. Click the OK button to return to the Dimension Storage Settings dialog box, and click the OK button there. Now we're going to go check our permissions.

    What we need is for the account that connects to the SQL Server database to have ALTER TRACE permissions. First let's look at the data source.

22. In BIDS, double-click on the `Adventure Works DW2008.ds` data source to open the designer.

23. Click the Impersonation Information tab.

24. The default option is Use the Service Account. This means SSAS will use the service account it's running under to connect to the data source. Click OK to close this dialog.

25. On the server running Analysis Services, open the Services applet:

   o Windows Server 2003: Right-click on My Computer, click Manage. In the Computer Management applet, open Services and Applications and click Services.

   o Windows Server 2008: Right-click on Computer and click Manage. In the Server Manager applet, open Configuration and then click Services.

26. Scroll down to find the SQL Server Analysis Services service. Right-click on it and select Properties.

27. Click the Log On tab (Figure 6-32).

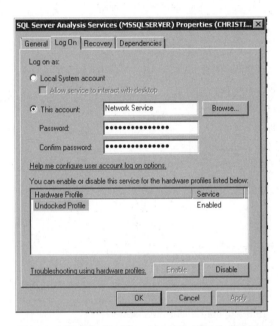

Figure 6-32. *The Log On properties for the SSAS service*

28. Note the account named; because the data source is using this account to access the data source, this is the account we have to provide for.

29. Close the dialog box and management applets.

30. Go back to SQL Server Management Studio.

31. Right-click on the server in the Object Explorer and select Properties.

32. In the Properties dialog box, select the Permissions page.

33. If you scroll down in the Logins or Roles pane, you should see NT AUTHORITY\NETWORK SERVICE (Figure 6-33).

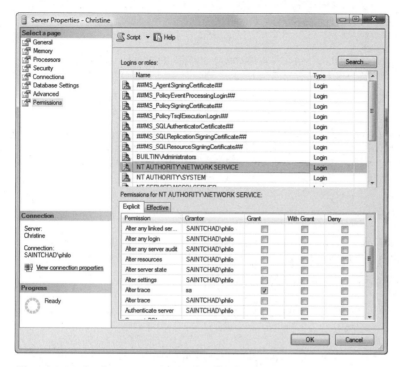

Figure 6-33. Setting properties on the database server

34. Select NT AUTHORITY\NETWORK SERVICE and then scroll down in the Permissions pane and select the Alter Trace check box.

35. Click the OK button.

36. Return to BIDS and process the dimension (with the new Automatic MOLAP storage setting). Click Yes to "Would you like to build and deploy the project first?"

37. Click Run when the Process Dimension dialog box shows up. After processing, click Close and then click Close again.

38. Click the Dimension Structure tab and then the Browser tab. You should get the warning at the bottom of the browser. Click Reconnect.

39. Let's edit the promotion again—back to SQL Server Management Studio.

40. Change LL Road Frame Sale to **All Road Frame Sale**. Remember to click the down arrow to write the change!

41. Back to BIDS.

42. In the Dimension menu, click Refresh.

43. Note that the promotion now indicates All Road Frame Sale and the members are re-sorted!

Our dimension will now automatically update with changes to the database.

UnknownMember

The UnknownMember is effectively a catchall where any measures that don't match members in this dimension will be assigned. You have the option as to whether to have an unknown member and whether it's visible. You can also give the member a specific name (for example, Unassigned Sales or Respondents Didn't Answer).

WriteEnabled

While most of the time OLAP cubes are used for read-only analysis, Analysis Services does have the ability to edit cubes and dimensions directly. However, only parent-child dimensions can be write-enabled.

We talked a lot about attributes earlier. Let's take a closer look at them.

Attributes

If dimension members are nouns, attributes are adjectives. Let's look back at our Promotion dimension (Figure 6-34). The dimension has attributes for the percentage discount, start and end date, maximum and minimum quantities, promotion category, type, and name. Each of these is amplifying information for each promotion in the dimension.

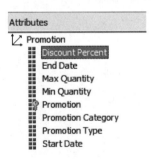

Figure 6-34. Attributes on the Promotion dimension

Attributes are often called *containers* for dimension members. I find the adjective concept easier to grasp. However, you can see how that approach works—take all the Promotion categories, and each one is something of a "bucket" for the promotions that have that category. However, not every attribute is really a "container"—for example, a Product dimension might have a Price attribute, but grouping by price ($12.95, $13.15, $14.75) wouldn't be that productive.

In addition to enabling drill-down or reporting, attributes are also useful for querying. For example, you can select for all blue bicycles, or use an MDX query like the following, to show sales by product for all products with a list price over $1,000.00. We'll dive into MDX in Chapter 9.

```
WITH
MEMBER Measures.[List Price] AS [Product].[List Price].CurrentMember.MemberValue

SELECT
{[Date].[Fiscal].[Fiscal Quarter].Members}
ON COLUMNS,

Filter([Product].[Product].Members, (Measures.[List Price]>1000))
ON ROWS

FROM [Adventure Works]
WHERE ([Measures].[Internet Sales Amount])
```

In the dimension structure, attributes are generally defined by dragging a field from the data source view over to the dimension (Figure 6-35).

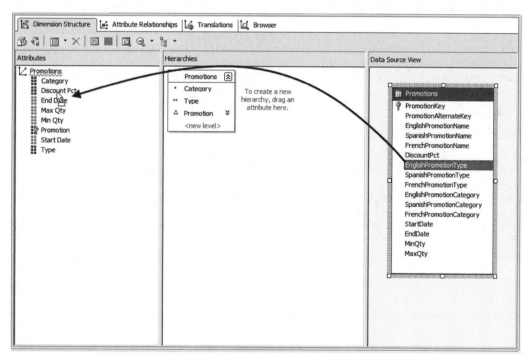

Figure 6-35. *Creating an attribute from the dimension data source view*

For the most part, a dimension consisting of members and attributes is similar to a database table with rows and columns. In fact, you can view the dimension with its attributes as a table: in the browser, selecting Member Properties from the Dimension menu will show all the attributes for the dimension in a table (Figure 6-36).

Current level: ▫ (All)	Category	Discount Pct	End Date	Max Qty	Min Qty	Start Date	Type
⊟ ● All							
● All Road Frame Sale	Reseller	0.35	8/15/2003 12:00:00 AM	0	0	7/1/2003 12:00:00 AM	Excess Inventory
● Half-Price Pedal Sale	Customer	0.5	9/15/2003 12:00:00 AM	0	0	8/15/2003 12:00:00 AM	Seasonal Discount
● Mountain Tire Sale 2009	Customer	0.5	8/30/2003 12:00:00 AM	0	0	6/15/2003 12:00:00 AM	Excess Inventory
● Mountain-100 Clearance Sale	Reseller	0.35	6/30/2002 12:00:00 AM	0	0	5/15/2002 12:00:00 AM	Discontinued Product
● Mountain-500 Silver Clearance Sale	Reseller	0.4	6/30/2004 12:00:00 AM	0	0	5/1/2004 12:00:00 AM	Discontinued Product
● No Discount	No Discount	0	12/31/2004 12:00:00 AM	0	0	6/1/2001 12:00:00 AM	No Discount
● Road-650 Overstock	Reseller	0.3	8/31/2002 12:00:00 AM	0	0	7/1/2002 12:00:00 AM	Excess Inventory
● Sport Helmet Discount-2002	Reseller	0.1	7/31/2002 12:00:00 AM	0	0	7/1/2002 12:00:00 AM	Seasonal Discount
● Sport Helmet Discount-2003	Reseller	0.15	7/31/2003 12:00:00 AM	0	0	7/1/2003 12:00:00 AM	Seasonal Discount
● Touring-1000 Promotion	Reseller	0.2	9/30/2003 12:00:00 AM	0	0	7/1/2003 12:00:00 AM	New Product
● Touring-3000 Promotion	Reseller	0.15	9/30/2003 12:00:00 AM	0	0	7/1/2003 12:00:00 AM	New Product
● Volume Discount 11 to 14	Reseller	0.02	6/30/2004 12:00:00 AM	14	11	7/1/2001 12:00:00 AM	Volume Discount
● Volume Discount 15 to 24	Reseller	0.05	6/30/2004 12:00:00 AM	24	15	7/1/2001 12:00:00 AM	Volume Discount
● Volume Discount 25 to 40	Reseller	0.1	6/30/2004 12:00:00 AM	40	25	7/1/2001 12:00:00 AM	Volume Discount
● Volume Discount 41 to 60	Reseller	0.15	6/30/2004 12:00:00 AM	60	41	7/1/2001 12:00:00 AM	Volume Discount
● Volume Discount over 60	Reseller	0.2	6/30/2004 12:00:00 AM	0	61	7/1/2001 12:00:00 AM	Volume Discount

Figure 6-36. *Showing all attribute values for a dimension*

To really get value out of the collection of attributes in a dimension, we want to build hierarchies and set attribute relationships. Because our users will generally use various attributes in different ways, defining attribute relationships aids the user in seeing relationships in the dimension structure, as well as optimizing performance by defining how attributes are related within the data and by business rule. Let's dig into attribute relationships.

Attribute Relationships

Every attribute in a dimension is related to the key attribute of the dimension, either directly or indirectly (via a relationship with another attribute). By default, when you create a dimension in BIDS, each attribute in the table with the key attribute (in a star schema this would be the only table in the dimension; in a snowflake schema it's the main dimension table) is directly related to the key attribute. The attribute in foreign-key tables bound to the foreign key is also directly related to the key attribute. Finally, attributes based on fields in the foreign-key tables are directly related to the foreign-key attribute (Figure 6-37).

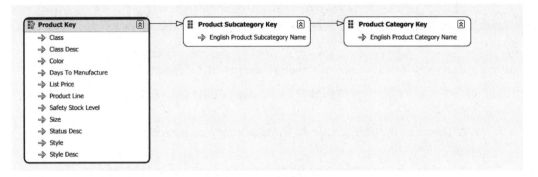

Figure 6-37. *Attribute relationships on a snowflake schema*

In Figure 6-37, we're looking at the default attribute relationships for a dimension built from the Product, Subcategory, and Category tables. Note that the English Product Subcategory Name and English Product Category Name are each related to the attributes bound to the foreign key in their table, which are related to the Product Key—the key attribute for the dimension.

Attribute relationships should always be built to represent *natural hierarchies*. A natural hierarchy is simply a one-to-many relationship. Be wary of creating "unnatural hierarchies" with attribute relationships, because these will significantly affect query performance and can even result in inaccurate results.

■ **Note** Also be sure to validate your data; even though a business rule may indicate that data is one-to-many, the actual data may violate the restriction. This is a potential problem with star schemas; with snowflakes, one-to-many relationships can be enforced with database constraints. If you have a star schema and all the attributes for a hierarchy are fields in a table, you could easily end up with many-to-many relationships in the data. So be sure to review your entire BI infrastructure and ensure that checks are in place to validate that the data will do what you think it does.

Building an attribute relationship in the dimension prompts Analysis Services to structure indexes to reflect that relationship. Another thing attribute relationships do is help SSAS understand which indexes *not* to build. For example, looking back at Figure 6-37, any relationship between category and product can be inferred from the relationships between product and subcategory, then subcategory and category. No explicit relationship needs to be built between product and category.

Attribute relationships can be rigid or flexible. A flexible relationship indicates that the attribute relationship can change over time. When an attribute relationship is flexible, aggregations under that relationship are dropped and recomputed during incremental updates. If the relationship is rigid, the relationships won't be recomputed. Exercise 6-4 presents an example of creating a dimension with attribute relationships.

■ **Caution** If the relationship between attributes with a rigid relationship actually changes, Analysis Services will throw an error during incremental processing.

Exercise 6-4. Create a Dimension with Attribute Relationships

The steps that follow walk you through the process of creating a dimension, and then of creating attribute relationships in that dimension. Let's create the Product dimension for our project.

1. Right-click on the Dimensions folder in the Solution Explorer and select New Dimension.

2. In the Dimension Wizard, if you see the Welcome to the Dimension Wizard message, click Next.

3. On the Select Creation Method page, select Use an Existing Table and click Next.

4. On the Specify Source Information page, select the Adventure Works DW2008 data source view, Product for the main table, and EnglishProductName for the Name column (Figure 6-38).

Figure 6-38. *Specifying the source information for the dimension*

5. Click the Next button.

6. The Related Tables page will detect that the Product Subcategory and Product Category tables are linked to the Product table. Leave them selected and click the Next button.

7. On the Select Dimension Attributes page, select attributes as in Table 6-2.

Table 6-2. Selecting Attributes

Attribute Name	Enable Browsing	Attribute Type
Product Key	Yes	Regular
Color	Yes	Regular
Safety Stock Level	No	Regular
Reorder Point	No	Regular
List Price	Yes	Regular
Size	Yes	Regular
Weight	Yes	Regular
Days To Manufacture	Yes	Regular
Product Line	Yes	Regular
Dealer Price	Yes	Regular
Class	Yes	Regular
Style	Yes	Regular
Model Name	Yes	Regular
Large Photo	No	Regular
English Description	No	Regular
Start Date	No	Slowly Changing Dimension—Date
End Date	No	Slowly Changing Dimension—End Date
Status	No	Slowly Changing Dimension—Status
Product Subcategory Key	Yes	Regular
Product Category Key	Yes	Regular

8. When you're finished, click the Next button.

9. Review the construction in the Completing the Wizard page, leave the name Product, and click Finish.

You should end up with a dimension that looks like Figure 6-39.

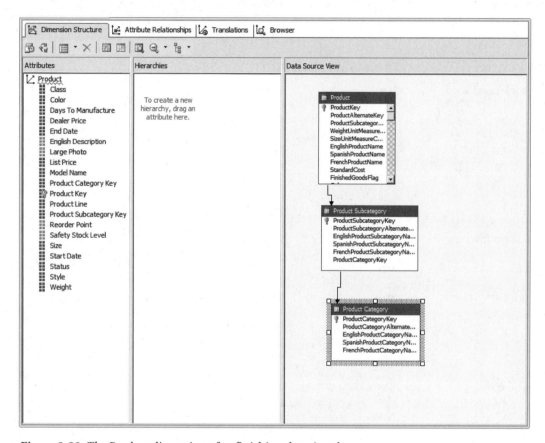

Figure 6-39. *The Product dimension after finishing the wizard*

10. We want to make the key values more descriptive and appropriate for end users. Click on the Product Key attribute.

11. Right-click on Product Key in the left-hand attributes pane, and select Rename.

12. Change the name to **Product**.

13. Check the properties for this attribute. The NameColumn property should be Product.EnglishProductName. This way, the user sees proper name values instead of key numbers in the list of Products.

14. Rename Product Category Key to **Category** and Product Subcategory Key to **Subcategory** the same way.

15. Change the NameColumn property for Category to **EnglishProductCategoryName**.

16. Change the NameColumn property for Subcategory to **EnglishProductSubcategoryName**.

17. Now let's create a hierarchy: drag the Category attribute to the Hierarchies pane. Then drag the Subcategory attribute to the <new level> area under it, and the Product attribute under that.

18. Click the hierarchy title bar and change the name to **Products** (Figure 6-40).

Figure 6-40. Products hierarchy

19. You should see a warning line (blue squiggle) under the Products title in the hierarchy. That's because there's no attribute relationship between the levels of the hierarchy. Click the Attribute Relationships tab.

20. Drag the Product attribute over the Subcategory attribute and drop it; then drag the Subcategory attribute over the Category attribute. The result should look like Figure 6-41.

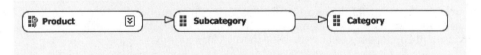

Figure 6-41. The Category hierarchy attribute relationship

21. Because a Product will always be in a given Subcategory, and a given Subcategory will always be in the same Category, those relationships can be rigid. Right-click on each relationship arrow, select Relationship Type, and then Rigid.

22. Note that the Attribute Relationships pane shows all relationships, most of which are to the key attribute of Product.

23. Click the Dimension Structure tab and then click the List Price attribute.

24. Change the properties for the attribute: DiscretizationBucketCount to 5, DiscretizationMethod to Automatic.

25. Process the dimension and then check it out in the browser. Note that the List Price hierarchy now consists of five ranges instead of a number of discrete values—that's the result of setting discretization buckets.

26. Look at the Products hierarchy and make sure you can navigate down categories, subcategories, and products.

27. Save the project; we'll keep using it.

Attribute Properties

Let's take a look at the properties for an attribute. Select an attribute and then view the Properties pane (right-click and then select Properties). Let's work down the Properties pane and look at some of the more significant properties. Here's the list:

AttributeHierarchyDisplayFolder: This is a free-text value. Any value you type in here will create a folder in the list of attribute hierarchies available (see Figure 6-42).

Figure 6-42. Creating display folders for grouping attributes

AttributeHierarchyEnabled, AttributeHierarchyVisible: The Enabled property indicates whether the attribute can be used at all for grouping measures. If an attribute is not visible but is enabled, it won't show up in client applications but can be used in MDX expressions.

DefaultMember: Indicates which member is selected when the dimension is selected.

OrderBy, OrderByAttribute: OrderBy indicates what to use to sort the attribute values (the attribute's key, name, or the key or name of another attribute). If you select AttributeKey or AttributeName, you will also have to enter the name of another attribute to use for sorting. We'll cover the properties in the "Parent-Child Dimensions" section that follows.

Parent-Child Dimensions

If you have to deal with employees in a company, or records in a file system, you're probably familiar with the concept of *recursive data structures*. Basically, in these structures, one of the fields annotates *parent* or *owner*, and the structure links to itself. Traversing a structure like this gives you a traditional tree or organizational chart type arrangement (see Figure 6-43).

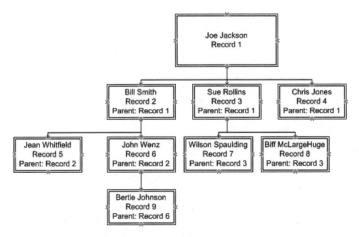

Figure 6-43. *A parent-child structure*

As you can tell, this kind of structure invites interesting challenges. However, Analysis Services handles it very easily. When you create a dimension that has a self-referring relationship, Analysis Services automatically creates a parent-child hierarchy in the dimension. That attribute is marked as a parent attribute type, and has an icon as seen in Figure 6-44.

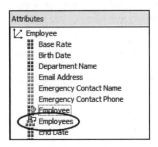

Figure 6-44. *A parent-child attribute in a dimension*

There are five specific properties relating to parent-child dimensions:

MembersWithData: Here you can select whether nonleaf data will be hidden or visible. The leaf members are the members that have no children (at the "bottom" of the hierarchy). Generally in an OLAP measure, all the values are at the leaf level; nonleaf members have values that are the result of aggregating the leaf values. For example, subcategory sales are the result of adding together the sales of all the products in the subcategory; nobody actually sells a subcategory. However, in parent-child dimensions, often nonleaf members will have values of their own. In a sales organization hierarchy, sales managers may also make sales. So how do you show both the rollup of the manager's subordinates and the manager's sales? If you indicate NonLeafDataVisible, SSAS deals with this by creating a "data member" that is a leaf member directly under the manager, containing the data for the manager.

MembersWithDataCaption: Here you can give a template for the data member used when you indicate that nonleaf data is visible in a hierarchy. You can type any string—an asterisk will be replaced with the name of the member above. See Figure 6-45.

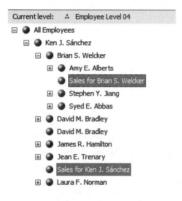

Figure 6-45. *Data members for nonleaf members in a parent-child hierarchy*

NamingTemplate: If you look back to Figure 6-43, note that there are four levels in the hierarchy. In a normal SSAS hierarchy, you would annotate each level by the attribute name (Categories, Subcategories, Products). However, in a parent-child hierarchy, there are no natural names—every level is Employees. So Analysis Services creates a dynamic name based on this template. You can enter multiple terms separated by semicolons. When you process the dimension, Analysis Services will assign the terms starting at the first level. When SSAS runs out of terms, it will keep using the last term but appending index numbers: 1, 2, 3, and so on.

RootMemberIf: If you think about the way this hierarchy is built, one thing to wonder is how to find where it starts. That's what this setting is about. You can establish the root member by setting its parent ID to itself, leaving the value blank, or leaving it undefined. With this setting, you can choose which method to use, or if all three are acceptable.

UnaryOperatorColumn: Remember our Account dimension? Each level had an indicator as to whether it should be added or subtracted to the aggregation. This property is where you indicate which column to use that holds those operators (see Figure 6-46).

AccountKey	ParentAccoun	AccountCodeA	ParentAccount	AccountDescription	AccountType	Operator
1		1		Balance Sheet		~
2	1	10	1	Assets	Assets	+
3	2	110	10	Current Assets	Assets	+
4	3	1110	110	Cash	Assets	+
5	3	1120	110	Receivables	Assets	+
6	5	1130	1120	Trade Receivables	Assets	+
7	5	1140	1120	Other Receivables	Assets	+
8	3	1150	110	Allowance for Bad Debt	Assets	+
9	3	1160	110	Inventory	Assets	+
10	9	1162	1160	Raw Materials	Assets	+
11	9	1164	1160	Work in Process	Assets	+

Figure 6-46. Native account table data showing the Operator column at far right

If you choose to use parent-child dimensions, be sure that your intended client applications handle them gracefully. You definitely want to keep them in the first round of requirements and test to verify that everything works as you expect. Now let's move on to our final special case for dimensions, and perhaps the most important—the Time dimension.

The Time Dimension

The Time dimension is a special case in OLAP technology. If you think about a calendar as a dimension, there are several facts about it that differentiate it from other dimensions:

- There are specific directions for *forward* (newer) and *back* (older).

- The members create a continuous range (1–31 January, 1–28 February, and so forth).

- When semiadditive measures calculate different dimensions in different ways, the Time dimension is usually the one dimension that is different from the others (inventory levels are added in other dimensions, and averaged in the Time dimension).

- Rolling averages over time have meaning (rolling averages by product do not).

- The concept of *to date* exists, which allows analysis of time periods not yet ended (year to date, quarter to date).

As a result, OLAP engines are designed to recognize Time dimensions and work with them. And SQL Server Analysis Services is no different. A basic Time dimension will be based on a table with date data— a date field, which will be a normal DateTime value, then a number of additional fields to enable Analysis Services to create hierarchies as well as perform various kinds of date math. The data source view table will then contain additional calculated fields to build out the full definition of a date model (see Figure 6-47).

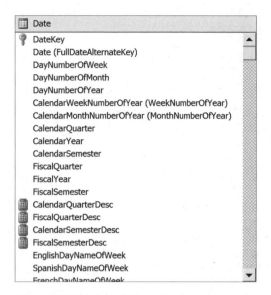

Figure 6-47. A date table in the data source view

Date dimensions can have multiple calendars. In addition to the standard calendar, you can also have the following:

Fiscal calendar: Many companies have financial reporting years that end on dates other than January 1. The fiscal calendar reflects the position in the fiscal year. For example, the US federal government has a fiscal year that starts on October 1. So October/November/December of 2008 is the first fiscal quarter of fiscal year 2009 (Q1FY09).

Reporting calendar: A reporting calendar follows a standard quarterly structure in which two months in the quarter have four weeks, and the other month has five. You can indicate which month has five weeks by selecting the month pattern (445, 454, or 544).

Manufacturing calendar: This calendar has thirteen "months" of four weeks each, divided into four quarters (three quarters with three months, one with four). You can set when the calendar starts as well as which quarter has the extra "month."

ISO 8601 calendar: This calendar follows the ISO standard calendar, which establishes a fixed number of seven-day weeks within a year, and establishes the start date of the year, based on the Gregorian calendar.

Being able to use and combine these calendars can provide a powerful new analytic capability to the end users, who may need to relate financial data to manufacturing data, or calendar data to reporting data.

You have several methods of creating time tables in BIDS:

Generate a time table in the data source: Using the new Dimension Wizard, this will create a Time dimension mapped to a data source view. If you generate the schema, it will also generate the table in the data source, optionally fill it with date data, and map it in the data source view.

Generate a time table on the server: This will generate the time table on the SSAS server, in the event you don't have permissions to create a table on the data source server.

Roll your own: Perhaps you already have a table populated with every date, or need to build your own for some other reason. It's possible, but painful.

■ **Note** There is no limit on the number of Time dimensions you can have in a cube.

The resulting structure of a Date dimension is shown in Figure 6-48.

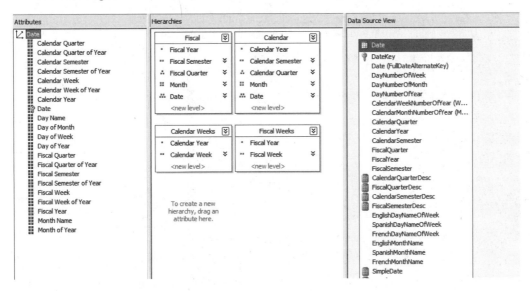

Figure 6-48. A Date dimension in the designer

The easiest way to understand dimensions is to create one, so let's embark on our last exercise. Exercise 6-5 shows how to create Date dimensions for use in analyzing time-based data.

Exercise 6-5. Create a Date Dimension

Follow the steps in this exercise to create a Date dimension. Such dimensions are useful in analyzing historical data that has been recorded over time.

1. Open the SSAS AdventureWorks project.

2. Right-click on the Dimensions folder and select New Dimension.

3. If you get the Welcome to the Dimension Wizard pane, click the Next button.

4. Select the option Generate a Time Table in the Data Source. Then click Next.

5. The calendar day options indicate how much of the dimension data source will be prepopulated with dates. The data in AdventureWorks goes from mid-2001 to mid-2004, so let's set the start date as 1/1/2001 and the end date as 12/31/2005.

■ **Note** As time goes on, you will need a process and mechanism in place to generate additional date data for the underlying table.

---

6. Leave the first day of the week as Sunday. In the time periods, select Year, Quarter, Month, Week, and Date. The screen should look like Figure 6-49.

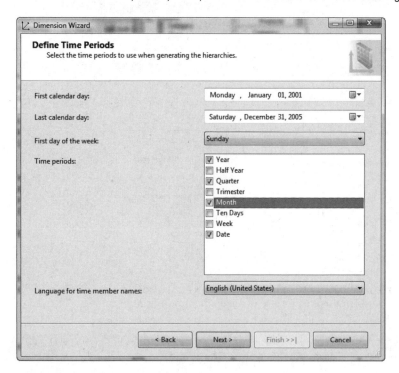

Figure 6-49. Defining the time periods

7. Click the Next button.

8. On the Select Calendars page, select the Regular Calendar and Fiscal Calendar check boxes. For the Fiscal Calendar, set the Start Day and Month to 1 October (see Figure 6-50).

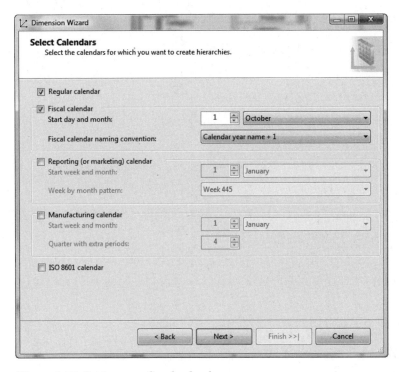

Figure 6-50. Setting up a fiscal calendar

9. Click the Next button.

10. Change the name to **Date**. Select the check box labeled Generate Schema Now near the bottom (Figure 6-51). This runs the wizard to create the data tables and data source view.

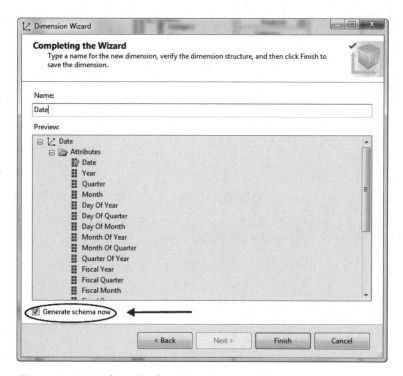

Figure 6-51. *Finishing the date wizard*

11. Click the Finish button.

12. The Schema Generation Wizard opens; if you're on the Welcome page, click the Next button.

13. On the Specify Target page, select Use Existing Data Source View and ensure that Adventure Works DW2008 is selected.

14. Click the Next button.

15. On the Subject Area Database Schema Options page, leave the defaults and click the Next button.

16. The Specify Naming Conventions page enables you to define naming conventions for the table that will be created in the database. Leave the defaults and click the Next button.

17. On the final page, click the Finish button.

18. Figure 6-52 shows the status page after a successful run. Note the warnings for the Promotion and Product dimensions. The wizard tries to generate a schema for every dimension in the database. Because those two dimensions are already bound to tables, they weren't touched.

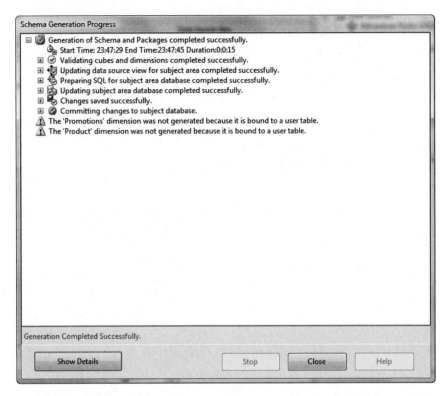

Figure 6-52. Finishing the schema generation wizard for the date table

19. Click the Close button.

Now you should see a Date dimension in your solution (Figure 6-53). We'll connect it up in the next chapter.

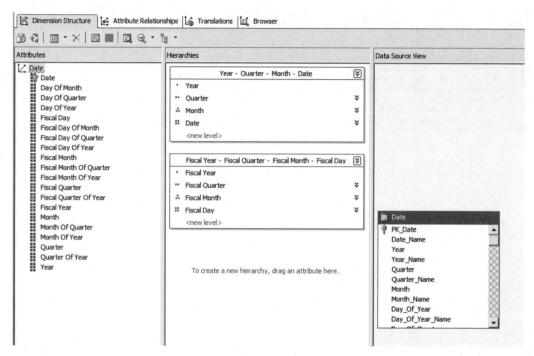

Figure 6-53. *The finished Date dimension*

> **20.** Save the solution.

Summary

So that's our tour of dimensions in SQL Server 2008 Analysis Services. Remember that defining your dimensional space is the key to developing a solid OLAP solution. Dimensions are like the tent poles, and in Chapter 7, we'll look at the fabric of the tent—that is to say, measures.

CHAPTER 7

■ ■ ■

Building a Cube

So we have a collection of dimensions, and a pile of data in our database. The question is—now what? To get value from our data and the work we've done so far, we need to assemble the dimensions into a cube. That's what we're going to do in this chapter. In addition to building a cube from our dimensions, we'll also look into calculated measures (deriving additional values from existing values), partitions (dividing data to ease management), and a deeper dive into aggregations and their design.

A cube in SQL Server Analysis Services is basically a collection of dimensions and measures (and some metadata.) Dimensions must be defined in the database before they can be used in the cube (which is why we did them first!). After we have a collection of dimensions, we can fill in the cube with fact data. After we have our fact data loaded, Analysis Services aggregates the data in accordance with the hierarchies in the dimensions.

Let's take a look at our "spreadsheet cube" from Chapter 2, shown again here in Figure 7-1. We have three *dimensions* shown: Product, Country, and Order Date. At various intersections we see dollar figures representing the sales corresponding to the dimension member values. For example, in 1997, $780 worth of Alice mutton was sold in Italy. Note that the numbers we're seeing most likely aren't the *leaf-level* values—they were probably aggregated from individual sales records.

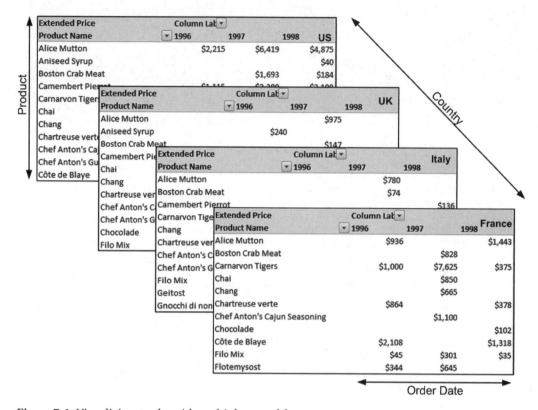

Figure 7-1. *Visualizing a cube with multiple spreadsheet pages*

For example, that $780 worth of Alice mutton may have been 30 orders at $26 each—we have no way of knowing from this representation of the cube. When you drill all the way down to the lowest member of each dimension, you're looking at the leaf-level values, which should be representative of the actual fact data the measures are built on.

Measures are fact data that's been combined in some way. This is the reason we're building cubes in the first place. Remember that the goal of OLAP is not to inspect rows and rows of individual records, but to review aggregated data. In this case, we don't want to see a list of 30 orders that each include Alice mutton; we want to know that $780 total was ordered from Italy in 1997. To get those totals, we want the data to be aggregated in some way. SSAS allows you to define how numeric data is aggregated in a number of ways (sum, count, maximum value, average, and so forth). By default, fact data will be summed, but a cube designer can select other methods of combining values.

In addition to actual numeric data, we often want values that are derived from existing numbers. For example, very often purchase orders will have the quantity ordered and the unit cost, but won't contain the extended price (total cost for the number of items ordered).

Another aspect to aggregations is choosing how much to calculate in advance. Consider a cube with dimensions for dates, a catalog of 5,000 products, 3,140 counties (or county equivalents—I'm looking at you, Louisiana and Alaska), and 15,000 customers. At the lowest level, there are 85,957,500,000,000 unique combinations…*per year!* We don't want Analysis Services calculating the totals for every single combination all the way up every hierarchy. The Aggregation Designer in BIDS lets us guide Analysis

Services as to how much to calculate in advance, with the remainder being figured out on the fly when a user makes a query.

We'll cover all these aspects of cubes in this chapter. Let's start with dimensions and how to build a cube.

Dimensions and Cubes

In Chapter 6, we created three dimensions: Promotions, Product, and Date. All of these are linked to the Reseller Sales table in the data source view, as shown in Figure 7-2 (with some tables excluded for clarity). The Reseller Sales table is going to be our fact table, the table containing all the detailed records we're interested in aggregating together. Product, Promotions, and Date are all dimension tables, effectively lookup tables for the ways we want to roll the reseller sales data together.

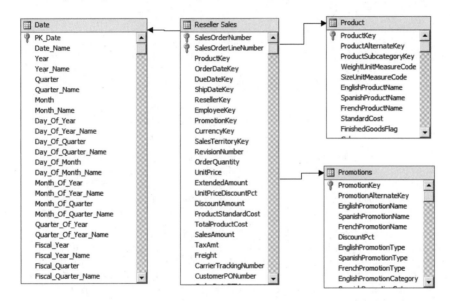

Figure 7-2. *Tables in the data source view*

The fields in our fact table consist of a primary key (SalesOrderNumber and SalesOrderLineNumber as a composite key), a collection of foreign keys, and our line-item detail data, most of which is numeric. When we build our cube, we'll create measures out of the numeric fields. Let's take a look at the mechanics of creating a cube.

■ **Note** Remember that measures will always be numeric. Even when you might want a measure similar to Return Reasons, what you'll really have is a dimension for the text of the reasons, and each measure value will be a count that's summed up. You'll look at this later in the chapter.

Creating Cubes

Creating a cube in BIDS is as simple as right-clicking on the Cubes folder in the Solution Explorer and selecting New Cube, as shown in Figure 7-3. This starts the Cube Wizard, which will walk you through the steps to create a new cube in the current database or database solution.

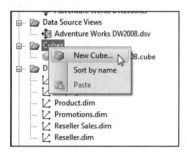

Figure 7-3. *Creating a new cube in BIDS*

When you start the wizard, you have three options for creating a new cube:

Use existing tables: This is the option you'll use most often, especially when you're starting out. This wizard will walk through selecting existing dimensions, generating new dimensions from existing tables, and creating measures from fact tables. We'll use this wizard in Exercise 7-1.

Create an empty cube: As indicated, this option simply creates a container. You can choose to bind it to a data source view or not. For more experienced administrators, you may end up using this option more often, as you use it to simply create a cube, and then add dimensions and measures manually. This is also the best choice if you have a cube for which all the dimensions are linked dimensions.

Generate tables in the data source: This wizard will walk you through creating skeleton dimensions and measures. When you finish the wizard, you'll have the option to generate the schema—creating the matching tables in a data source. This is a good way to start if you know what you want the cube to look like, and have to pull data together from multiple data sources. Design the cube, and use the wizard to create a matching staging database. Then you can design SSIS transformations to load data from the data sources into the staging database.

Using Measure Group Tables

If you're creating a cube from existing tables, the next step is to select the tables that contain measures. This page of the wizard is shown in Figure 7-4. Generally, you're interested only in tables that contain fact data—the numbers and details of the transactions you're trying to track. However, you might also want measures based on data in dimension tables. For example, our Product table has a field for Standard Cost. We may want to use that in some analysis on our data later, perhaps as part of a profit margin calculation. In that case, we would include the Product table as a measure group table so we could include the StandardCost field as a measure.

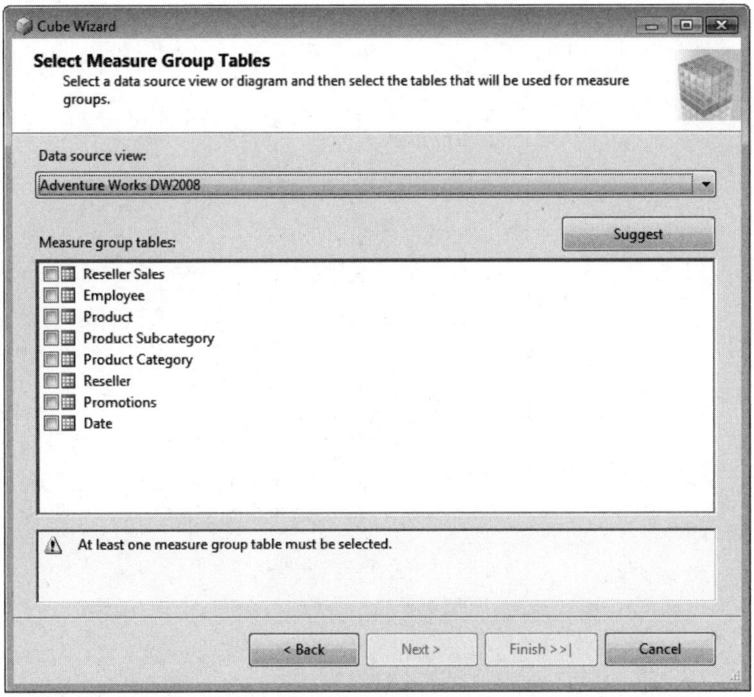

Figure 7-4. *Selecting measure group tables*

Also note the Suggest button—if you click this, BIDS will analyze the tables in the current data source view and suggest the tables that are likely to contain fact data of interest. After you've selected the tables you want to build measures from, the next page of the wizard presents the fields in those tables that are numeric. You can then select the fields you want to build measures from, as shown in Figure 7-5.

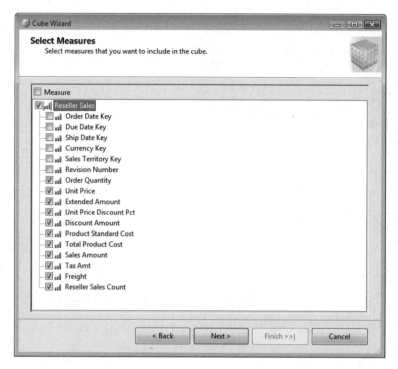

Figure 7-5. *Selecting measures*

Generally, the list of proposed measures will be overinclusive, especially because foreign keys are often numeric. Deselect any fields you don't want aggregated as measures—foreign keys, counters, document numbers, and so on. Note the last field in the list, Reseller Sales Count. The wizard always adds a measure that is simply a count of the number of records. Essentially this is a field whose value is 1 in every record. When you slice a cube by dimensions, the ones will be added for each record, giving a count of the number of sales corresponding with the selection. Figure 7-6 shows a breakdown of reseller sales by country and year for AdventureWorks.

Reseller Sales Count	Year				
Row Labels	CY 2001	CY 2002	CY 2003	CY 2004	Grand Total
Australia			3,009	1,939	4,948
Canada	2,404	12,727	18,801	7,829	41,761
France		2,680	7,715	3,953	14,348
Germany			4,480	2,900	7,380
United Kingdom		2,443	7,060	3,690	13,193
United States	8,431	40,391	59,107	24,819	132,748
Grand Total	**10,835**	**58,241**	**100,172**	**45,130**	**214,378**

Figure 7-6. *Spreadsheet showing counts of reseller orders*

Selecting Dimensions

The next step in the wizard is to select the dimensions in the solution that already exist, as shown in Figure 7-7. (Linked dimensions won't be displayed.) All the dimensions will be selected by default; you just need to deselect the dimensions you don't want included in the cube. After you've selected the dimensions, the final wizard page will confirm your configuration and then create the cube.

Figure 7-7. *Selecting dimensions*

When the wizard creates the cube, these dimensions will be included and the wizard will create associations for them with a "best guess" as to the type of association (see "Defining Dimension Usage" later in this chapter for more information). In Exercise 7-1, you'll use the Cube Wizard to create a cube for analyzing reseller sales data by using the dimensions we built in Chapter 6.

Exercise 7-1. Create a Cube

In this exercise, you're going to take the dimensions you created in Chapter 6 and use them to create a cube based on the data you have in the AdventureWorks relational database.

1. In the Solution Explorer (press Ctrl+Alt+L if it's not showing), right-click on the Cubes folder and select New Cube. This starts the Cube Wizard (see Figure 7-8).

Figure 7-8. *Starting the Cube Wizard*

2. Click the Next button.

3. On the Select Creation Method page, leave the default Use Existing Tables selected and click the Next button.

4. In the Select Measure Group Tables page, ensure that Adventure Works DW2008 is selected as the data source view (it should be, unless you've created another one on your own), and select Reseller Sales as the measure group table, as shown in Figure 7-9.

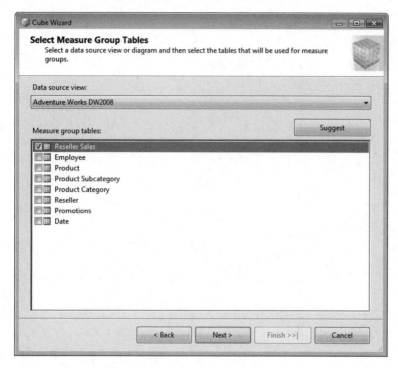

Figure 7-9. Selecting tables that contain measures

5. Click the Next button.

6. In the Select Measures page, all the fields will be selected by default: Deselect Order Date Key, Due Date Key, Ship Date Key, Currency Key, Sales Territory Key, and Revision Number, as shown in Figure 7-10.

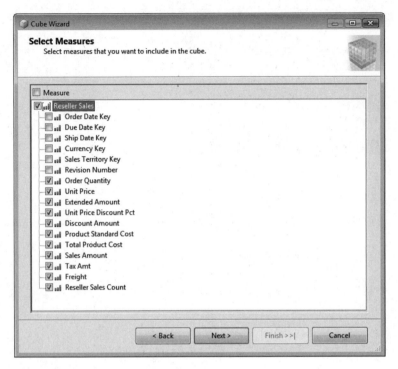

Figure 7-10. *Selecting measures for our cube*

7. Click the Next button.

8. On the Select Existing Dimensions page, ensure that Promotions, Date, and Product are selected, and then click the Next button.

9. On the Select New Dimensions page, leave all three dimensions and their hierarchies selected, as shown in Figure 7-11. Click the Next button.

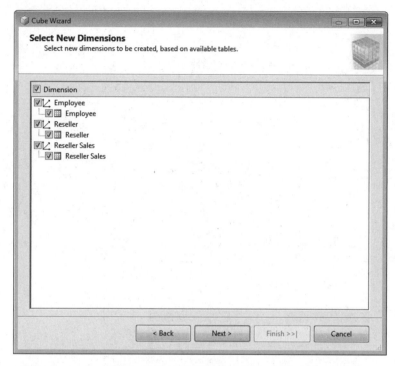

Figure 7-11. Creating new dimensions from existing linked tables

10. On the Completing the Wizard page, review the cube and click Finish. You should end up with a cube similar to what's shown in Figure 7-12. We'll use this to continue our discussion on cubes.

Figure 7-12. *A finished cube*

11. Save the solution.

Now that we have a cube, we know that we have a collection of fact data as measures, derived from the fields in the Reseller Sales table, and a number of dimensions. However, how does Analysis Services know how the measures and dimensions are related? We've seen some very complex schemas with dimensions related to fact tables through other dimensions, dimensions not related to fact tables, many-to-many dimensions, and so on. How does SSAS keep all this straight? Let's take a look at setting dimension usage.

Defining Dimension Usage

In our newly created cube, let's look at the Dimension Usage tab. Each dimension is assigned a relationship to each measure group in the cube. The usage for the cube we just created is shown in Figure 7-13. Each dimension is listed, linked to the Reseller Sales measure group that was created by the wizard.

Figure 7-13. *The Dimension Usage tab*

If you click the [-] button on a relationship, such as the Product relationship highlighted in Figure 7-13, you'll be presented with options for the relationship, as shown in Figure 7-14. The top section enables you to select the type of relationship between the dimension and the measure group; the lower section will change depending on the type of relationship selected.

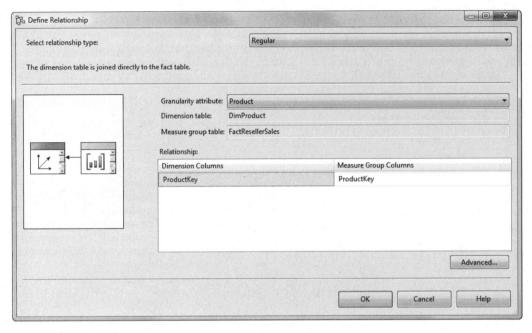

Figure 7-14. Defining a dimension relationship

The options available for relationship type are as follows:

No Relationship: There's no relationship between the dimension and the measure group. This is useful when you have many dimensions and many measure groups in a single cube, as in the AdventureWorks reference cube. Obviously, if there's no relationship, there won't be any settings to change.

Regular: This is a standard *one-to-many* reference relationship with a foreign key in the measure group and a primary key in the dimension. If you select this relationship type, you'll have to choose the attribute from the dimension that defines how the measure group is linked to it. (For example, perhaps we have a measure group regarding overhead costs for product categories; then in the Product dimension, you would select the Category granularity attribute.)

■ **Note** If you select a dimension attribute other than the key for the granularity attribute, you must make sure all the other attributes in the dimension are also related to your key so that the server can aggregate data properly.

Fact: This indicates that the dimension was created from the fact table it's related to. There are no settings here to change.

Referenced: A referenced relationship indicates an indirect relationship through another table. For example, if we created an explicit dimension for product subcategories and related that dimension to reseller sales, we would have to indicate the indirect relationship through the Products table.

Many-to-Many: If we wanted to define a relationship between customers and products, we would have to use a many-to-many relationship (one customer can be related to many products; one product can be related to many customers). To define the relationship, you'll need to select the intermediate measure group and then the tables necessary to make the connection.

Data Mining: A relationship necessary for a data-mining dimension. We'll cover this in more depth in Chapter 13.

This may all seem like a hassle, but it becomes very important to have this kind of control when we start looking at more-complex cubes, such as the AdventureWorks cube. A portion of the dimension usage table from AdventureWorks is shown in Figure 7-15. Note the dimensions that are marked as "no relation" with various measure groups. For example, the Sales Reasons group is associated with only the Sales Reason and Internet Sales Order Details dimensions.

Dimensions	Internet Sales	Internet Orders	Internet Customers	Sales Reasons	Reseller Sales
Date	Date	Date	Date		Date
Date (Ship Date)	Date	Date	Date		Date
Date (Delivery Date)	Date	Date	Date		Date
Customer	Customer	Customer	Customer		
Reseller					Reseller
Geography					Reseller
Employee					Employee
Promotion	Promotion	Promotion	Promotion		Promotion
Product	Product	Product	Product		Product
Sales Territory	Sales Territory Region	Sales Territory Region	Sales Territory Region		Sales Territory Region
Sales Reason	Sales Reasons	Sales Reasons	Sales Reasons	Sales Reason	
Internet Sales Order Details	Internet Sales Order	Internet Sales Order	Internet Sales Order	Internet Sales Order	
Reseller Sales Order Details					Reseller Sales Order
Sales Summary Order De…					
Source Currency	Source Currency Code	Source Currency Code	Source Currency Code		Source Currency Code
Destination Currency	Exchange Rates				Exchange Rates
Sales Channel					
Organization					
Department					

Figure 7-15. The dimension usage table for the Adventure Works cube

Having these measure groups and dimensions in a single place can make both development and maintenance easier, and it also provides end users an ability to combine data in a single report (for example, Internet Sales vs. Reseller Sales). If these measure groups were in separate cubes, many tools couldn't handle mapping to them both. We've talked about measure groups a lot. Let's dig more deeply into them.

Measures and Measure Groups

We've covered the concept of measures quite a few times; measures are the numbers our users are after to analyze. They're numbers. In OLAP, they're generally aggregated in some way so that we can look at various breakdowns of values. In BIDS, you can find all the measures and measure groups in a cube on the left side of the Cube Structure tab, as shown in Figure 7-16.

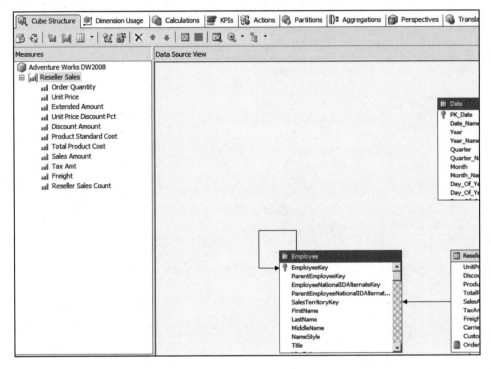

Figure 7-16. The Measures pane in BIDS

Measures

Measures are our numbers. Dimensions are around the edge; measures are in the middle, and where we're focused. Measures consist of our transactional data, and the aggregated values as we roll them up by dimension. A measure generally corresponds to a single column of data in a fact table (the table itself relates to a measure group, covered in a few pages). Take a look at Figure 7-17, showing sales data by country and territory across the top, and product categories and product lines down the left.

Row Labels	Australia		Australia Total	Canada			Canada Total	Grand Total
	Queensland	Victoria		Alberta	Ontario	Quebec		
⊟ Accessories		$9,233.09	$9,233.09	$13,411.50	$57,047.88	$15,313.69	$85,773.08	$95,006.17
⊞ Accessory		$9,233.09	$9,233.09	$13,411.50	$57,047.88	$15,313.69	$85,773.08	$95,006.17
⊟ Bikes	$35,086.45	$314,875.54	$349,961.99	$1,062,605.27	$4,956,175.27	$2,483,295.86	$8,502,076.41	$8,852,038.40
⊞ Mountain				$982,769.05	$1,841,210.42	$948,025.49	$3,772,004.96	$3,772,004.96
⊞ Road				$71,703.09	$2,940,401.67	$1,150,792.63	$4,162,897.40	$4,162,897.40
⊞ Touring	$35,086.45	$314,875.54	$349,961.99	$8,133.13	$174,563.18	$384,477.74	$567,174.04	$917,136.04
⊟ Clothing		$12,223.55	$12,223.55	$45,792.47	$185,127.53	$52,469.44	$283,389.45	$295,613.00
⊞ Accessory		$12,223.55	$12,223.55	$31,214.06	$152,371.57	$33,035.10	$216,620.72	$228,844.28
⊞ Mountain				$14,196.14	$31,423.78	$18,824.82	$64,444.74	$64,444.74
⊞ Road				$382.28	$1,332.18	$609.52	$2,323.98	$2,323.98
⊟ Components	$3,985.03	$18,058.65	$22,043.68	$274,217.17	$980,837.27	$440,786.33	$1,695,840.77	$1,717,884.45
⊞ Components	$973.30	$7,742.12	$8,715.41	$19,703.28	$49,619.33	$21,514.19	$90,836.80	$99,552.21
⊞ Mountain				$234,130.56	$455,327.85	$222,567.39	$912,025.81	$912,025.81
⊞ Road				$16,369.16	$461,736.58	$124,825.67	$602,931.41	$602,931.41
⊞ Touring	$3,011.73	$10,316.53	$13,328.26	$4,014.18	$14,153.50	$71,879.07	$90,046.75	$103,375.01
Grand Total	$39,071.48	$354,390.84	$393,462.31	$1,396,026.42	$6,179,187.96	$2,991,865.33	$10,567,079.70	$10,960,542.02

Figure 7-17. Facts and aggregations

Let's say that our fact data is reported at the territory level by product line—the numbers on a white background. Those are our facts at the leaf level. They are summed by Analysis Services to produce the totals by country and by product category (gray background), and the grand totals by country and by product category (white numbers on a dark gray background). The $10 million figure in the lower right is the grand total—the summation of every fact in the cube, also the result of the (All) member on every dimension.

Native measures are generally the result of a single field; however, more-complex measures can be generated by creating a calculated measure. For example, if a record contains fields for unit cost and quantity ordered, you could have the subtotal calculated by creating a measure for unit cost × quantity. You'll take a closer look at calculated measures later in the chapter.

One of the big things we want from Analysis Services is combining numerical data. Although we generally think about simply adding the numbers together, there are other ways to combine them. Selecting a measure in the Measures pane gives us access to the properties for the measure, the first of which is AggregateFunction. The only truly additive functions (aggregated the same way along every dimension) are Sum and Count. No matter which way you slice the data, these will return the total (either adding the values or counting the records) of the child members.

There are two nonadditive aggregations: DistinctCount and None. These do not aggregate numbers. None simply performs no aggregation—if a combination of dimension members doesn't return a distinct value from the fact table, a null is returned. DistinctCount is unique in that its value must be calculated from the selection of dimension members every time—there aren't subtotals to "roll up" because distinct values may collide.

The other aggregate functions are all semiadditive; they can be aggregated along some dimensions, but not others. As mentioned previously, an inventory measure can be summed along a geographic dimension (adding inventory stock levels for different locations), but not along the Time dimension (you can't take values of 15 widgets in the warehouse in July, 20 widgets in August, and 25 widgets in September, and add them to get a result of 60 widgets—the value in September should be 25). The aggregate functions of Min/Max, ByAccount, AverageOfChildren, First/LastChild, and FirstNonEmpty are all semiadditive aggregation functions.

The DataType property is generally set to Inherited, where it will pick up the data type of the underlying measure field. You can, however, set it specifically to a data type. The data type must match the data type of the fact table field, with the exception that if the aggregate function is set to count or distinct count, the data type will reflect the integer nature of the count value.

DisplayFolder gives you a way to organize measures when you have a large number of them. This is a free-form text field. Any value you enter here will generate a folder in the measure group with the measure inside it. Folders with the same name will be combined in a measure group. See Figure 7-18 for an example.

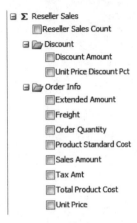

Figure 7-18. *Display folders for organizing measures*

If you need to perform a calculation to create the value of the fact data at the leaf level, you can use MDX in MeasureExpression to be evaluated before any data is aggregated. Consider a quota system in which selling widgets to customers in certain states gets an additional 10 percent bonus credit for the salesperson. To calculate the credit correctly, you have to evaluate the sale at the record (that this specific product was sold to one of those specific customers). Then the values can be added normally.

FormatString is very important . The format code you place here governs how the value is rendered by client applications (Excel, Reporting Services, and so forth). If you drop down the selector, you can see preformatted format strings for numeric values, currency, and dates. However, this is a free-form field, and you can enter any format string by using standard Microsoft formatting conventions.

That sums up the fundamentals of measures. Of course, there's more we can do with measures— we'll look at calculated measures later in the chapter, and KPIs and actions in Chapter 11. We'll dig into partitions and aggregations in Chapter 12. For now, let's take a look at how we deal with groups of measures.

Measure Groups

A *measure group* is a collection of measures. (Sorry.) Specifically, they are the OLAP representation of the underlying fact table that contains the data we're reporting on. In addition, they also represent the aggregations that result from rolling up our measure data. A measure group represents all the measures from a single fact table. The measures within are based on the fields in that table or calculated view in the data source view. If you create a new measure group (in BIDS, Cube menu → New Measure Group), you'll be asked which table to base it on. Similarly, if you create a new measure, the New Measure dialog box will prompt you for a source table and then offer the fields in that table to select from. The table you select will dictate which measure group will contain the new measure.

As you've already seen, measure groups are the containers used to associate dimensions in a cube. BIDS also uses measure groups as a starting point for partitions and aggregations. By default, partitions are set by measure group. However, they can be further divided by using a query to split the data in a measure group (for example, breaking down sales measure data by year). Aggregations are set by partition, so of course by default they'll also be set by measure group. We'll look at partitions and aggregations later in this chapter.

In general, measure groups are just containers. Let's take a look at some of the properties of a measure group and how we can use them, and then we'll dig into measures themselves. The properties for a measure group are broken down into the incredibly descriptive groups of Basic, Advanced, Configurable, Storage, and Misc. Well, let's not pay any attention to the groupings and just walk through them:

AggregationPrefix: This property is a leftover from the SQL Server 2000 days, when it was used to set a prefix on tables created for aggregations. It's deprecated and will probably be gone in the next version of SQL Server.

ErrorConfiguration: Here you can set specific responses for errors that occur when processing the cube. The configuration set here will be used as the error configuration by any new partitions created for the measure group.

EstimatedRows: You can enter the estimated number of rows per partition here for predictive calculations. (This number will also be used as the default on new partitions.)

IgnoreUnrelatedDimensions: When this is set to true, any dimensions that are not associated with the measure group will be forced to the All member when they are included in a query against the cube. For an unrelated dimension, the measure group will have the same value for every member, but forcing it back to All makes it clear to the end user what's happening.

ProcessingMode: Another property that's actually a default setting for new partitions. In this case, the options are Regular or LazyAggregations. The Regular setting means that data for the cube won't be available for processing until all the aggregations have been calculated. With a setting of LazyAggregations, users can query the cube after the data is loaded and before all the aggregations are complete.

Type: Similar to dimension types, there are several options here that help enable special business intelligence features in Analysis Services.

StorageLocation: This determines where the cache files for the measure group will be stored. If you click the selector button [...] you'll have a list of locations specified on the server where you can place the storage files.

DataAggregation: This setting dictates how data can be aggregated for the measure group.

ProactiveCaching: Setting up caching here will set the default proactive caching setting for any partitions created from the measure group. The dialog should look familiar from when we worked on proactive caching with dimensions in Chapter 6.

StorageMode: Operates the same as ProactiveCaching, explained in the preceding list item.

One final aspect to measures we want to look at are calculated measures—how we can derive values from existing values in a fact table.

Calculated Measures

Calculated measures are actually a subset of "calculated dimension members." However, we want to focus on calculated measures, because this is where you'll do most of your calculating.

We'll start by looking at the Calculations tab of the cube designer, as shown in Figure 7-19. The organizer is at the top left, calculation tools are in the lower left, and the calculation designer is the right-hand pane.

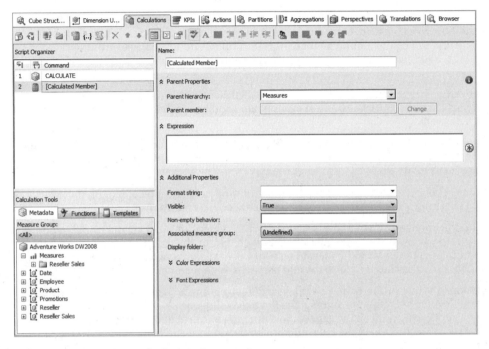

Figure 7-19. Designing calculations in our cube

The script organizer lists all the script sections in the current cube. Script sections can be calculations, named sets, or script commands. There is one script command that is there by default, and that's the CALCULATE statement. This is the command that instructs the Analysis Services server to generate all the aggregations for the cube.

■ **Warning** You can end up in an interesting place by accident as a result of the CALCULATE statement. If you create a new calculation and fill in a few fields, and then switch to the script view, you won't be able to switch back to the designer, because there will be a syntax error in the script as a result of the unfinished fields. You may get frustrated and just try a Ctrl+Alt+Delete. The next time you process the cube, it will run fine (no errors), but you'll have no aggregated values in your cube (the CALCULATE statement didn't run). The way to fix this is to go back to the Calculations tab and enter CALCULATE for the script command, and then reprocess.

You can create a new calculated member by choosing New Calculated Member from the Cube menu. This will create a new calculation and open the designer. The calculation will have a default name and be set to the Measures hierarchy. Note that you also have a Format string, and can set the measure group and a display folder.

The Expression box seems small, but it will automatically grow as you type. Expressions must be well-formed MDX. We'll dig into MDX in depth in Chapter 9, but we'll look at some lightweight examples here. Note that the Calculation Tools section in the lower left has the cube structure available. You can drag and drop measures, dimensions, and hierarchies from here. There's also a Functions tab, which lists all the MDX functions you may need (and then some!).

The simplest type of calculated member we've referred to previously is figuring out the total for a line item from the quantity and unit cost. This is some very simple math; we can open the Reseller Sales folder in the Metadata tab, and drag Order Quantity to the Expression box, type an asterisk (*), and then drag over Unit Price, which gives us this:

```
[Measures].[Order Quantity]*[Measures].[Unit Price]
```

Note that BIDS inserted the parent hierarchy name ([Measures]) for us. So if we name the calculation [Line Item Total] (standard SQL syntax—you must use square brackets around item names that have spaces), and set the Format string to Currency, we can process the cube and see results similar to Figure 7-20.

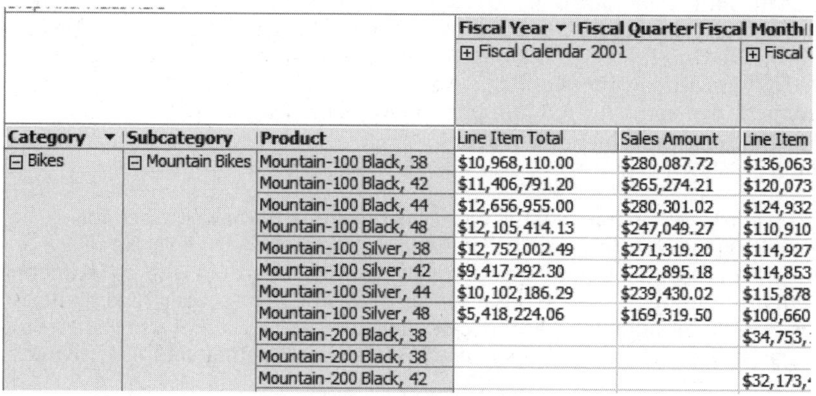

			Fiscal Year ▾	Fiscal Quarter	Fiscal Month
			⊞ Fiscal Calendar 2001		⊞ Fiscal (
Category ▾	Subcategory	Product	Line Item Total	Sales Amount	Line Item
⊟ Bikes	⊟ Mountain Bikes	Mountain-100 Black, 38	$10,968,110.00	$280,087.72	$136,063
		Mountain-100 Black, 42	$11,406,791.20	$265,274.21	$120,073
		Mountain-100 Black, 44	$12,656,955.00	$280,301.02	$124,932
		Mountain-100 Black, 48	$12,105,414.13	$247,049.27	$110,910
		Mountain-100 Silver, 38	$12,752,002.49	$271,319.20	$114,927
		Mountain-100 Silver, 42	$9,417,292.30	$222,895.18	$114,853
		Mountain-100 Silver, 44	$10,102,186.29	$239,430.02	$115,878
		Mountain-100 Silver, 48	$5,418,224.06	$169,319.50	$100,660
		Mountain-200 Black, 38			$34,753,
		Mountain-200 Black, 38			
		Mountain-200 Black, 42			$32,173,

Figure 7-20. Calculating the line-item total

Hold on, that can't be right—why are the totals from our calculated measure so much higher than the total sales amount? Let's take a look at the order quantity and unit price, shown in Figure 7-21.

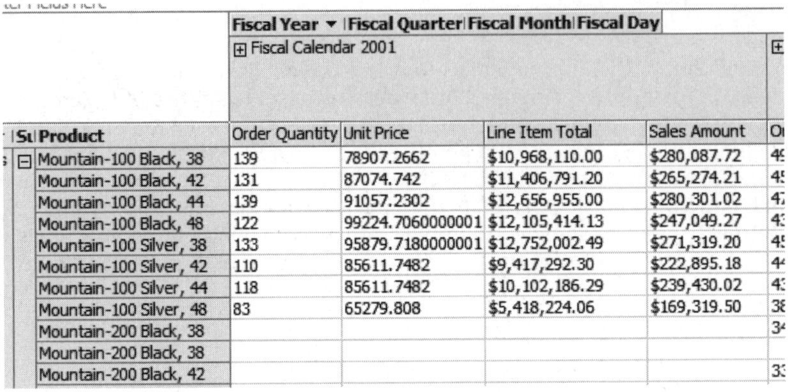

SET FIELDS HERE

	Product	Fiscal Year ▾	Fiscal Quarter	Fiscal Month	Fiscal Day	
		⊞ Fiscal Calendar 2001				⊞
		Order Quantity	Unit Price	Line Item Total	Sales Amount	O
⊟	Mountain-100 Black, 38	139	78907.2662	$10,968,110.00	$280,087.72	49
	Mountain-100 Black, 42	131	87074.742	$11,406,791.20	$265,274.21	45
	Mountain-100 Black, 44	139	91057.2302	$12,656,955.00	$280,301.02	47
	Mountain-100 Black, 48	122	99224.7060000001	$12,105,414.13	$247,049.27	43
	Mountain-100 Silver, 38	133	95879.7180000001	$12,752,002.49	$271,319.20	45
	Mountain-100 Silver, 42	110	85611.7482	$9,417,292.30	$222,895.18	44
	Mountain-100 Silver, 44	118	85611.7482	$10,102,186.29	$239,430.02	43
	Mountain-100 Silver, 48	83	65279.808	$5,418,224.06	$169,319.50	38
	Mountain-200 Black, 38					34
	Mountain-200 Black, 38					
	Mountain-200 Black, 42					33

Figure 7-21. Adding additional data

If we multiply the Order Quantity shown here by the Unit Price, we can see that it comes out to the Line Item Total. So it looks like the Line Item Total is being calculated at whatever level we're at, and that doesn't make sense for this type of calculation. (You can't multiply the total number of items bought for a period of time by the total of all the Unit Prices—remember we left Unit Price as a sum.) This value should probably be calculated as a measure expression.

Instead, let's try calculating the average order amount per sale. We'll use this formula:

```
[Measures].[Extended Amount]/[Measures].[Reseller Sales Count]
```

This type of calculation will work well no matter what type of aggregation we have. In fact, for averages we actually want to calculate them considering all child data, no matter what level we're at. Consider two classrooms, one with 100 students, and the other with 10. The classroom with 100 students scores an average of 95 percent on an exam, while the classroom with 10 students scores a 75 percent. To figure the average score across the whole student body, you can't just average 75 percent and 95 percent to get 85 percent; you have to go back to the original data and add 110 scores together, and then divide by 110 (the answer is 93 percent).

And this is a beautiful example of what makes Analysis Services so powerful. When we calculate an average based on two measures, it will produce the total of the first divided by the total of the second, based on the measures used and members selected. If we deploy, process, and check the browser, we'll see the numbers in Figure 7-22.

			Fiscal Year ▾	Fiscal Quarter	Fiscal Month	Fiscal Day	
			⊞ Fiscal Calendar 2001		⊞ Fiscal Calendar 2002		
Category ▾	Subcategory	Pr	Sales Amount	Average Order Amount	Sales Amount	Average Order Amount	
⊟ Bikes	⊞ Mountain Bikes		$1,975,676.13	$5,833.75	$9,401,805.19	$5,304.42	
	⊞ Road Bikes		$945,805.16	$1,533.08	$9,333,541.15	$2,087.37	
	⊞ Touring Bikes						
	Total		$2,921,481.29	$3,058.11	$18,735,346.34	$3,007.91	
⊞ Components			$248,561.74	$735.39	$2,853,256.99	$744.61	
⊟ Clothing	⊞ Bib-Shorts				$66,564.89	$241.37	
	⊞ Caps		$1,168.60	$18.00	$7,844.82	$22.61	
	⊞ Gloves				$51,170.46	$122.46	
	⊞ Jerseys		$12,384.66	$82.61	$90,705.96	$110.74	
	⊞ Shorts				$26,127.43	$160.78	
	⊞ Socks		$1,498.79	$28.72	$5,074.60	$28.71	
	⊞ Tights				$66,320.47	$238.75	
	⊞ Vests						
	Total		$15,052.05	$56.28	$313,808.63	$126.64	
⊞ Accessories			$8,538.89	$50.83	$72,635.28	$75.28	
Grand Total			$3,193,633.97	$1,847.25	$21,975,047.24	$1,628.96	

Figure 7-22. *Calculating average sales amount*

We can have more-complex calculations as well. This calculation in the AdventureWorks cube is as follows:

```
Case
    When IsEmpty( [Measures].[Reseller Sales-Sales Amount] )
    Then 0
    Else ( [Product].[Product Categories].CurrentMember,
           [Measures].[Reseller Sales-Sales Amount]) /
         ( [Product].[Product Categories].[(All)].[All],
           [Measures].[Reseller Sales-Sales Amount] )
End
```

This will return for any cell associated with a product the ratio of sales for that product (or group of products) as compared to the sales for all products. You can see what this would look like in Figure 7-23. Note the CASE statement—if there is no measure from the reseller sales amount, this calculation returns a zero (for example, if you're measuring Internet sales). This highlights that calculations run against all measures in a cube, so if you're creating a calculation that is specific to a measure, you will need to exclude it from other measures.

After we're sure we're in our Reseller Sales measure, we want to calculate the sales for the currently selected group against all sales. The first half is an MDX expression indicating to use the currently selected member, while the second half indicates using all members (the grand total). Note also how the percentages don't break down across geography—only across the product hierarchy. However, every individual product, subcategory, and category is compared to the total.

Reseller Ratio to All Products	Column Labels ⌄							
	⊟ Australia		Australia Total	⊟ Canada			Canada Total	Grand Total
Row Labels ⌄	⊞ Queensland	⊞ Victoria		⊞ Alberta	⊞ Ontario	⊞ Quebec		
⊟ Accessories		2.61%	2.35%	0.96%	0.92%	0.51%	0.81%	0.87%
⊞ Bike Racks		1.38%	1.24%	0.34%	0.26%	0.22%	0.26%	0.29%
⊞ Bottles and Cages		0.03%	0.03%	0.01%	0.01%	0.01%	0.01%	0.01%
⊞ Cleaners		0.04%	0.04%	0.02%	0.01%	0.01%	0.01%	0.01%
⊞ Helmets		0.70%	0.63%	0.40%	0.49%	0.22%	0.40%	0.41%
⊞ Hydration Packs		0.45%	0.40%	0.15%	0.09%	0.04%	0.08%	0.09%
⊞ Locks				0.02%	0.04%	0.01%	0.03%	0.03%
⊞ Pumps				0.02%	0.03%	0.01%	0.02%	0.02%
⊞ Tires and Tubes				0.00%	0.00%	0.00%	0.00%	0.00%
⊞ Bikes	89.80%	88.85%	88.94%	76.12%	80.21%	83.00%	80.46%	80.76%
⊞ Clothing		3.45%	3.11%	3.28%	3.00%	1.75%	2.68%	2.70%
⊞ Components	10.20%	5.10%	5.60%	19.64%	15.87%	14.73%	16.05%	15.67%
Grand Total	100.00%	100.00%	100.00%	100.00%	100.00%	100.00%	100.00%	100.00%

Figure 7-23. Percentage of sales by product

If we just wanted to see products compared to other products in their subcategory, and subcategories compared to others in their category, we could simply change the [All] to .Parent. We'll dig into MDX more in Chapter 9.

In Exercise 7-2, let's build a calculated measure just so you can be sure you know what you're doing.

Exercise 7-2. Create a Calculated Measure

In this exercise, you'll create a calculated measure in our SSAS AdventureWorks cube and set some properties to take a look at how it all fits together.

1. Open the SSAS AdventureWorks project if you don't already have it open.

2. Double-click on the Adventure Works DW2008.cube to open it.

3. Click the Calculations tab.

4. From the Cube menu, select New Calculated Member to create a new calculation and open the designer.

5. Name the calculation **[Average Order Amount]** (you must include the square brackets).

6. The Parent hierarchy should already be set to Measures.

7. For the Expression, click the Metadata tab in the Calculation Tools section in the lower left. Open the measures, then Reseller Sales. Drag Extended Amount to the Expression window (Figure 7-24).

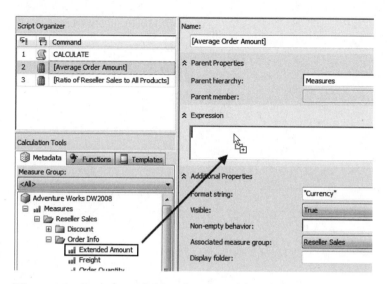

Figure 7-24. *Dragging a field to the Expression box*

8. Type a slash (/) after [Measures].[Extended Amount]. Then drag Reseller Sales Count over after the slash. The Expression window should look like Figure 7-25.

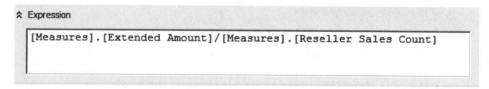

Figure 7-25. *The calculation expression*

9. Change the Format string to **$#,##0.00;($#,##0.00)**.

■ **Tip** Don't use *Currency*—Excel doesn't seem to recognize it.

---

10. Visible should be True, and Associated Measure Group should be Reseller Sales.

11. Open up the Color Expressions section by clicking the two down-arrows next to the header.

12. Set the Fore Color text to the following:

```
IIF([Measures].[Average Order Amount]>500, 16711680, 255)
```

This translates to "If this value is greater than 500, set the text color to blue (16711680); otherwise set it to red (255)." The numbers are populated if you use the color picker button.

The final Calculation page should look like Figure 7-26.

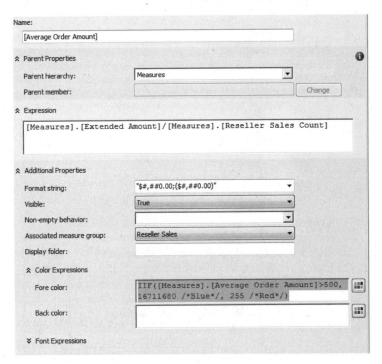

Figure 7-26. The calculated measure

13. Deploy and process the cube. Then view the measure either in the BIDS browser or Excel. Both will properly render the formatting and font colors, as shown in Figure 7-27.

Category ▼	Subcategory ▼	Pr	Fiscal Year ▼ Fiscal Quarter Fiscal Month Fiscal Day		
			⊞ Fiscal Calendar 2001	⊞ Fiscal Calendar 2002	⊞ Fiscal Calendar 2003
			Average Order Amount	Average Order Amount	Average Order Amount
⊟ Bikes	⊞ Mountain Bikes		$5,833.75	$5,304.42	$3,122.45
	⊞ Road Bikes		$1,533.08	$2,087.37	$2,443.70
	⊞ Touring Bikes				$2,424.13
	Total		$3,058.11	$3,007.91	$2,655.83
⊞ Components			$735.39	$744.61	$649.51
⊟ Clothing	⊞ Bib-Shorts			$241.37	$209.45
	⊞ Caps		$18.00	$22.61	$29.21
	⊞ Gloves			$122.46	$98.87
	⊞ Jerseys		$82.61	$110.74	$164.72
	⊞ Shorts			$160.78	$209.68
	⊞ Socks		$28.72	$28.71	$33.22
	⊞ Tights			$238.75	$197.63
	⊞ Vests				$283.49
	Total		$56.28	$126.64	$152.98
⊞ Accessories			$50.83	$75.28	$111.25
Grand Total			$1,847.25	$1,628.96	$1,234.67

Figure 7-27. *Using the calculated measure*

14. Save the solution.

Summary

For the most important part of our cubes, it may seem like this went more quickly than the previous chapter. As I mentioned, generally most of the work is in creating the dimensions. After that's done, generating the cube can go fairly quickly. However, we still have more work to do. Although you've deployed things a few times, you really don't have a strong grasp of what *deploy* and *process* really mean. And that's what Chapter 8 is about.

CHAPTER 8

■ ■ ■

Deploying and Processing

We've discussed deploying and processing cubes a few times, but it's been a very rote "push this button, hope it works" process for you. In this chapter, we're going to discuss what *deploying* actually means and the various ways of doing it. We'll talk a bit about synchronizing databases between servers—similar to mirroring for transactional databases. Finally, we'll talk about processing the various Analysis Services objects and how to schedule processing should you need to.

Deploying a Project

After you've designed an OLAP solution, consisting of data sources, data source views, dimensions, and cubes, as well as data-mining structures, roles, and assemblies if you created them, you need to deploy the project to a server before you can actually do anything with it. When you deploy a project, it creates (or updates) an Analysis Services database. Then all the subordinate objects are created in the database.

Generally, you will deploy a project from BIDS to a development server. If everything passes there, you can use the Deployment Wizard to deploy the project to testing and production servers. You could also write code to use Analysis Management Objects (AMO) to manage deployment of SSAS objects. This becomes particularly compelling when there's a series of complex operations necessary in deployment, such as seeding confidential information or connecting a number of disparate systems.

Project Properties

As you saw in earlier exercises, you need to adjust the project properties before you can deploy the project. If nothing else, you'll have to change the deployment server name, because the default value is *localhost*. To change the project properties, right-click on the solution name and select Properties (Figure 8-1).

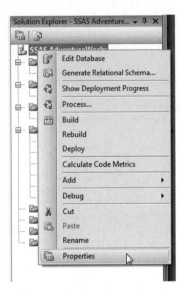

Figure 8-1. Opening the project properties

After you have the properties dialog box open (Figure 8-2), you'll have access to all the properties that govern deployment of the project to a server. You can choose from multiple configurations by using the Configuration Manager at the top. Each configuration is simply a collection of settings, so you can specify which server to deploy to, whether the database should be processed, and so forth.

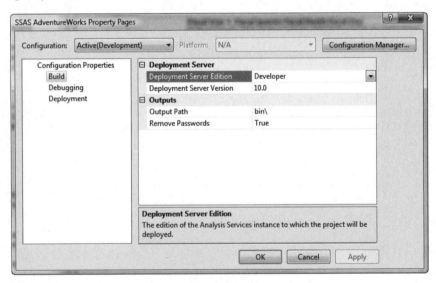

Figure 8-2. The project properties dialog box

Following is a list describing each of the properties available from the dialog box in Figure 8-2. The list is divided into three parts, one for each tab in the dialog:

Build Tab Properties

- **Deployment Server (Edition, Version):** Indicates the SSAS edition and version of the server the database will be deployed to. This will run verifications on the project you're trying to deploy to verify all the features you're trying to use are in fact available on the target server. For example, perspectives are available only on the Enterprise Edition of SQL Server Analysis Services. So if you're using perspectives and you set the target server edition to Standard, you will get an error regarding the use of perspectives when you try to deploy.

- **Output Path:** When you build the solution, a number of structures are generated (for example, XML manifests). The output path indicates where these structures should be copied to.

- **Remove Passwords:** This option will remove any hard-coded passwords from the data sources you're using as they're deployed. If you are hard-coding passwords, you may not want those stored with the project. Alternatively, you may work against a development database, but ultimately deploy to test or production, where you don't know the passwords and a DBA must go in to enter these separately (via SSMS).

Debugging Tab Properties

- **Start Object:** When you start debugging a solution, this is the object that will be started by BIDS. The Currently Active Object option means that whatever object is open and has the focus will be where debugging starts. (This is for debugging .NET stored procedures.)

Deployment Tab Properties

- **Processing Option:** This option indicates whether BIDS should process the database after it's deployed. You may instead choose to deploy a database solution and have the processing job run during a later maintenance window.

- **Transactional Deployment:** On occasion you'll have errors when you process a database object (we'll look at this later in the chapter).

- **Deployment Mode:** The options here are Deploy Changes Only or Deploy All—simply an option to deploy the full project or just the parts that have changed. For very large projects, you may want to deploy only the changes, but deploying the full project ensures that what's on the server matches what you've been working on.

- **Server:** This is simply the server name you will deploy to (the default value is *localhost*).

- **Database:** If you need the database to have a different name on the server, you can change it here. The default value is the name of the project.

Deployment Methods

Let's take a closer look at the methods we have of deploying a database. The methods available include the following:

AMO Automation: The word *automation* here is a little bit misleading, considering how much work we've been doing in BIDS to date. The concept here is that you can use AMO to create database objects on the server. So in theory, you could use code to create dimensions, build cubes, add members, and so on. Consider a large multinational company that has millions of sales every day—their sales cubes would be huge. You might set up partitions so that the current quarter's data is processed daily, but all historical data is processed only weekly. Instead of setting up new partitions each quarter, you could have an automated process to move the just-completed quarter's data into the current year, and then create a new partition for the current quarter to be processed daily. The work on the server to make the changes to the database would be effectively considered "deploying" the new partition.

Backup/Restore: This is an easy way to move a cube intact. If you need a cube available on a system but would never need to actually process it on the system (perhaps a demo system that doesn't need to be absolutely current), then you could back up the cube from the production system and restore it to the demo system. So long as the source data wasn't needed again, the cube would function on the demo system as if it were "live."

Deployment Wizard: SSAS client tools include a deployment wizard that can be scripted to deploy databases. When you run the wizard, you'll be prompted for an Analysis Services database file (*.asdatabase). You'll find this in the folder you specified in the project properties. After you've selected the file, walking through the wizard should seem fairly familiar; you've seen all the deployment options before. When you finish the wizard, it will deploy the database and supporting objects. You will also have the ability to save the deployment as a script that you can run to repeat the deployment.

Synchronize Database Wizard: This wizard runs similarly to the import/export wizard in SQL Server. Running the Synchronize Database Wizard copies the cube data and database metadata from one server to another. This is best used to move databases between development, testing, and production servers.

XMLA Scripts: Finally, you can deploy a database (or database objects) by generating XML for Analysis (XMLA) scripts (either from existing objects or by creating your own) and running these scripts on the server. XMLA scripts can be run from SSMS, an SSIS task, or executed via code.

Now that you have an understanding of the basics of deploying databases in Analysis Services, we'll dig a bit into using the Deployment Wizard and synchronizing databases. Then we'll take a look at processing databases, cubes, and other database objects.

Using the Deployment Wizard

The SQL Server Analysis Services Deployment Wizard is a stand-alone application. With the Deployment Wizard, you can take the database file generated when you build an SSAS database from BIDS and either deploy the database to a target server or generate a deployment script in XMLA.

Running the Wizard

The Deployment Wizard is installed with the SQL Server client tools, and can be found via Start → All Programs → Microsoft SQL Server 2008 → Analysis Services. It runs just like a standard wizard; when you launch it, you'll get a welcome screen (Figure 8-3).

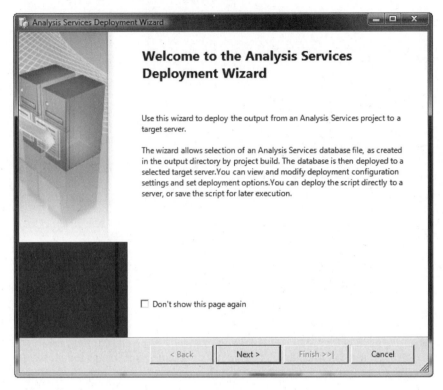

Figure 8-3. The welcome screen for the Deployment Wizard

As I mentioned earlier, the wizard will ask for the location of the asdatabase file generated by BIDS when you build an Analysis Services solution. You'll find this file in the Build location directory (generally \bin on the project directory). Next you'll be prompted for the server and database names, then partition and role options (Figure 8-4). The options for both roles and partitions cover the same general idea: do you want the deployed database to overwrite or add to existing roles and partitions?

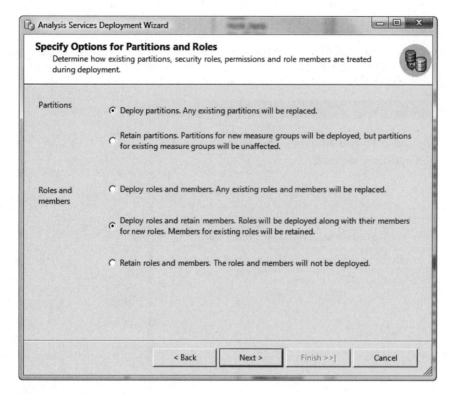

Figure 8-4. *Role and partition options*

The next page of the wizard, shown in Figure 8-5, enables you to specify some key configuration properties for the objects in the database. On this page, you can edit the connection strings for each data source, as well as the impersonation information, and indicate the file locations for log files and storage files.

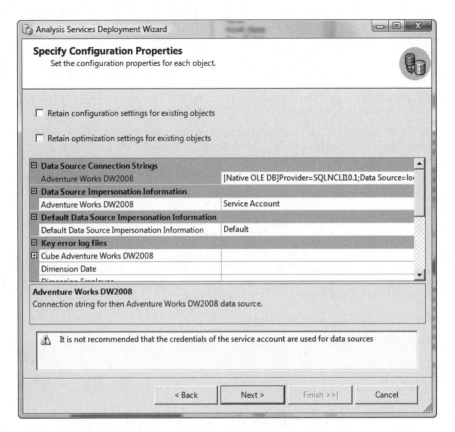

Figure 8-5. *Database object properties*

After verifying the configuration properties, you can specify the processing method, writeback options, and whether processing should be wrapped in a transaction. Then the final page will verify that you want to deploy the solution, and ask whether you want to generate a script (there's no way to generate a script at this point without actually going through the deployment). The script will be generated in the location you indicate in XMLA format.

The two great options for the Deployment Wizard are to create an easily repeatable method for developers to deploy their databases (just run the script, instead of having to click through the wizard every time), and for DBAs to have a way to deploy a database without needing BIDS installed.

After building the solution, you'll have a number of files that are also the input files for the Deployment Wizard. Each will have the same name as the solution, and an extension indicating the contents. In addition, the wizard will generate a file ending in Script.xmla.

Input Files

The configuration files for an SSAS solution are XML files named after the solution with extensions indicating their use:

asdatabase: This file is effectively a manifest file for the solution. This file lists all the objects in the database.

deploymenttargets: This simply lists the server and database names where the objects in the asdatabase file should be created.

configsettings: This file gives specific information about the objects and connection information about data connections. Settings in this file override any information in the asdatabase file.

deploymentoptions: This file contains settings from the deployment dialog box indicating whether the deployment is transactional, processing options, and so forth.

These files are what you'll need to run a deployment script generated with the Deployment Wizard, as you'll examine in the next section.

Deployment Scripts

The deployment script is an XML representation of the entire SSAS database (in other words, it has no dependencies on the input files). As a result, you can use the script to deploy and process the database simply by executing the script against the desired Analysis Server.

After you have the XMLA script, you have several options to execute it:

- Execute it from code via an AMO connection.

- Run it via PowerShell.

- Execute it in Management Studio.

- Run it with an Analysis Services Execute DDL Task in SQL Server Integration Services.

You can automate the generation of the deployment script, either via command-line execution of the Deployment Wizard or by creating the deployment script with another process (after all, it's just XML). So again we can appreciate the options for automating creation, deployment, and processing of Analysis Services databases and their contents.

After a database is deployed, we may want to replicate it on another server. Analysis Services provides a synchronization capability to enable copying an active database from one server to another.

Synchronizing SSAS Databases

You can synchronize two Analysis Services databases, which is effectively an import operation from one Analysis Server to another. The Synchronize Database Wizard is hidden in SQL Server Management Studio. To find it, connect to the SSAS instance you will be importing to, and then right-click the Databases folder and select Synchronize, as shown in Figure 8-6.

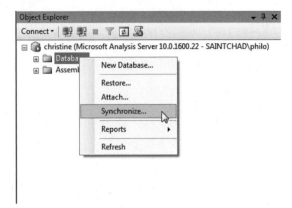

Figure 8-6. Synchronizing SSAS databases

Walking through the wizard is pretty straightforward. First you select a source server and database. The destination server is the server you launched the wizard from (Figure 8-7). The tricky thing about working with the wizard is that you can't synchronize a database to the same server, so you need two Analysis Servers to run the wizard.

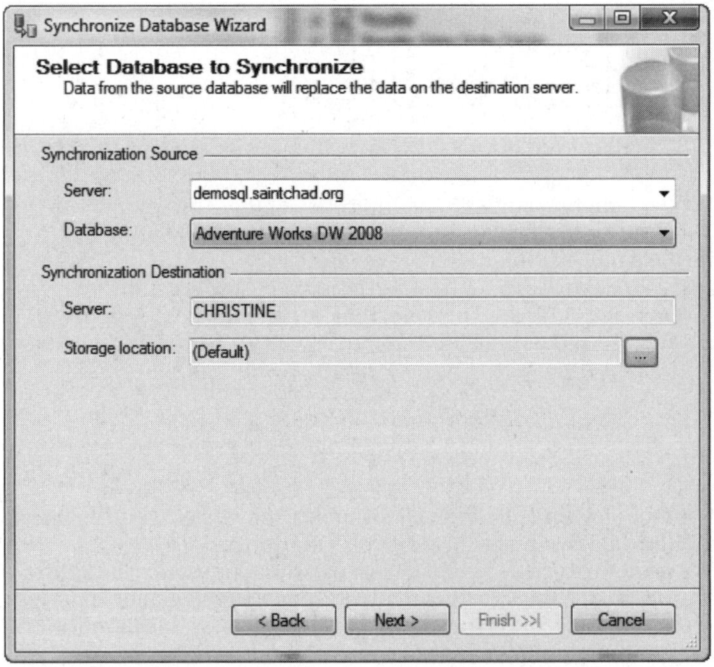

Figure 8-7. Selecting the source server and database

After you've selected a source database, the wizard will poll the database and fetch a list of partitions, as shown in Figure 8-8. This is informative only—you can't change any of the locations (though you can after the database is deployed).

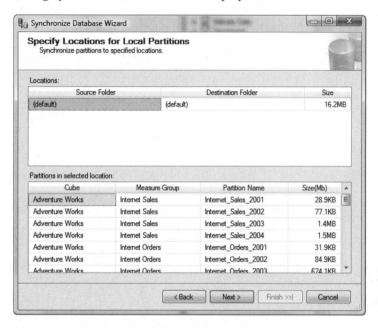

Figure 8-8. Specifying partition locations

Finally, you'll have the option to sync roles and members, and indicate whether you want to run the synchronization now or generate an XMLA script to a file. Again, you can then use the file using the methods described previously to create the database wherever you may need it.

We've covered various ways of deploying databases and database objects, basically getting our structures in place on an Analysis Services server. After they're there, they're just metadata. If we want working cubes, we need to process the database and its objects. And that's what we're looking at in the next section.

Processing

We've deployed our project, with data source views mapped to various tables, dimensions, and cubes. However, these are just the structures of the data that make up our cube. Our dimensions have no members, and our cube has no data. We need to process the database if we want our users to be able to browse the data. Also, after the cube is processed, if the structure of an object changes, or data in the object needs to be updated, or the aggregation design is changed, we need to reprocess one or more structures to reflect the changes. (I'll refer to both as *processing*.)

What Processing Does for Us

Overall, processing a database (or individual cubes or dimensions) consists of loading the fact data, and then running the scripts for aggregation, calculated members, named sets, and so on. Let's take a look at a notional example that will produce output similar to that shown in Figure 8-9. For this example, we'll imagine that the individual fact record is the reported sales for the month by product subcategory.

| Row Labels | Bikes | | Bikes Total | Components | | | Components Total | Clothing | | | | Clothing Total | Grand Total |
	Mountain Bikes	Road Bikes		Handlebars	Forks	Wheels		Caps	Gloves	Jerseys	Shorts		
Calendar 2002	9331011.485	10803179.02	20134190.51	53650.5315	49709.8888	452549.7191	555910.1394	9466.7441	90897.0838	110845.6436	49383.768	260593.2395	20950693.89
Quarter 1, 2002	2182568.556	1695038.228	3877606.784					921.2951		9517.332		10438.6271	3888045.411
January 2002	345553.98	341737.6376	687291.6176					129.6625		1268.9776		1398.6401	688690.2577
February 2002	1011972.012	802402.3117	1814374.324					501.1886		4470.262		4971.4506	1819345.774
March 2002	825042.564	550898.2789	1375940.843					290.444		3778.0924		4068.5364	1380009.379
Quarter 2, 2002	1976637.914	1905150.104	3881788.018					1479.1897		16931.2362		18410.4259	3900198.444
April 2002	392263.842	421688.8465	813952.6885					248.952		2941.7208		3190.6728	817143.3613
May 2002	1198331.462	884953.6246	2083285.087					638.9767		8048.393		8687.3697	2091972.457
June 2002	386042.61	598507.6325	984550.2425					591.261		5941.1224		6532.3834	991082.6259
Quarter 3, 2002	2811937.438	3863388.982	6675326.42	35341.0863	26166.7828	288627.8321	350135.7012	3990.6653	52536.8767	48901.7598	26207.2314	131636.5332	7157098.655
July 2002	714159.555	1103498.761	1817658.316	8670.3255	4638.636	69261.5904	82570.5519	1093.6594	14859.0661	13854.4418	8368.605	38175.7723	1938404.64
August 2002	1080316.709	1613246.035	2693562.744	14312.218	11035.716	124441.9453	149789.8793	1751.9129	19964.8671	20093.5774	8766.9386	50577.296	2893929.919
September 2002	1017461.174	1146644.187	2164105.361	12358.5428	10492.4308	94924.2964	117775.27	1145.093	17712.9435	14953.7406	9071.6878	42883.4649	2324764.096
Quarter 4, 2002	2359867.577	3339601.709	5699469.286	18309.4452	23543.106	163921.887	205774.4382	3075.594	38360.2071	35495.3156	23176.5366	100107.6533	6005351.378
October 2002	577273.9378	890091.319	1467365.257	5158.3881	4096.452	43782	53036.8401	798.721	10421.3911	9537.5182	6766.872	27524.5023	1547926.599
November 2002	979397.2685	1516305.173	2495702.442	7304.9396	9529.674	71073.444	87908.0576	1159.7012	14939.9993	13699.19	8674.554	38473.4445	2622083.944
December 2002	803196.3708	933205.2167	1736401.588	5846.1175	9916.98	49066.443	64829.5405	1117.1718	12998.8167	12258.6074	7735.1106	34109.7065	1835340.835
Grand Total	9331011.485	10803179.02	20134190.51	53650.5315	49709.8888	452549.7191	555910.1394	9466.7441	90897.0838	110845.6436	49383.768	260593.2395	20950693.89

Figure 8-9. A cube showing purchase data by month and by product subcategory

When we process the cube, first the dimensions are processed. Analysis Services does this by going through each attribute and running a SELECT query to return the members of the attribute. For example, in the Products dimension, for the Subcategory attribute, SSAS runs the following query:

```
SELECT
  DISTINCT
[dbo_DimProductSubcategory].[ProductSubcategoryKey] AS ↩
[dbo_DimProductSubcategoryProductSubcategoryKey0_0],
[dbo_DimProductSubcategory].[EnglishProductSubcategoryName] AS ↩
[dbo_DimProductSubcategoryEnglishProductSubcategoryName0_1],
[dbo_DimProductSubcategory].[ProductCategoryKey] AS ↩
[dbo_DimProductSubcategoryProductCategoryKey0_2]
  FROM [dbo].[DimProductSubcategory] AS [dbo_DimProductSubcategory]
```

This returns results similar to those shown in Table 8-1.

Table 8-1. *The Subcategory Attribute*

ProductSubcategoryKey0	EnglishProductSubcategoryName0	ProductCategoryKey0
1	Mountain Bikes	1
2	Road Bikes	1
3	Touring Bikes	1
4	Handlebars	2
5	Bottom Brackets	2
6	Brakes	2
7	Chains	2

This process continues through each dimension attribute. After all the attributes in a hierarchy are populated, the hierarchy is constructed. Analysis Services also turns the tabular data into unique names. For example, the HL Road Handlebars product may have a unique name of [Components].[Handlebars].[HL Road Handlebars] denoting its full hierarchy. After all the dimensions are completely processed, SSAS starts processing the cube by assembling the dimensions, which can be visualized similar to Figure 8-10.

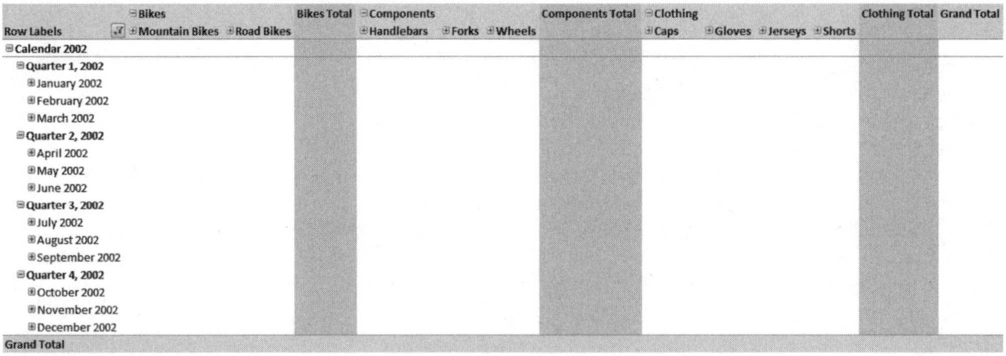

Figure 8-10. *A "cube" with the dimensions in place*

The next step is to load the fact data—the direct transactional data we pull from our database. This will look similar to what you see in Figure 8-11.

| Row Labels | Bikes | | Bikes Total | Components | | | Components Total | Clothing | | | | Clothing Total | Grand Total |
	Mountain Bikes	Road Bikes		Handlebars	Forks	Wheels		Caps	Gloves	Jerseys	Shorts		
Calendar 2002													
Quarter 1, 2002													
January 2002	345553.98	341737.6376						129.6625		1268.9776			
February 2002	1011972.012	802402.3117						501.1886		4470.262			
March 2002	825042.564	550898.2789						290.444		3778.0924			
Quarter 2, 2002													
April 2002	392263.842	421688.8465						248.952		2941.7208			
May 2002	1198331.462	884953.6246						638.9767		8048.393			
June 2002	386042.61	598507.6325						591.261		5941.1224			
Quarter 3, 2002													
July 2002	714159.555	1103498.761		8670.3255	4638.636	69261.5904		1093.6594	14859.0661	13854.4418	8368.605		
August 2002	1080316.709	1613246.035		14312.218	11035.716	124441.9453		1751.9129	19964.8671	20093.5774	8766.9386		
September 2002	1017461.174	1146644.187		12358.5428	10492.4308	94924.2964		1145.093	17712.9435	14953.7406	9071.6878		
Quarter 4, 2002													
October 2002	577273.9378	890091.319		5158.3881	4096.452	43782		798.721	10421.3911	9537.5182	6766.872		
November 2002	979397.2685	1516305.173		7304.9396	9529.674	71073.444		1159.7012	14939.9993	13699.19	8674.554		
December 2002	803196.3708	933205.2167		5846.1175	9916.98	49066.443		1117.1718	12998.8167	12258.6074	7735.1106		
Grand Total													

Figure 8-11. A "cube" with just the fact data loaded

The next step is the CALCULATE statement. After the fact data is loaded, Analysis Services runs all the scripts in the cube. The first script, the default MDX script, has automatically included in it the CALCULATE statement. CALCULATE effectively directs every cell in the cube to aggregate from the cells below it. (Cells are created based on the structure of the dimensions and their hierarchies.)

■ **Caution** If you ever have a cube that processes without errors but shows no aggregate values, check for the presence of the CALCULATE statement in the default script. There are (very few) reasons to delete the statement or move it, but for most cubes it should be the first statement in the default script. You can "lose" it if you get frustrated with your scripts, open the script window, and just press Ctrl+A, then Delete. Ask me how I know this.

Each subtotal and total is aggregated from the granular data beneath it, similar to the indicators shown in Figure 8-12. After the CALCULATE statement has executed, we'll have our fully populated cube, as shown (again) in Figure 8-12.

| Row Labels | Bikes | | Bikes Total | Components | | | Components Total | Clothing | | | | Clothing Total | Grand Total |
	Mountain Bikes	Road Bikes		Handlebars	Forks	Wheels		Caps	Gloves	Jerseys	Shorts		
Calendar 2002	9331011.485	10803179.02	20134190.51	53650.5315	49709.8888	452549.7191	555910.1394	9466.7441	90897.0838	110845.6436	49383.768	260593.2395	20950693.89
Quarter 1, 2002	2182568.556	1695038.228	3877606.784					921.2951		9517.332		10418.6271	3888045.411
January 2002	345559.98	341737.6376	687291.6176					129.6625		1268.9776		1398.6401	688690.2577
February 2002	1011972.012	802402.3117	1814374.324					501.1886		4470.262		4971.4506	1819345.774
March 2002	825042.564	550898.2789	1375940.843					290.444		3778.0924		4068.5364	1380009.379
Quarter 2, 2002	1976637.914	1905150.104	3881788.018					1479.1897		16931.2362		18410.4259	3900198.444
April 2002	392263.842	421688.8465	813952.6885					248.952		2941.7208		3190.6728	817143.3613
May 2002	1198331.462	884953.6246	2083285.087					638.9767		8048.393		8687.3697	2091972.457
June 2002	386042.61	598507.6325	984550.2425					591.261		5941.1224		6532.3834	991082.6259
Quarter 3, 2002	2811937.438	3863388.982	6675326.42	35341.0863	26166.7828	288627.8321	350135.7012	3990.6653	52536.8767	48901.7598	26207.2314	131636.5332	7157098.655
July 2002	714159.555	1103498.316	1817658.316	8670.3255	4638.636	69261.5904	82570.5519	1093.6594	14859.0661	13854.4418	8368.605	38175.7723	1938404.64
August 2002	1080316.709	1613246.035	2693562.744	14312.218	11035.716	124441.9453	149789.8793	1751.9129	19964.8671	20093.5774	8766.9386	50577.296	2893929.919
September 2002	1017461.174	1146644.187	2164105.361	12358.5428	10492.4308	94924.2964	117775.27	1145.093	17712.9435	14953.7406	9071.6878	42843.4649	2324764.096
Quarter 4, 2002	2359867.577	3339601.709	5699469.286	18309.4452	23543.106	163921.887	205774.4382	3075.594	38360.2071	35495.3156	23176.5366	100107.6533	6005351.378
October 2002	577273.9378	890091.319	1467365.257	5158.3881	4096.452	43782	53036.8401	798.721	10421.3911	9537.5182	6766.872	27524.5023	1547926.599
November 2002	979397.2685	1516305.173	2495702.442	7304.9396	9529.674	71073.444	87908.0576	1159.7012	14939.9993	13699.19	8674.554	38473.4445	2622083.944
December 2002	803196.3708	933205.2167	1736401.588	5846.1175	9916.98	49066.443	64829.5405	1117.1718	12998.8167	12258.6074	7735.1106	34109.7065	1835340.835
Grand Total	9331011.485	10803179.02	20134190.51	53650.5315	49709.8888	452549.7191	555910.1394	9466.7441	90897.0838	110845.6436	49383.768	260593.2395	20950693.89

Figure 8-12. Aggregating fact data and completing the cube

After the CALCULATE statement, Analysis Services will go on to execute any other scripts, such as calculated members, named sets, and other scripts. Note that because calculated members are calculated after the CALCULATE statement, they are not aggregated (calculations must provide for all levels of a hierarchy, if necessary).

In addition to the database, cubes, dimensions, and measure groups, the server will process (as necessary) partitions, mining models, and mining structures. Processing any higher-level object will force processing of subordinate objects (for example, processing a database will cause processing of everything within it).

AVAILABILITY WHEN PROCESSING

One thing to be concerned about when processing objects in Analysis Services is, what happens to users who are currently using cubes? Luckily, SSAS is pretty smart about this. While the objects are being processed, the existing objects (if they've already been processed) are left in place and accessible to users. The processing job is wrapped in a transaction. If the processing fails, the transaction is rolled back and the deployed solution is left in place.

If the processing job succeeds, the affected objects are locked (and unavailable to end users) while they are replaced. However, any query that is blocked is queued, and will be run as soon as the cube is fully available again.

How to Initiate Processing from BIDS

There are various ways to invoke processing an object in Analysis Services—via BIDS, Management Studio, XMLA, through AMO with code, or as we covered previously in the chapter, as the final action in a deployment. Let's look at using BIDS first. Then we'll get into the other tools.

In BIDS, you can initiate processing on the database (solution), cubes, dimensions, or mining structures in one of three ways:

- Right-click on the object in the Solution Explorer and select Process from the context menu. (Note that you can select multiple objects by using the Ctrl key and then process them all.)

- With the object open, click the Process button (⟳) on the toolbar.

- With the object open, select the appropriate object menu (Dimension, Cube, Mining Model) and click Process.

Any of these methods will open the Process dialog box, as shown in Figure 8-13. All objects use the same dialog.

Figure 8-13. The Process dialog box

The Object list indicates all the objects that you have selected to be processed; you can remove objects from the list by selecting them and clicking the Remove button.

There are several options for processing the objects in the Object list. You can select the options from a drop-down list in the Process Options column, as shown in Figure 8-14. Each object can be selected separately.

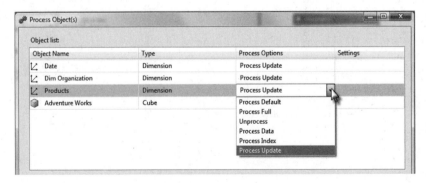

Figure 8-14. Selecting processing options

The processing options have the following effects:

Process Default: Checks the status of an object, and if the object is unprocessed, it is fully processed; if the object is partially processed, the processing is finished.

Process Full: Drops all data and performs full processing on the object and all objects it contains. This is necessary when there has been a structural change in the object.

Unprocess: Drops all data in the objects and subordinate objects. The data is not reloaded. Generally, this is a step in troubleshooting, because this will also delete all the binary data store files.

Process Data: Will drop all data and then reload the data without building aggregations or indexes (see the following Process Index item).

Process Index: Creates or rebuilds indexes and aggregations on processed partitions.

Process Update: (Only supported for dimensions.) Will force a reread of data and update of dimension attributes. Will also add new members to a dimension if they have been created in the underlying data.

After you've set the processing options for the objects you've selected, you can see what the effect of processing them will be by clicking the Impact Analysis button to open the Impact Analysis dialog box, as shown in Figure 8-15.

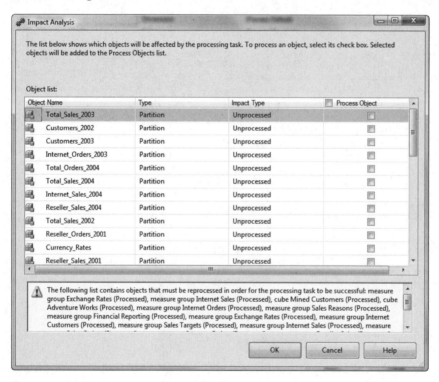

Figure 8-15. The Impact Analysis dialog box

The Object list shows objects affected by all the objects you've selected to process and their end state after you run the processing. In Figure 8-15, you can see that processing the AdventureWorks cube with a setting of Structure Only will leave all its partitions in an unprocessed state. Depending on the processing you've selected, you may or may not see a list of impacted objects in the Object list. For example, running a cube on Process Full will not show any objects in the list, because all subordinate objects are automatically processed. The yellow alert box at the bottom of the dialog alerts you to objects that will automatically be processed along with the objects you've selected.

The Change Settings button in the dialog box will open the Change Settings dialog, as shown in Figure 8-16. In this dialog, you can set whether processing tasks should be performed in parallel or sequentially (with a governing transaction), whether to use existing tables for writeback storage or always create new ones, and whether SSAS should simply always process any objects affected by the current objects and their processing settings.

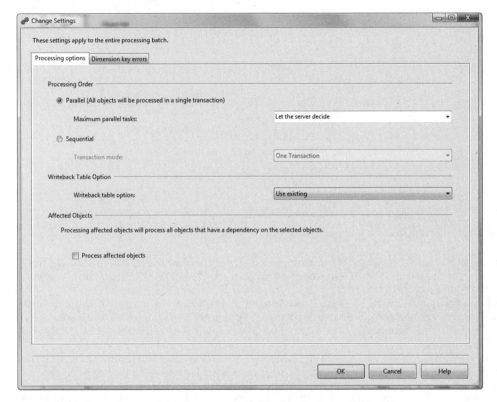

Figure 8-16. The Processing Options tab of the Change Settings dialog box

The other tab in the dialog, shown in Figure 8-17, is for how you want dimension key errors handled. This is a very important area, because dimension key errors are a critical aspect of dealing with processing cubes and databases. When a cube is processed, the fact tables are joined to the dimension tables. Now remember that relationships may be defined in the data source view, so they are not enforced on the source tables. As a result, there's a strong chance of having data in the fact tables that doesn't map to a key in a dimension table.

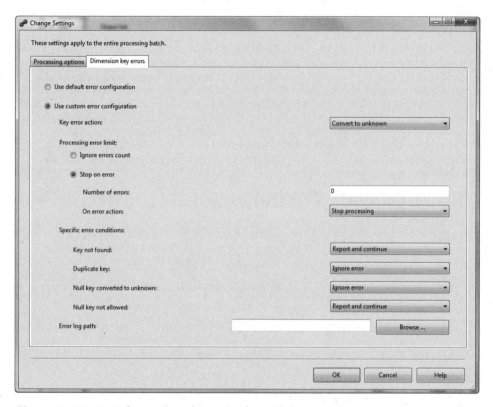

Figure 8-17. *Settings for errors in dimension keys while processing*

Your first option is whether to accept the default error configuration, which can be specified at the measure group level or the partition level. Alternatively, you can select to define the error configuration here for this processing run.

The first option is what to do when a record has a key that doesn't match any key in the dimension table. You can either have SSAS convert the value to flag it as "unknown," in which case it will appear in the Unknown member of the dimension, or you can select to delete the record from the cube if it doesn't match a dimension member.

The next option allows you to stop processing if there are more than a certain number of errors. You can either ignore the error count—essentially process no matter what happens—or indicate that the object(s) should stop processing (or just stop logging) after a certain number of errors. Finally, there's a section where you can indicate actions based on the type of error, and if you select Report and Continue for any of them, you can indicate a path to the error log where they should be logged as they happen.

Finally, you're ready to process the objects you've selected. Click the Run button to open the Process Progress dialog box, as shown in Figure 8-18 for the AdventureWorks cube. You can see the cube and its measure groups listed first, and then the related cube Mined Customers and its measure groups as well.

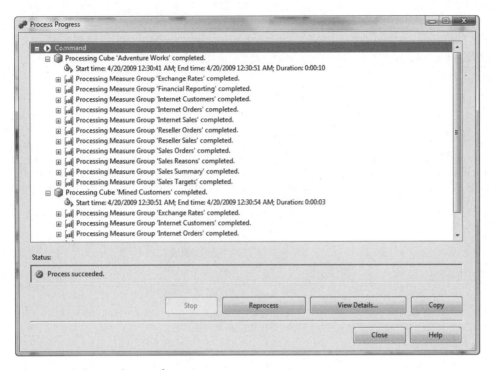

Figure 8-18. *Processing a cube*

If you open up a measure group, you'll see the processing information for that group. In the case of Reseller Sales for AdventureWorks, the four partitions are processed as shown in Figure 8-19. Under each partition is the processing data for that partition, and then the SQL queries used to build the partition. You can highlight the query, copy it, and run the query in SSMS. This is a great troubleshooting tool when you're getting errors during processing.

Figure 8-19. *Details of processing a measure group*

Note that from the Process Progress dialog box, you can also stop the processing while it's in progress, reprocess the selected objects, and view the details or copy either specific lines or errors. If you get errors in processing, often you'll see a long list of errors as SSAS hits several of them simultaneously. The error that's truly indicative of what's going on is often at the bottom of the list, but you may have to hunt around to find it.

■ **Tip** When you're first building a cube, I often find the first time it's processed is when you find all your authentication errors. If you get a large batch of errors, look for authentication problems in the error reports. Fix those first and then try again.

Processing from SQL Server Management Studio

Processing from SSMS is essentially the same as processing from BIDS, except that you can't multiselect objects from the Object Explorer pane. You *can* multiselect from the Object Details pane (View menu → Object Explorer Details), but then you can only multiselect from objects of the same type by selecting the object type folder (for example, Cubes or Dimensions).

The Process dialog will look somewhat different, as shown in Figure 8-20, but has all the same functionality. Everything else we've discussed remains the same.

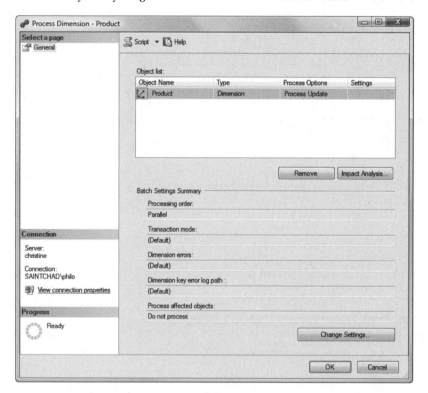

Figure 8-20. The Process object dialog box in SSMS

Processing via XMLA

Everything you do in BIDS and SSMS is actually executed against Analysis Services with XMLA, so obviously you can simply execute a cube manually the same way. At its simplest, an XMLA script to process a database will look like this:

```
<Process xmlns="http://schemas.microsoft.com/analysisservices/2003/engine">
  <Object>
    <DatabaseID>Adventure Works DW</DatabaseID>
  </Object>
  <Type>ProcessFull</Type>
  <WriteBackTableCreation>UseExisting</WriteBackTableCreation>
</Process>
```

Pretty straightforward, right? Of course, XMLA provides all the options you need to perform all the functions and options we've covered so far in this section. The benefit of using XMLA is that you can generate and run the script from an automated process, and you can also perform more-complex actions (process one dimension and use the results to select a table that serves as the data source for another dimension, for example). For more information about processing SSAS objects with XMLA, see the TechNet article at http://technet.microsoft.com/en-us/library/ms187199.aspx.

Processing with Analysis Management Objects (AMO)

Finally, if you don't relish learning yet another dialect of XML, you can process Analysis Services objects by using AMO from a .NET language. In AMO, after you have a reference to the object you want to process, it's simply a matter of calling the Process() method. The Process() method is inherited for each object type, so it has the same signature. It's overloaded, so you can call it with no parameters, or one to five parameters covering various processing options.

For more information on processing SSAS objects via AMO, see the Books Online article also available in TechNet at http://technet.microsoft.com/en-us/library/ms345091.aspx.

Scheduling OLAP Maintenance

We know that if we're using MOLAP, the data and aggregations in cubes (and dimensions and partitions) are static and don't reflect the underlying data until refreshed. In general, there are two approaches to consider, depending on whether you are using a staging database or you have a data source view to connect directly to live systems.

If you're using a staging database, the best approach is to process the OLAP database or cubes from the ETL package that is moving the data into the staging database. For example, if you are using SQL Server Integration Services to extract and clean data before loading it into your staging database, you could add an Analysis Services Processing Task at the end. The SSAS Processing Task in Integration Services gives you a task editor, as shown in Figure 8-21. The options in the dialog should look pretty familiar. In this way, whenever the staging database is loaded with new data, the cubes (or partitions) will be automatically processed to reflect the new data available.

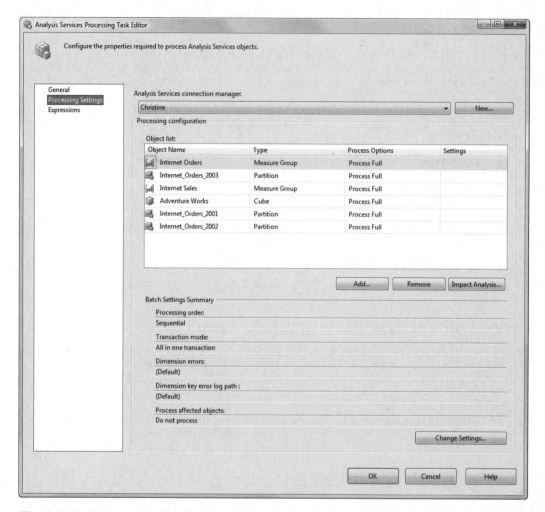

Figure 8-21. Setting options for the AS Processing Task in Integration Services

On the other hand, perhaps you don't want the related objects processed whenever you load new data. Maybe you're loading data in an ongoing manner through the week but want to process cubes only over the weekend (the business isn't that time-sensitive). Or if you're not using a staging database, and the cube links directly to the live data, absent ROLAP or proactive caching you may not get an indication of when there's new data. Finally, you may be linked to a transactional database that's very busy during working hours, but you just need to process the cubes every night. In that case, you can schedule cube processing by using the SQL Server Agent.

The SQL Server Agent is a feature of SQL Server that can schedule and run various jobs, including processing SSAS objects. Connect to a SQL Server instance that has the agent installed and running, and you'll see it at the bottom of the Object Explorer. If you open the agent in the tree view, you'll see a folder containing all the jobs for that agent, as shown in Figure 8-22.

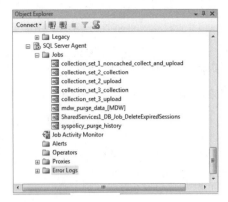

Figure 8-22. *SQL Server Agent in SSMS*

From here you can right-click on the Jobs folder and select New Job. Within the New Job Step dialog box, you can create a number of steps; the one we are most interested in is processing our SSAS objects. You can do this with a type of SQL Server Analysis Services command. The downside is that the command text is just a big text box and has to be in XMLA format (Figure 8-23). However, after you have your steps set up properly, you can also add scheduling, alerts, notifications, and so on to make the job more robust.

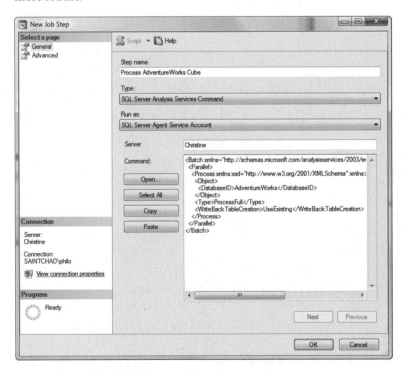

Figure 8-23. *A SQL Agent job step to process a cube*

Summary

Hopefully, what happens behind the curtains when you process a cube is now a little bit less of a mystery. You've looked at deploying solutions, processing cubes and objects, and a little bit of scheduling and maintenance.

Over the past eight chapters, I've often mentioned MDX queries as a way of selecting data from OLAP solutions. Let's take a look at how MDX works in Chapter 9.

CHAPTER 9

■■■

MDX

So, after eight chapters of referring to Multidimensional Expressions (MDX), you finally get to find out what those expressions actually are. Just as SQL is a structured query language for returning tabular results from relational databases, MDX is a query language for returning multidimensional data sets from multidimensional data sources.

MDX was introduced by Microsoft with SQL Server OLAP Services 7.0 in 1997. Since then it has been adopted by a number of vendors and has become the de facto OLAP query language. The latest specification was published in 1999, but Microsoft continues to add improvements.

Why the Need?

For an idea of a comparison between SQL and MDX, let's look at two examples before we dive into the details. We want a breakdown of reseller sales by year and by product subcategory for the category of Bikes. To start with, we can use this query:

```
Use AdventureWorks2008
Go

Select
    Datepart(YYYY, Sales.SalesOrderHeader.ShipDate) As OrderYear,
    Sales.SalesOrderHeader.TotalDue,
    Production.ProductCategory.Name As Category,
    Production.ProductSubcategory.Name As Subcategory
From
    Sales.SalesOrderDetail Inner Join
    Sales.SalesOrderHeader On
    Sales.SalesOrderDetail.SalesOrderID = Sales.SalesOrderHeader.SalesOrderID Inner Join
    Production.Product On
    Sales.SalesOrderDetail.ProductID = Production.Product.ProductID Inner Join
    Production.ProductSubcategory On
    Production.Product.ProductSubcategoryID =
    Production.ProductSubcategory.ProductSubcategoryID Inner Join
    Production.ProductCategory On
    Production.ProductSubcategory.ProductCategoryID =
    Production.ProductCategory.ProductCategoryID
Where
    Production.ProductCategory.Name = 'Bikes'
```

However, this query gives us just a tabular answer, as shown in Table 9-1. What we want is a pivot table. So we can either rely on the client to pivot the data for us or do it on the server. Because this data set in AdventureWorks is 40,000 rows, we're bringing a lot of data across the wire just to aggregate it on the client side.

Table 9-1. Tabular Results from the SQL Query

Order Year	Total Due	Subcategory	Category
2001	38331.9613	Road Bikes	Bikes
2001	45187.5136	Road Bikes	Bikes
2001	3953.9884	Road Bikes	Bikes
2001	3953.9884	Road Bikes	Bikes

Luckily, the PIVOT command was introduced with SQL Server 2005, so we can write a SQL query to have the crosstab result set created on the server. The query to produce a pivoted set of results is shown here:

```
Select
    Subcategory,
    [2001] As CY2001,
    [2002] As CY2002,
    [2003] As CY2003,
    [2004] As CY2004
From
(
  Select
    DatePart(YYYY, Sales.SalesOrderHeader.ShipDate) As OrderYear,
    Production.ProductSubcategory.Name As Subcategory,
    Cast(Sales.SalesOrderDetail.LineTotal As Money) As ExtAmount
  From
    Sales.SalesOrderDetail Inner Join
    Sales.SalesOrderHeader On
    Sales.SalesOrderDetail.SalesOrderID =
    Sales.SalesOrderHeader.SalesOrderID Inner Join
    Production.Product On
    Sales.SalesOrderDetail.ProductID = Production.Product.ProductID Inner Join
    Production.ProductSubcategory On
    Production.Product.ProductSubcategoryID =
    Production.ProductSubcategory.ProductSubcategoryID Inner Join
    Production.ProductCategory On
    Production.ProductSubcategory.ProductCategoryID =
    Production.ProductCategory.ProductCategoryID
  Where
    ProductCategory.Name = 'Bikes' And
    ProductSubcategory.Name In ('Mountain Bikes','Touring Bikes')
) P
    Pivot
    (
```

```
        Sum(ExtAmount)
        For OrderYear
        In ([2001], [2002], [2003], [2004]
    )
) As Pvt
```

Wow! This query is quite complex. The results of the preceding pivot query are displayed in Table 9-2, and include all four years of sales totals for two subcategories belonging to the Bikes category.

Table 9-2. Results from SQL Query, Using PIVOT

Subcategory	CY2001	CY2002	CY2003	CY2004
Mountain Bikes	5104234.8587	10739209.2384	12739978.3739	7862021.4699
Touring Bikes	NULL	NULL	6711882.202	7584409.0678

However, if we've built a cube for the sales data, we can write an MDX query to produce similar results. The MDX will look like this:

```
Select
    Non Empty [Ship Date].[Calendar Year].Children On Columns,
    Non Empty {Product.Subcategory.[Mountain Bikes], Product.Subcategory.[Touring Bikes]}
        On Rows
From
    [Adventure Works]
Where
    (Product.Category.[Bikes], Measures.[Sales Amount])
```

Far simpler! And when you start to learn your way around MDX, you'll find it much easier to understand and write than the pivot query. In addition, as you'll see, MDX offers a lot more power for various dimensional approaches to looking at data. Let's dive in.

Tuples and Sets

Even though we can work with numerous dimensions (and we will), for visualizing how MDX operates, we'll use a cube (three dimensions) to explain the basic concepts. Just as SQL queries work in cells, rows, and columns, the basic units in MDX that we work with are members, tuples, and sets.

You should be fairly familiar with members by now—a *member* is an individual value in an attribute of a dimension. Members can be leaf members (the lowest level of a hierarchy), parent members (upper levels of a hierarchy), or an All member (the top level of an attribute encompassing all the lower members). That's pretty straightforward; but we're going to have to dig more deeply into understanding tuples and sets.

Notation

Tuples and sets are deeply connected to the problem of selecting members from dimensions. First we need to look at how we indicate which dimensions and members we're talking about. It's actually pretty straightforward—dimensions, hierarchies, attributes, and members are simply laid out in a breadcrumb manner, separated by dots.

For example, to select the LL Fork product from the Product dimension, you would use the following:

```
[Product].[Product].[LL Fork]
```

That's the Product dimension, the Product attribute hierarchy, and the LL Fork member. The square brackets are required only if the name contains numbers, spaces, or other special characters, or it's a keyword. For the preceding example, this will also work:

```
Product.Product.[LL Fork]
```

If you have a hierarchy, you will also have to specify the level from which you are selecting the member. So if you're selecting the LL Fork from the Product level of the Product Categories hierarchy, it will look like this:

```
[Product].[Product Categories].[Product].[LL Fork]
```

I do want to point out that in a hierarchy you don't "walk down" the selections. For example, note that we didn't say anything about subcategories in the preceding example. We didn't say anything about categories, either; it's just that the hierarchy is named Product Categories, so it can be a bit misleading.

Finally, we're using member names for our annotations here, mostly because they're easier to read. However, please recognize that unless you have a constraint in the underlying data source, there is no guarantee that member names will be unique. This may seem somewhat puzzling until you consider this example:

```
[Date].[Calendar].[Month].[November]
```

■ **Note** In AdventureWorks, the member names for months are actually formatted as *November 2003*, but that's simply in how you format the name. *November* is a perfectly valid option too.

If your months are formatted in this way, that member would return an error as being ambiguous. So how do we point to that month? In MDX you also have the option of using the member key to select a member; this is indicated by prefixing the member with an ampersand. For example, LL Fork in the Product dimension will have a member key of 391. To find this member key, review the Properties pane of the Product attribute belonging to the Product dimension. The KeyColumns property lists the ProductKey field of the Product table as your member key. Figure 9-1 shows the BIDS environment with these panes displayed.

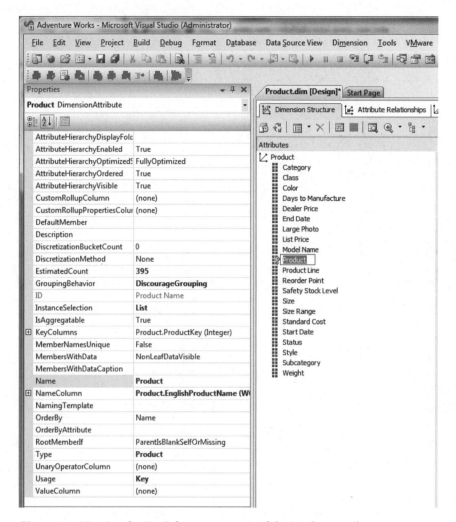

Figure 9-1. *Viewing the KeyColumns property of the Product attribute*

Using this value, selecting LL Fork (Product Key 391) from the AdventureWorks cube looks like this:

```
[Product].[Product Categories].[Product].&[391]
```

If you're comfortable that the member name is unique, you can replace &[391] with [LL Fork], and the query will run just fine.

Tuples

When we were building reports in the BIDS browser or Excel, we were actually building up and executing MDX queries. Let's take a look at our notional cube again, shown in Figure 9-2. We have a basic cube with three dimensions: Product, Region, and Date. Each dimension has four members. Finally, let's say the cube has a single measure: Sales Amount. Simple enough.

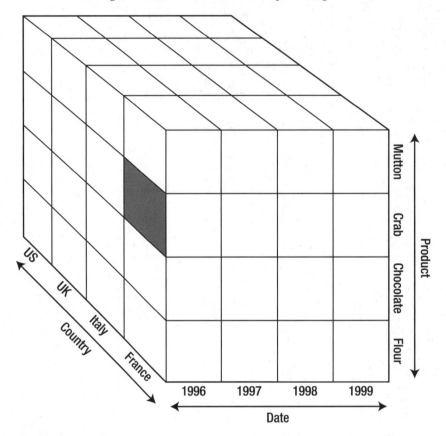

Figure 9-2. *A cube*

Now let's say we want to see the amount of crab sold in France in 1996. We select the appropriate members of each dimension ([Products].[Crab], [Dates].[1996], [Region].[France]), giving us the result shown in Figure 9-3. In selecting a member from each dimension in the cube, we've isolated a single cell (and therefore a single value). The highlighted area in Figure 9-3 is a tuple.

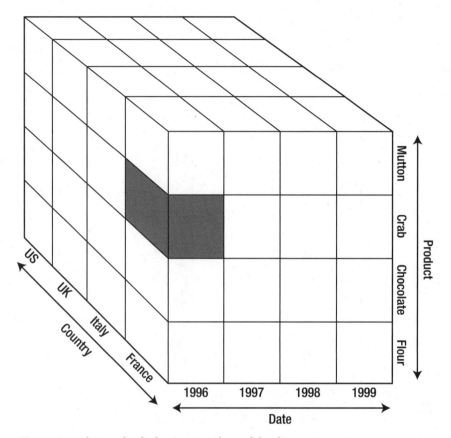

Figure 9-3. *The result of selecting members of the dimensions*

You may be thinking, "So a tuple is a cell?" Not quite. A *tuple* is the result of selecting a single member from each of the dimensions in a cube, resulting in a single cell (value) in the cube. "But wait," you say, looking at the AdventureWorks cube, "so I have to go and select a member from every single dimension every time?" Yes, but you don't have to do it by hand. Let's say you want the value of sales for all products in Italy in 1996. That would look like the cube in Figure 9-4. So—is the highlighted section a tuple?

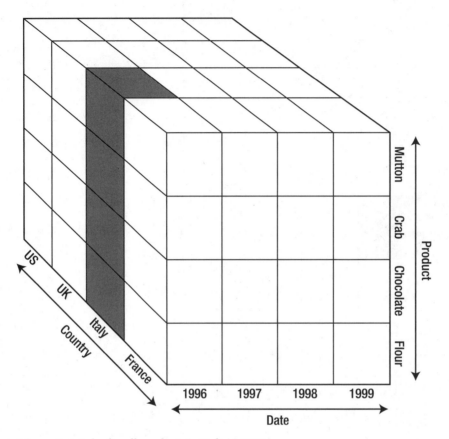

Figure 9-4. *Sales for all products in Italy in 1996*

Yes! It's a tuple. "But we didn't select a member from the Products dimension," you may complain. Ah, but we did. Remember that every dimension *has a default member*. You can designate the default member in the Dimension Designer in BIDS for each attribute. If you don't designate a default member, Analysis Services will choose an arbitrary member from the highest level of the attribute hierarchy.

Now here's the neat part: if you indicate that an attribute can be aggregated (set IsAggregatable to true), then SSAS will automatically create a level above all the members with an (All) member, consisting of the aggregation of all members in that dimension. Because this is the top level, and has a single member, then if you don't set a default member, SSAS uses the (All) member as the default. So when we select just Italy from the notational cube, Analysis Services is providing the default member for every other dimension in the cube.

■ **Tip** This also points out a little something about dimension design: be sure your default member makes sense!

A note about annotation: tuples are always surrounded by parentheses. If you look at Figure 9-3 again, that tuple is annotated:

```
([Products].[Crab], [Dates].[1996], [Region].[France])
```

Generally, because a tuple will be a collection of members, the parentheses are pretty easy to remember. Where this will trip you up is when you use a single member to denote a tuple. For example, let's say you wanted to see the total sales for 2003. "That's easy," you would say. "The tuple would be [Date].[Calendar Year].[2003]!" No—that is the member. The *tuple* would be the following:

```
([Date].[Calendar Year].[2003])
```

I have a colleague whose first attempt at troubleshooting MDX queries would be to just wrap everything in parentheses. …

■ **Note** So by now you're probably dying to know if *tuple* is generally pronounced to rhyme with *couple* or with *pupil*. The answer is—there isn't a preferred pronunciation. It's a personal preference (and often the subject of fairly robust debates). Personally, I alternate.

Now remember that I said a tuple represents a single cell, or value, in the cube. You may look back at Figure 9-4 and say, "But there are four cells in the selection." This is true. But if you select Italy from the Region dimension and 1996 from the Date dimension, you'll get a single value—the sum of sales for all products. This often causes consternation when you're trying to design a report; you want a breakdown by product, but instead you get a single value (if you're using a bar chart, you'll get one bar, which we often call *the big box*, as shown in Figure 9-5).

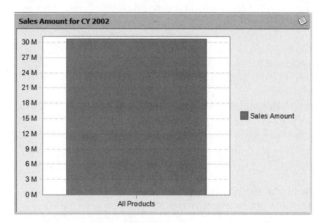

Figure 9-5. A bar chart with one bar

So what do we do if we want to see a bar for each product? Each bar is represented by a tuple, and the four together are a *set*, which we'll look at in the next section.

Sets

A *set* is a collection of zero or more tuples. That's it. Now let's look at that bar chart again, the one with one big box. If we want the chart to show a bar for each product, we need to select the tuple for each product. The tuples for the four products are annotated like this:

```
{([Products].[Mutton], [Dates].[1996], [Region].[France]),
([Products].[Crab], [Dates].[1996], [Region].[France]),
([Products].[Chocolate], [Dates].[1996], [Region].[France]),
([Products].[Flour], [Dates].[1996], [Region].[France])}
```

This is a set—a collection of tuples with the same dimensionality. The tuples are separated by commas, and the set is set off with curly braces. *Dimensionality* refers to the idea that the tuples must each have the same dimensions, in the same order, and use the same hierarchy. You can use members from different levels in the same hierarchy:

```
{([Products].[Categories].[Category].[Bikes]),
([Products].[Categories].[Product].[Racer 150])}
```

But not from different hierarchies:

```
{([Products].[Categories].[Category].[Bikes]), ([Products].[Style].[Mens])}
```

You can also create a set by using the range operator (:). To do this, simply indicate the first and last members of a contiguous range, for example:

```
{[Date].[Calendar].[Month].[March 2002]:
[Date].[Calendar].[Month].[June 2002]}
```

This indicates a set consisting of four members: March 2002, April 2002, May 2002, and June 2002.

Finally, remember that a set consists of zero or more tuples, including zero and one. So yes, a tuple can be a set, or a set can contain no tuples at all (which is an empty set). This would generally be about dynamically generating sets to be used in functions. Consider the Average function, designed to return the average value of a set of tuples passed in by another function. The passing function may return a single tuple, or none for given criteria. We can either accept a set with one tuple, or we've got to write logic to read the set first and branch if the collection of tuples has only one tuple, or none. It's easier to just accept a set with a single tuple or an empty set.

Okay, hopefully you've got a loose grasp on what a tuple is, and what a set is, because we're going to start applying the concepts.

MDX Queries

You saw an example of an MDX query at the beginning of the chapter. Similar to SQL, it's a SELECT statement, but while SQL queries return table sets, MDX queries return hierarchical row sets. MDX statements are simply fragments of MDX designed to return a scalar value, tuple, set, or other object. Let's start by looking at MDX queries.

As we've seen, an MDX query looks somewhat like a SQL query—a SELECT statement with FROM and WHERE clauses. However, even though it looks like a SQL query, it's significantly different. For one thing, with a SQL query you select fields, and the records returned are rows. An MDX query selects cells, and the multidimensional results are organized along multiple axes, just like the cubes we've looked at (though not limited to three—an MDX query can have up to 128 axes!).

Also, whereas SQL queries can join multiple tables and/or views, MDX queries are limited to a single cube. In addition, you may be used to writing SQL queries from just about anywhere and getting a rowset in return that you can parse. The results of MDX queries, on the other hand, aren't easily human readable, and generally need a front-end application designed to work with them (you'll look at OLAP front ends in Chapter 14).

Using SQL Server Management Studio

We'll be using SQL Server Management Studio (SSMS) to run the queries in this chapter. It's a good tool for testing MDX expressions and queries; the only significant limitation is that it can work with only two axes. Most MDX UI tools will either error out if you try to work with more than two axes (like SSMS), expand the axes as Cartesian joins (as Reporting Services does), or turn any dimension beyond two into a *slicer*, or filter selector, as Excel and ProClarity do.

You'll find SSMS in the Start menu under Microsoft SQL Server 2008. Click to open it. When it opens, you'll be greeted with a connection dialog box. Select Analysis Services for the Server type, and then enter the server name (**localhost** works if it's running locally). Click the Connect button.

You should have the Object Explorer pane on the left (Press F8 if it's not there). Open out the server node and then the Databases folder and the databases underneath. Select the database you want to work with (Adventure Works DW 2008) and then click the New Query button in the menu strip (or press Ctrl+N). This will open an MDX query window, with a cube browser down the left side.

SELECT

A basic MDX query consists of a SELECT statement and a FROM statement. The SELECT statement is the meat of the query; you define the axes here. This is easiest to understand with a few examples. Open SSMS (see the preceding sidebar). You can enter MDX queries and then click the Execute button to run them.

■ **Tip** When you click the Execute button (or press F5), SSMS will execute all the text in the window. You can have multiple queries in the window, and SSMS will execute each in turn. However, if you highlight one query, SSMS will execute only the highlighted text.

SSMS with an MDX query open is shown in Figure 9-6.

Figure 9-6. *SQL Server Management Studio (SSMS) with an MDX query open*

To start, let's look at a basic MDX query:

```
SELECT {[Measures].[Reseller Sales Amount]} ON COLUMNS ,
       {[Product].[Category]} ON ROWS
FROM [Adventure Works]
```

The results of this query are shown in Table 9-3. We've selected a measure member for the columns—this creates the column heading and values. Then we selected a dimension member for the rows, giving us the All Products member by default. Finally, we indicated which cube to run the query on. Note that both `[Measures].[Reseller Sales Amount]` and `[Product].[Category]` are in curly braces—the `SELECT` statement expects a set expression for each axis. (And even though we have just a single tuple in the first axis, a set can be a single tuple!)

Table 9-3. *MDX Query Results*

--	Reseller Sales Amount
All Products	$80,450,596.98

What if we want a list of the product categories? The first answer is that we can simply list them, as shown in this query:

```
SELECT { [Measures].[Reseller Sales Amount] } ON COLUMNS ,
        {[Product].[Category].[Accessories],
        [Product].[Category].[Bikes],
        [Product].[Category].[Clothing],
        [Product].[Category].[Components]} ON ROWS
FROM [Adventure Works]
```

This will return the results shown in Table 9-4. This is fairly straightforward: we've selected the Reseller Sales Amount as our column (header and values), and the set consisting of each category member from the Category hierarchy in the Product dimension.

Table 9-4. *MDX Query Returning Multiple Product Categories*

--	Reseller Sales Amount
Accessories	$571,297.93
Bikes	$66,302,381.56
Clothing	$1,777,840.84
Components	$11,799,076.66

We're not restricted to a single column, either. Let's say we want to compare Reseller Sales Amount to the cost of freight. We can simply add the Reseller Freight Cost measure to the set we've selected for Columns, as shown next. This produces the result shown in Table 9-5.

```
SELECT { [Measures].[Reseller Sales Amount],
[Measures].[Reseller Freight Cost]} ON COLUMNS ,
        {[Product].[Category].[Accessories],
        [Product].[Category].[Bikes],
        [Product].[Category].[Clothing],
        [Product].[Category].[Components]} ON ROWS
FROM [Adventure Works]
```

Table 9-5. *MDX Query Showing Two Measures as Columns*

--	Reseller Sales Amount	Reseller Freight Cost
Accessories	$571,297.93	$14,282.52
Bikes	$66,302,381.56	$1,657,560.05
Clothing	$1,777,840.84	$44,446.19
Components	$11,799,076.66	$294,977.15

Well now we come to an interesting question: how do we use MDX to produce a pivot table? We've done a number of examples showing a breakdown of a value by one dimension in rows, and another dimension in columns. Can we show Reseller Sales Amount by Categories and Years? Sure we can—or I wouldn't have put the question in the book.

Let's try adjusting the query slightly, as shown next. The results are in Table 9-6.

```
SELECT { [Date].[Fiscal Year].[FY 2002],
        [Date].[Fiscal Year].[FY 2003]} ON COLUMNS ,
        {[Product].[Category].[Accessories],
        [Product].[Category].[Bikes]} ON ROWS
FROM [Adventure Works]
```

***Table 9-6.** MDX Query Using Dimensions for Columns and Rows*

--	FY 2002	FY 2003
Accessories	$36,814.85	$124,433.35
Bikes	$15,018,534.07	$22,417,419.69

We see the selected fiscal years across the column headers—that's good. And we see the categories we chose as row headers—also good. But what are those dollar amounts? A little investigation would show these are the values for the Reseller Sales Amount, but where does that come from? If we check the cube properties in BIDS, we'll find that there is a property DefaultMeasure, set to Reseller Sales Amount. Well that makes sense.

But how do we show a measure other than the Reseller Sales Amount? We can't add it to either the ROWS or COLUMNS set, because measures don't have the same dimensionality as the other tuples in the set. (See how this is starting to make sense?)

WHERE

What we can do is use a WHERE clause. The WHERE clause in an MDX query works just like the WHERE clause in a SQL query. It operates to restrict the query's results. In this case, we can use the WHERE clause to select what measure we want to return. So we end up with a query as shown here:

```
SELECT { [Date].[Fiscal Year].[FY 2002],
        [Date].[Fiscal Year].[FY 2003]} ON COLUMNS ,
        {[Product].[Category].[Accessories],
v[Product].[Category].[Bikes]} ON ROWS
FROM [Adventure Works]
WHERE ([Measures].[Reseller Freight Cost])
```

In addition to selecting the measure we want to look at, we can also use the WHERE clause to limit query results in other ways. The following query will show results similar to the previous one, but with the measure restricted to sales in the United States:

```
SELECT { [Date].[Fiscal Year].[FY 2002],
        [Date].[Fiscal Year].[FY 2003]} ON COLUMNS ,
        {[Product].[Category].[Accessories],
```

```
        [Product].[Category].[Bikes]} ON ROWS
FROM [Adventure Works]
WHERE ([Measures].[Reseller Freight Cost],
       [Geography].[Country].[United States])
```

This use of the WHERE clause is also referred to as a *slicer*, because it slices the cube. When looking at an MDX query, remember that the WHERE clause is always evaluated first, followed by the remainder of the query (similar to SQL). Although these two examples seem very different (one selecting a measure, the other slicing to a specific dimension member), they're not. Measures in OLAP are just members of the [Measures] dimension. In that light, both selectors in the second query ([Reseller Freight Cost] and [United States]) are selecting a single member of their respective dimensions.

Axes

Another way to deal with selecting measures is to simply add another axis, because OLAP is multidimensional. After all, our "cube" concept started with three dimensions. So why are we limited to two dimensions (rows and columns) in our query? We're not. MDX queries can have up to 128 axes, and the first five are named: COLUMNS, ROWS, PAGES, SECTIONS, CHAPTERS. Beyond those five, then you simply indicate the axis with an ordinal (for example, ON AXIS 0).

Now, if you try to run an MDX query with three axes in SSMS you'll get an error; SSMS will tell you "Results cannot be displayed for cellsets with more than two axes." In *Fast Track to MDX* by Mark Whitehorn, Robert Zare, and Mosha Pasumansky (Springer, 2005), the authors quote ProClarity's vice-president of Research and Development regarding ProClarity's experience with users and more than two dimensions. The short answer is that flat screens and spreadsheets can show only two dimensions, so additional dimensions have to be represented in other ways (slicers, pages, and so forth). This can result in a user getting an initial result that looks incorrect.

Let's say you want a report of sales by product, by year, and by country. You arrange products in rows, years in columns, and countries as pages. Now because there are thousands of products, you don't want to list them all, especially if most of them will show empty data rows. Well, an MDX query won't return products that have no sales data. What gets returned when a specific product has sales in the United States but not in France? A row will be returned for that product, as it has a sales value. But on the page for US data, we'll see the product listed with no values.

The obvious question from a user is "Why is this product listed with no values, but other products I know exist aren't listed?" The display gets fairly confusing fairly quickly.

MDX Functions

You saw how to create a grid with multiple member values. However, having to list all the members in a large hierarchy will get unwieldy very quickly. In addition, if members change, we could end up with invalidated queries. However, MDX offers functions that we can use to get what we're looking for. MDX functions work just as functions in any language: they take parameters and return an object. The return value can be a scalar value (number), a set, a tuple, or other object.

In this case, either the Members function or the Children function will work to give us what we are looking for. Let's compare the two. The following query produces the results in Table 9-7.

```
SELECT { [Measures].[Reseller Sales Amount] } ON COLUMNS ,
        {[Product].[Category].Members} ON ROWS
FROM [Adventure Works]
```

Table 9-7. *MDX Query Using Members Function*

--	Reseller Sales Amount
All Products	$80,450,596.98
Accessories	$571,297.93
Bikes	$66,302,381.56
Clothing	$1,777,840.84
Components	$11,799,076.66

Now let's look at a query using the Children function, and the results shown in Table 9-8.

```
SELECT { [Measures].[Reseller Sales Amount] } ON COLUMNS ,
        {[Product].[Category].Children} ON ROWS
FROM [Adventure Works]
```

Table 9-8. *MDX Query Using Children Function*

--	Reseller Sales Amount
Accessories	$571,297.93
Bikes	$66,302,381.56
Clothing	$1,777,840.84
Components	$11,799,076.66

It's pretty easy to see the difference: the Members function returns the All member, while the Children function doesn't. If you try this on the [Product Categories] hierarchy, you'll see the extreme difference, because Members returns all the members from the [Categories] level, the [Subcategories] level, and the [Products] level, as shown in Figure 9-7. Note that we have All Products, then Accessories (category), then Bike Racks (subcategory), and then finally the products in that subcategory. On the other hand, Figure 9-8 shows selecting all the children under the Mountain Bikes subcategory.

	Reseller Sales Amount
All Products	$80,450,596.98
Accessories	$571,297.93
Bike Racks	$197,736.16
Hitch Rack - 4-Bike	$197,736.16
Bike Stands	(null)
All-Purpose Bike Stand	(null)
Bottles and Cages	$7,476.60
Mountain Bottle Cage	(null)
Road Bottle Cage	(null)
Water Bottle - 30 oz.	$7,476.60
Cleaners	$11,188.37
Bike Wash - Dissolver	$11,188.37
Fenders	(null)
Fender Set - Mountain	(null)

Figure 9-7. *Selecting a hierarchy Members collection*

	Reseller Sales Amount
Mountain-100 Black, 38	$1,174,622.74
Mountain-100 Black, 42	$1,102,848.18
Mountain-100 Black, 44	$1,163,352.98
Mountain-100 Black, 48	$1,041,901.60
Mountain-100 Silver, 38	$1,094,669.28
Mountain-100 Silver, 42	$1,043,695.27
Mountain-100 Silver, 44	$1,050,610.85
Mountain-100 Silver, 48	$897,257.36
Mountain-200 Black, 38	$1,471,078.72

Figure 9-8. *Selecting all the products under the Bikes category*

We'll take a closer look at moving up and down hierarchies later in the chapter.

There's another difference between SQL and MDX that I haven't mentioned yet. In SQL we learn that we cannot depend on the order of the records returned; each record should be considered independent. However, in the OLAP world result sets are governed by dimensions, and our dimension members are always in a specific order. This means that concepts such as *previous, next, first,* and *last* all have meaning.

First of all, at any given time in MDX, you have a *current member*. As we are working with the cube and the queries, we consider that we are "in" a specific cell or tuple, and so for each dimension we have a specific member we are working with (or the default member if none are selected).

The CurrentMember function operates on a hierarchy ([Product Categories] in this case) and returns a tuple representing the current member. If we run this, we get the result shown in Figure 9-9. Remember, a tuple is also a set.

	Reseller Sales Amount
All Products	$80,450,596.98

Figure 9-9. *The Reseller Sales Amount query using a* `CurrentMember` *function*

For example, let's say for a given cell you want the value from the cell before it (most commonly to calculate change over time, but you may also want to produce a report of change from one tooling station to the next, or from one promotion to the next). Generally, the dimension where we have the most interest in finding the previous or next member is the Time dimension. We frequently want to compare values for a specific period to the value for the preceding period (in sales there's even a term for this—*year over year growth*).

Let's look at an MDX query for year over year growth:

```
WITH
        MEMBER [Measures].[YOY Growth] AS
                ([Date].[Fiscal Quarter].CurrentMember,
[Measures].[Reseller Sales Amount])-
                ([Date].[Fiscal Quarter].PrevMember,
[Measures].[Reseller Sales Amount])
SELECT NONEMPTY([Date].[Fiscal Quarter].Children *
{[Measures].[Reseller Sales Amount],
[Measures].[YOY Growth]}) ON COLUMNS ,
        NONEMPTY([Product].[Model Name].Children) ON ROWS
FROM [Adventure Works]
WHERE ([Geography].[Country].[United States])
```

There are a number of new features we're using here. First let's look at what the results would look like in the MDX Designer in SSMS, shown in Figure 9-10.

	Q1 FY 2004	Q1 FY 2004	Q2 FY 2004	Q2 FY 2004	Q3 FY 2004
	Reseller Sales Amount	YOY Growth	Reseller Sales Amount	YOY Growth	Reseller Sales Amount
Bike Wash	$1,971.19	1971.1905	$1,407.90	-563.2923	$808.14
Cable Lock	(null)	-2570.52	(null)	(null)	(null)
Chain	$1,785.17	1785.168	$1,280.23	-504.9394	$886.51
Classic Vest	$35,855.00	35855.0022	$28,725.51	-7129.4942	$17,076.70
Cycling Cap	$2,484.16	726.8316	$1,790.63	-693.5283	$1,005.55
Front Brakes	$9,995.28	9995.2806	$6,070.50	-3924.7806	$4,473.00
Front Derailleur	$6,532.39	6532.386	$5,214.93	-1317.456	$3,513.22
Full-Finger Gloves	(null)	-22220.5676	(null)	(null)	(null)
Half-Finger Gloves	$9,512.15	4131.1998	$7,095.78	-2416.3721	$2,365.73
Hitch Rack - 4-Bike	$32,445.52	32445.516	$24,937.10	-7508.412	$12,600.00

Figure 9-10. *Results of the MDX query*

Well the first thing we run into is the `WITH` statement—what's that? In an MDX query, we use the `WITH` statement to define a *query-scoped calculated measure*. (The alternative is a *session-scoped calculated measure* and is defined with the `CREATE MEMBER` statement, which we won't cover here.) In this case, we have created a new measure, [`YOY Growth`], and defined it.

The definition follows the AS keyword, and creates our YOY measure as the difference between two tuples based on the [Fiscal Quarter] dimension and the [Reseller Sales Amount] measure. In the tuples, we use the CurrentMember and PrevMember functions. As you might guess, they return the current member of the hierarchy and the previous member of the hierarchy, respectively.

Look at this statement:

```
([Date].[Fiscal Quarter].CurrentMember,
[Measures].[Reseller Sales Amount])
-
([Date].[Fiscal Quarter].PrevMember,
[Measures].[Reseller Sales Amount])
```

Note the two parenthetical operators, which are identical except for the operator at the end. Each one defines a tuple based on all the current dimension members, except for the [Date] dimension, where we are taking the current or previous member of the [Fiscal Quarter] hierarchy, and the [Reseller Sales Amount] member of the [Measures] dimension.

As the calculated measure is used in the query, for each cell calculated, the Analysis Services parser determines the current member for the hierarchy, and creates the appropriate tuple to find the value of the cell. Then the previous member is found, and the value for *that* tuple is identified. Finally, the two values are subtracted to create the value returned for the specific cell.

Our next new feature in the query is the NONEMPTY function. Consider a query to return all the Internet sales for customers on July 1, 2001:

```
SELECT [Measures].[Internet Sales Amount] ON 0,
([Customer].[Customer].[Customer].MEMBERS,
    [Date].[Calendar].[Date].&[20010701]) ON 1
FROM [Adventure Works]
```

If you run this query, you'll get a list of all 18,485 customers, most of whom didn't buy anything on that particular day (and so will have a value of (null) in the cell). Instead, let's try using NONEMPTY in the query:

```
SELECT [Measures].[Internet Sales Amount] ON 0,
NONEMPTY(
    [Customer].[Customer].[Customer].MEMBERS,
    {([Date].[Calendar].[Date].&[20010701],
    [Measures].[Internet Sales Amount])}
) ON 1
FROM [Adventure Works]
```

The results here are shown in Table 9-9. Note that now we have just the five customers who made purchases on July 1. The way that NONEMPTY operates is to return all the tuples in a specified set that aren't empty. (Okay, that was probably obvious.) Where it becomes more powerful is when you specify two sets—then NONEMPTY will return the set of tuples from the first set that are empty based on a cross product with the second set.

Table 9-9. Using the NONEMPTY Function

--	Internet Sales Amount
Christy Zhu	$8,139.29
Cole A. Watson	$4,118.26
Rachael M. Martinez	$3,399.99
Ruben Prasad	$2,994.09
Sydney S. Wright	$4,631.11

Let's consider two dimensions, [Geography] and [Parts]. The [Geography] dimension has a hierarchy including [Region]. Now we want to create a report showing parts purchased by country for the countries in the North America region (Canada, Mexico, United States). If you look at the data in Figure 9-11, note that only five of the products out of eleven had sales in the countries we're interested in (shaded).

	Australia	Canada	France	Germany	Mexico	UK	USA
Bike Racks	$####						
Bike Stands		$####	$####			$####	
Brakes	$####	$####	$####	$####	$####		
Chains		$####	$####	$####	$####	$####	
Fenders							
Forks			$####	$####			
Helmets		$####					
Jerseys							
Pedals				$####	$####		
Saddles							

Figure 9-11. Using NONEMPTY

So we want to use the NONEMPTY function:

```
NONEMPTY(
[Parts].Members,
{
([Geography].[Region].[North America].Members, [Measures].[Sales Amount])
})
```

This will return the list of members of the Parts dimension that have values in the North America region for the Sales Amount measure, and will evaluate to {[Bike Stands], [Brakes], [Chains], [Helmets], [Pedals]}.

The second NONEMPTY function has just one argument:

```
NONEMPTY([Product].[Model Name].Children) ON ROWS
```

This will evaluate and return a set of members of the Model Name hierarchy that have values in the current context of the cube (taking into account default members and measures as well as the definition in the query).

Categories of Functions

Now that you know about functions, it's time to look at the different categories that are available to you. MDX offers functions relating to hierarchies, to aggregations, and to time.

Tree Functions

We've learned the importance of structure in OLAP dimensions—that's why we have hierarchies. Now we'll get into how to take advantage of those hierarchies in MDX. In SQL we learned to never presume what order records are in. If we needed a specific order, we had to ensure it in a query or view. In Analysis Services we define the order in the dimension (or there's a default order that doesn't change).

What this means is that we can operate on dimension members to find the next member or previous member. We can move up to parents or down to children. We do these with a collection of functions that operate on a member to get the appropriate "relative." For example, run the following query in SSMS:

```
SELECT [Measures].[Internet Sales Amount] ON 0,
[Date].[Calendar].[Month].[March 2002].Parent ON 1
FROM [Adventure Works]
```

This should return the Internet Sales for Q1, Calendar Year 2002, the parent of March 2002 in the Calendar hierarchy. In the preceding query, Parent is a function that operates on a member and returns the member above it in the hierarchy. If you execute Parent on the topmost member of a hierarchy, then SSAS will return a null.

You can "chain" Parent functions—for example, .Parent.Parent to get the "grandparent," or two levels up the hierarchy. But this becomes quickly painful. Instead there is the Ancestor() function to move up a dimensional hierarchy, as shown in the following query.

```
SELECT [Measures].[Reseller Sales Amount] ON 0,
Ancestor([Product].[Product Categories].[Product].[Chain],
    [Product].[Product Categories].[Subcategory]) ON 1
FROM [Adventure Works]
```

Ancestor() takes two arguments. The first is the dimension member to operate on, and the second is the hierarchy level to move up to. In the preceding query, we return the Subcategory for the [Chain] member of products. Ancestor() can also take a numeric value instead of a hierarchy level. In that case, the result returned is the specified number of steps *up* from the member in the first argument.

■ **Note** The second argument for Ancestor() is the *level name*, not a member name.

Now that we've moved up the tree from a given member, let's look at how we move down. If you think about a tree, you should quickly realize that while moving *up* a tree always gives us a specific member, moving *down* a tree is going to give us a collection of members. So it is that while the functions to move up return members, the functions to move down return sets. If you look at Figure 9-12, the ancestor of, for example, the Touring Bikes subcategory is the Bikes category (single member). The ancestor of the Road-150 product is the Road Bikes subcategory (single member). On the other hand, the descendants of the Bikes category are the Mountain Bikes, Road Bikes, and Touring Bikes subcategories (set of members).

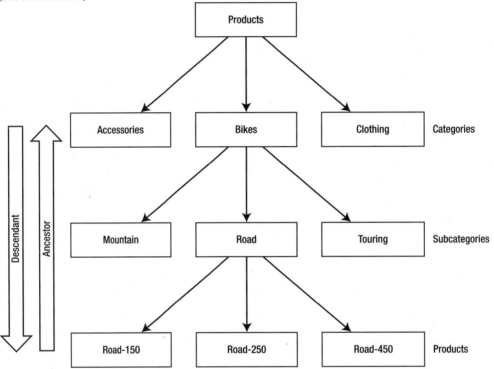

Figure 9-12. Ancestors vs. descendants in a tree

In the same vein, analogous to the .Parent function, we have .Children. However, as you have probably figured out, while .Parent returns a member, .Children returns a set. Try the following query:

```
SELECT [Measures].[Reseller Sales Amount] ON 0,
([Product].[Product Categories].[Bikes].Children) ON 1
FROM [Adventure Works]
```

You should get a breakdown of the reseller sales by the three subcategories under Bikes (Mountain Bikes, Road Bikes, Touring Bikes).

■ **Note** If you want to move down a hierarchy but return a single member, you can use the `.FirstChild` or `.LastChild` operators to return a single member from the next level in the hierarchy.

Aggregate Functions

One of the major reasons we've wanted to do the OLAP thing is to work with aggregated data. In all the examples you've seen to date, all the individual values are the result of adding the subordinate values (when we look at sales for Bikes in June 2003, we're adding all the individual bicycle sales for every model together for each day in June).

Let's look at an example:

```
WITH
    MEMBER [Measures].[Avg Sales] AS
        AVG({[Product].[Product Categories].CurrentMember.Children},
            [Measures].[Reseller Sales Amount])
SELECT
    NONEMPTY([Date].[Fiscal Quarter].Children *
    {[Measures].[Reseller Sales Amount],
    [Measures].[Avg Sales]}) ON COLUMNS ,
    NONEMPTY([Product].[Product Categories].Children) ON ROWS
FROM [Adventure Works]
WHERE ([Geography].[Country].[United States])
```

In this query, we've created a calculated measure using the `AVG()` function, which will average the value in the second argument across the members indicated in the first argument. In this case we've used `CurrentMember` to indicate to use the current selected member of the Product Categories hierarchy, and then take the children of that member. In the `SELECT` statement, we then use the cross product between the set of fiscal quarters and the set consisting of the Reseller Sales Amount measure and our calculated measure. (This produces the output that displays the two measures for each quarter.) Part of the output is shown in Figure 9-13.

	Q1 FY 2002	Q1 FY 2002	Q2 FY 2002	Q2 FY 2002	Q3 F
	Reseller Sales Amount	Avg Sales	Reseller Sales Amount	Avg Sales	Reseller S
Accessories	$6,479.87	$6,479.87	$8,607.94	$8,607.94	$3,
Bikes	$2,346,155.56	$1,173,077.78	$3,678,471.79	$1,839,235.89	$3,14
Clothing	$12,034.33	$4,011.44	$14,428.68	$4,809.56	$9,
Components	$190,981.36	$95,490.68	$294,916.32	$147,458.16	$132

Figure 9-13. The output of the AVG() query

We can use Excel to examine the data underlying our grid here. Connect Excel to the AdventureWorks cube and create a pivot table with Categories down rows, Fiscal Quarters across the columns, and add a filter for the United States. You can then drill down into the subcategories, as shown in Figure 9-14, and do the math yourself to check the averages.

	A	B	C	D	E
1	Geography	United States ▼			
2					
3	Reseller Sales Amount	Column Labels ▼			
4		⊟ FY 2002			
5		⊟ H1 FY 2002		⊟ H2 FY 2002	
6	Row Labels ▼	⊞ Q1 FY 2002	⊞ Q2 FY 2002	⊞ Q3 FY 2002	⊞ Q4 FY 2002
7	⊞ Accessories	$6,479.87	$8,607.94	$3,694.13	$9,144.48
8	⊟ Bikes	$2,346,155.56	$3,678,471.79	$3,148,390.07	$3,036,593.11
9	⊞ Mountain Bikes	$1,644,627.10	$2,223,571.21	$1,832,934.59	$1,600,710.16
10	⊞ Road Bikes	$701,528.46	$1,454,900.58	$1,315,455.49	$1,435,882.95
11	⊞ Touring Bikes				
12	⊞ Clothing	$12,034.33	$14,428.68	$9,165.23	$15,368.97
13	⊞ Components	$190,981.36	$294,916.32	$132,181.13	$302,021.98
14	Grand Total	$2,555,651.12	$3,996,424.72	$3,293,430.56	$3,363,128.54

Figure 9-14. Verifying the averages from the AVG() query

Another aggregate function that is very useful, and leverages the capabilities of SSAS well, is TOPCOUNT(). Often when you're doing analysis on business data, you'll see the 80/20 rule in action: 80 percent of your sales are in 20 percent of the regions, products, styles, and so forth. So you end up with charts that look like the one in Figure 9-15.

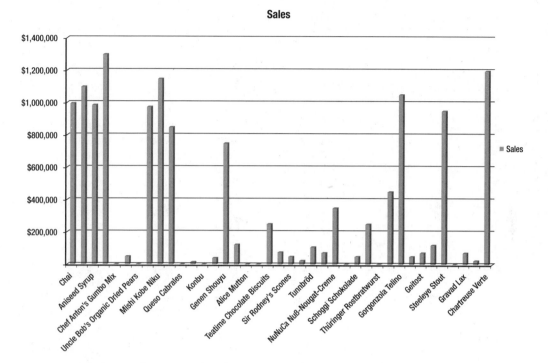

Figure 9-15. *Sales across a range of products*

What we're probably interested in is our top performers; for example, how did the top ten products sell? In T-SQL we can use a TOP function after sorting by the field we're interested in; in MDX we have TOPCOUNT().

Let's adjust our MDX query with the [Avg Sales] measure from before:

```
WITH
    MEMBER [Measures].[Avg Sales] AS
        AVG({[Product].[Product Categories].CurrentMember.Children},
            [Measures].[Reseller Sales Amount])
SELECT NONEMPTY([Date].[Fiscal Year].Children *
    {[Measures].[Reseller Sales Amount],
    [Measures].[Avg Sales]}) ON COLUMNS ,
    NONEMPTY([Product].[Product Categories].Subcategory.Members) ON ROWS
FROM [Adventure Works]
WHERE ([Geography].[Country].[United States])
```

First, I've changed the time dimension from fiscal quarters to fiscal years, just to make the results easier to see. Second, I've changed the SELECT statement to select Subcategory.Members from the [Product Categories] hierarchy. Note that I've changed .Children to .Members. Although the [Product Categories] hierarchy has a member named Members, the [Subcategory] level doesn't, so we use Children.

This query will return results as shown in Figure 9-16. If we just look at FY 2004, the average sales range from $462 to $362,000. In fact, more than 98 percent of our sales are concentrated in the top ten items.

	FY 2002	FY 2002	FY 2003	FY 2003	FY 2004	FY 2004
	Reseller Sales Amount	Avg Sales	Reseller Sales Amount	Avg Sales	Reseller Sales Amount	Avg Sales
Bike Racks	(null)	(null)	(null)	(null)	$94,320.52	$94,320.52
Bottles and Cages	(null)	(null)	(null)	(null)	$3,741.49	$3,741.49
Cleaners	(null)	(null)	(null)	(null)	$5,611.43	$5,611.43
Helmets	$27,926.42	$9,308.81	$61,271.31	$20,423.77	$61,150.61	$20,383.54
Hydration Packs	(null)	(null)	(null)	(null)	$30,201.49	$30,201.49
Locks	(null)	(null)	$10,229.98	$10,229.98	(null)	(null)
Pumps	(null)	(null)	$8,600.31	$8,600.31	(null)	(null)
Tires and Tubes	(null)	(null)	(null)	(null)	$461.66	$461.66
Mountain Bikes	$7,301,843.06	$912,730.38	$6,382,156.72	$638,215.67	$5,526,442.20	$276,322.11
Road Bikes	$4,907,767.48	$223,080.34	$9,356,485.02	$374,259.40	$6,520,506.29	$362,250.35
Touring Bikes	(null)	(null)	(null)	(null)	$4,837,550.96	$219,888.68
Bib-Shorts	(null)	(null)	$106,325.41	$35,441.80	(null)	(null)
Caps	$3,758.84	$3,758.84	$7,465.17	$7,465.17	$7,054.30	$7,054.30
Gloves	(null)	(null)	$104,178.17	$17,363.03	$24,178.42	$8,059.47
Jerseys	$41,675.62	$13,891.87	$85,470.15	$28,490.05	$190,604.57	$31,767.43
Shorts	(null)	(null)	$51,716.77	$17,238.92	$156,556.33	$52,185.44

Figure 9-16. Results of the Average Sales query across all subcategories

If we want to focus our business on the highest performers, we want to see just those top performers. So let's take a look at the query necessary:

```
WITH
    MEMBER [Measures].[Avg Sales] AS
        AVG({[Product].[Product Categories].CurrentMember.Children},
        [Measures].[Reseller Sales Amount])
SELECT
    NONEMPTY([Date].[Fiscal Year].Children *
    {[Measures].[Reseller Sales Amount],
    [Measures].[Avg Sales]}) ON COLUMNS ,
    TOPCOUNT(
        [Product].[Product Categories].Subcategory.Members,
        10,
        [Measures].[Reseller Sales Amount]) ON ROWS
FROM [Adventure Works]
WHERE ([Geography].[Country].[United States])
```

The only thing I've changed here is to add the TOPCOUNT() function to the rows declaration; this will return the top ten subcategories based on reseller sales. You'll notice that in several years the top sellers have a (null) for the sales amount—so why are they in the list? The trick to TOPCOUNT() is that it selects based on the set *as specified for the query*. In this case, we get the AdventureWorks cube sliced by the US geography, and the top subcategories are evaluated from that. After we have those top ten, *those* subcategories are listed by fiscal year.

Okay, now let's wrap up our tour of MDX queries with one of the main reasons we really want to use OLAP: time functions.

Time Functions

Much of the analysis we want to do in OLAP is time based. We want to see how data this month compares to the previous month, or perhaps to the same month last year (when looking at holiday sales, you want to compare December to December). Another comparison we often want to make is year to date, or YTD (if it's August, we want to compare this year's performance to the similar period last year, for example, January to August).

These queries in SQL can run from tricky to downright messy. You end up doing a lot of relative date math in the SQL language, and will probably end up doing a lot of table scans. In an OLAP solution, however, we're simply slicing the cube based on various criteria—what Analysis Services was designed to do.

So let's take a look at some of these approaches. We'll write queries to compare performance from month to month, to compare a month to the same month the year before, and to compare performance year to date.

We'll start with this query:

```
Select
    Nonempty([Date].[Fiscal].[Month].Members) On Columns,
    TopCount([Product].[Product Categories].Subcategory.Members,
    10,
    [Measures].[Reseller Sales Amount]) On Rows
From
    [Adventure Works]
Where
    ([Geography].[Country].[United States], [Date].[Fiscal Year].[FY 2004])
```

This will give us the reseller sales for the top ten product subcategories for the 12 months in fiscal 2004.

The first thing we want to look at is adding a measure for growth month to month:

```
WITH
    MEMBER [Measures].[Growth] AS
    ([Date].[Fiscal].CurrentMember,[Measures].[Reseller Sales Amount]) -
    ([Date].[Fiscal].PrevMember, [Measures].[Reseller Sales Amount])
SELECT NONEMPTY([Date].[Fiscal].[Month].Members
    * {[Measures].[Reseller Sales Amount], [Measures].[Growth]})
    ON COLUMNS ,
    TOPCOUNT([Product].[Product Categories].Subcategory.Members,
      10,
      [Measures].[Reseller Sales Amount])
    ON ROWS
FROM [Adventure Works]
WHERE ([Geography].[Country].[United States], [Date].[Fiscal Year].&[2004])
```

We've added a calculated measure ([Measures].[Growth]) that uses the .CurrentMember and .PrevMember functions on the [Date].[Fiscal] hierarchy. (Note that you can't use the member functions on [Date].[Fiscal].[Month], as you may be tempted to—you can use them only on a hierarchy, not a level.) This will give us a result as shown in Figure 9-17.

	July 2003	July 2003	August 2003	August 2003	September 2003	September 2003
	Reseller Sales Amount	Growth	Reseller Sales Amount	Growth	Reseller Sales Amount	Growth
Road Bikes	$552,109.24	-62240.7274	$763,628.50	211519.2598	$575,531.55	-188096.9547
Mountain Bikes	$368,086.87	-27183.0994	$682,742.05	314655.1813	$522,222.00	-160520.0457
Touring Bikes	$379,339.15	379339.1469	$287,751.84	-91587.3043	$552,774.64	265022.7963
Mountain Frames	$100,664.69	-4871.5102	$211,653.53	110988.8394	$184,368.86	-27284.6769
Road Frames	$122,400.71	13058.6889	$179,865.49	57464.7737	$164,378.71	-15486.78
Touring Frames	$106,469.68	106469.6835	$72,821.19	-33648.4897	$135,276.01	62454.8191
Jerseys	$22,493.82	15664.7452	$24,472.80	1978.9754	$24,265.22	-207.5768
Shorts	$11,365.67	6218.5272	$21,279.40	9913.7336	$16,329.84	-4949.56
Vests	$11,325.13	11325.1299	$12,967.25	1642.1163	$11,562.63	-1404.6201
Cranksets	$5,912.79	5912.79	$13,794.16	7881.3698	$19,536.59	5742.4342

Figure 9-17. Showing sales change month to month

Of course, the numbers look awful. Let's add a FORMAT() statement to our calculated measure:

```
WITH
    MEMBER [Measures].[Growth] AS
      Format((([Date].[Fiscal].CurrentMember,
      [Measures].[Reseller Sales Amount]) -
      ([Date].[Fiscal].PrevMember,
      [Measures].[Reseller Sales Amount]), "$#,##0.00")
```

Now we'll get results that look like Figure 9-18.

	July 2003	July 2003	August 2003	August 2003
	Reseller Sales Amount	Growth	Reseller Sales Amount	Growth
Road Bikes	$552,109.24	-$62,240.73	$763,628.50	$211,519.26
Mountain Bikes	$368,086.87	-$27,183.10	$682,742.05	$314,655.18
Touring Bikes	$379,339.15	$379,339.15	$287,751.84	-$91,587.30

Figure 9-18. Formatting our calculated measure

By formatting the measure in the MDX, we find that when we use any well-behaved front end, we get the same, expected, formatting.

Our bike sales are probably seasonal (mostly sold in the spring and summer, with a spike at Christmastime), so comparing month to month may not make sense. Let's take a look at comparing each month with the same month the previous year. For this, we just need to change our calculated measure as shown:

```
WITH
    MEMBER [Measures].[Growth] AS
      Format((([Date].[Fiscal].CurrentMember,
      [Measures].[Reseller Sales Amount]) -
      (ParallelPeriod([Date].[Fiscal].[Fiscal Year], 1,
```

```
        [Date].[Fiscal].CurrentMember),
    [Measures].[Reseller Sales Amount]), "$#,##0.00")
```

In this case, we're using the ParallelPeriod() function. This function is targeted toward time hierarchies; given a member, a level, and an index, the function will return the corresponding member that lags by the index count.

■ **Note** The index in ParallelPeriod() is given as a positive integer, but counts *backward*.

The easiest way to understand this is to look at the illustration in Figure 9-19.

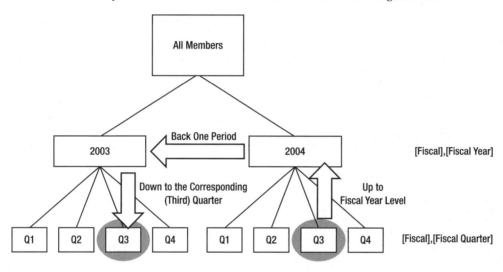

Figure 9-19. Understanding how ParallelPeriod() works

Our final exercise here is to figure out how our revenues compare to last year if it's August 15. We can't compare to last year's full-year numbers, because that's comparing eight months of performance to twelve. Instead, we need last year's revenues through August 15. First, of course, we need to calculate the year-to-date sales for our *current* year. For this we use the PeriodsToDate() function, as shown here:

```
SUM(PeriodsToDate([Date].[Fiscal].[Fiscal Year],
    [Date].[Fiscal].CurrentMember),
    [Measures].[Reseller Sales Amount])
```

The PeriodsToDate() function takes two arguments: a hierarchy level and a member. The function will return a set of members from the beginning of the period at the level specified up to the member specified. You can see this is pretty generic; we could get all the days in the current quarter, all the quarters in the current decade, or whatever.

■ **Note** There is a YTD() function in MDX, which is a shortcut for PeriodsToDate. The only problem is that it will work with only calendar years, so here we're using the more abstract function.

Now let's calculate the YTD for last year. We'll do this by using ParallelPeriod() to get the matching time period last year, and then running the PeriodsToDate() function to return the set of members from last year up to that member. Here we go:

```
WITH
    MEMBER [Measures].[CurrentYTD] AS
      SUM(PeriodsToDate([Date].[Fiscal].[Fiscal Year],
        [Date].[Fiscal].CurrentMember),
        [Measures].[Reseller Sales Amount])

    MEMBER [Measures].[PreviousYTD] AS
      SUM(PeriodsToDate([Date].[Fiscal].[Fiscal Year],
        ParallelPeriod([Date].[Fiscal].[Fiscal Year],
          1,
          [Date].[Fiscal].CurrentMember)),
        [Measures].[Reseller Sales Amount])

SELECT [Date].[Fiscal].[Month].Members
    * {[Measures].[CurrentYTD],
    [Measures].[PreviousYTD]}
    ON COLUMNS ,
    TOPCOUNT([Product].[Product Categories].Subcategory.Members,
      10,
      [Measures].[Reseller Sales Amount])
    ON ROWS
FROM [Adventure Works]
WHERE ([Geography].[Country].[United States],
    [Date].[Fiscal Year].&[2004])
```

This returns the results shown in Figure 9-20.

	July 2003	July 2003	August 2003	August 2003	September 2003	September 2003	
	CurrentYTD	PreviousYTD	CurrentYTD	PreviousYTD	CurrentYTD	PreviousYTD	
Road Bikes	$552,109.24	$731,519.63	$1,315,737.75	$1,803,278.07	$1,891,269.30	$2,603,505.78	
Mountain Bikes	$368,086.87	$466,129.19	$1,050,828.91	$1,261,511.74	$1,573,050.91	$1,897,076.12	
Touring Bikes	$379,339.15	(null)	$667,090.99	(null)	$1,219,865.63	(null)	
Mountain Frames	$100,664.69	$94,552.34	$312,318.23	$280,771.92	$496,687.08	$425,282.15	
Road Frames	$122,400.71	$179,547.51	$302,266.20	$439,426.49	$466,644.90	$654,979.36	
Touring Frames	$106,469.68	(null)	$179,290.88	(null)	$314,566.89	(null)	
Jerseys	$22,493.82	$9,245.30	$46,966.62	$20,299.74	$71,231.85	$30,645.08	
Shorts	$11,365.67	$4,457.55	$32,645.07	$10,407.91	$48,974.91	$17,496.48	
Vests	$11,325.13	(null)	$24,292.38	(null)	$35,855.00	(null)	
Cranksets	$5,912.79	(null)	$19,706.95	(null)	$39,243.54	(null)	

Figure 9-20. *The results of the YTD queries*

Note that even though I've removed the FORMAT() statement, we've retained the formatting—an artifact of summing formatted values. You could still put the statement back if you wanted to ensure how the numbers would be represented.

Summary

That's a high-speed tour of MDX, and hopefully now you know enough to be dangerous. Just as T-SQL is far richer than just the basic SELECT statement, MDX goes much deeper than what we've covered here. My goal was to give you some idea of using MDX against a cube to frame your understanding of MDX in general. In the next chapter, we're going to use MDX in some more advanced features of SQL Server Analysis Services.

■ ■ ■

CHAPTER 10

Cube Features

This chapter is something of a grab bag of features that build on the foundation of an OLAP cube to provide additional usability for end users. The first section, "Business Intelligence," includes those enhancements that can be added with the Business Intelligence Wizard, including *time intelligence* (period-to-date, rolling average, and so forth), *account intelligence* (asset, liability, income, expense), the *unary operator* (to indicate how values should be aggregated up a hierarchy), *attribute ordering*, and *currency conversion*.

Then you'll work with some of the tools available via the Calculations tab. You'll spend some time on *calculated members*, building measures that derive from other measures. A common example of a calculated measure is an average function, dividing a total value by the number of units (for instance, test scores divided by the number of students). This will also take us into the script organizer, where you'll look at some of the functions available there, and the template manager.

Named sets are also found on the Calculations tab, and provide a way of presenting a predefined collection of dimension members. You can either build a fixed set that is a common subset of members (for example, a group of location members that are manufacturing plants or retail stores), or a dynamic set that is determined on demand ("top sellers this year").

In the final section, you will look at four other areas of the Cube Designer. *Key performance indicators (KPIs)* provide a way to encapsulate business performance. A lot of the analysis performed on analytic data is to determine how a business is performing (or why it isn't performing). A key performance indicator is a quick, visual way to evaluate performance against a specified goal. Analysis Services enables you to define KPIs in a cube so that everyone is using the same metrics.

Actions are an interesting and, in my opinion, underused feature of Analysis Services. Actions are passed through to a client that is capable of implementing them. They are forwarded to the user, usually in a context menu, providing some kind of amplifying information regarding the selected data. For example, an action can launch a browser to a specific URL, or it may execute an MDX statement.

I showed you a little bit about *perspectives* earlier in the book. When you consider how vast a robust cube can get, working through all the various dimensions and measures can be painful. SSAS provides a way to filter all the features of a cube based on various functions. For example, a finance user will see only those dimensions, measures, actions, and so on that concern financial functions.

Translations provide a way to enable cubes for multilingual use. In the Translations tab, you can create a set of translated terms and phrases for the visible features of the cube.

So let's dig in and take a look at what we can do with our cubes.

Business Intelligence

You can add business intelligence features to a cube or a dimension by invoking the Business Intelligence Wizard, as shown in Figure 10-1. First, you need to either select your cube in the Solution Explorer, or ensure that your cube is open and selected in the Cube Designer. Invoke the wizard from the

BIDS menu bar by clicking Cube ➤ Add Business Intelligence. From the wizard you can select the enhancement you want to add. It's as simple as selecting an option and clicking the Next button.

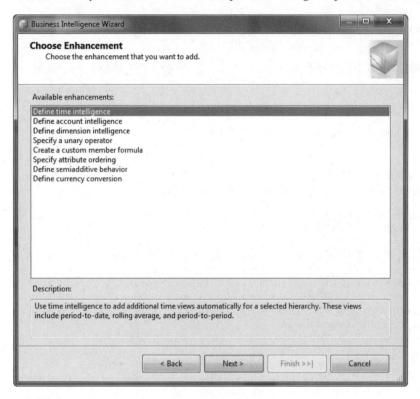

Figure 10-1. *The Business Intelligence Wizard*

The wizard makes changes to the objects involved and their metadata to support the feature you're adding. If it's necessary to select a dimension, measure, or other object, the wizard will prompt you as appropriate.

Time Intelligence

The Time Intelligence Enhancement enables you to select a specific time hierarchy, and then select from a list of time-based calculations to add to that hierarchy. The calculations offered will depend on the hierarchy selected (for example, you can't add Year Over Year Growth to a fiscal-quarter hierarchy). The following list shows a few of the time calculations available to you via the wizard:

> **Period to date**: These will give you shortcut calculations for year to date, semester to date, quarter, month, and 12 months to date. These will return the aggregation of a measure value from the beginning of the time period to the current date.

Moving averages: These will calculate an average for each date for the time period prior. For example, a three-month moving average will calculate the average from January through March for April, February through April for May, and so on.

Periodic growth: These provide either actual value or percentage change from one time period to another—for example, year over year. This calculates the year-to-date value for this year and the previous year and performs the comparison between the two of them.

After selecting the calculations, you will have to select the measures you want the new time views applied to, as shown in Figure 10-2. Because every dimension on a measure increases the processing time when processing the cube (to calculate and preload aggregations), we want to be cautious when adding these calculated measures, and add them to the measures only where we truly need them for reporting.

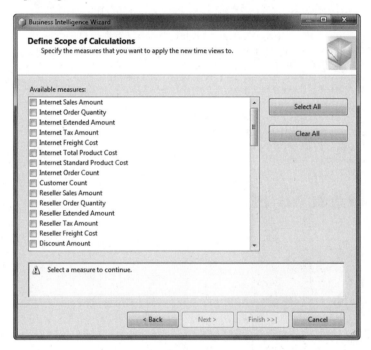

Figure 10-2. Selecting measures for time intelligence

After you select the measures and click Next, the final page of the wizard presents a summary of the changes to be made to the cube by the wizard. After you review these, click Finish, and the wizard makes the necessary changes to the cube's metadata:

- Create the named calculation in the time dimension in the data source view (the underlying data source isn't affected).

- Create a dimension attribute in the time dimension mapped to the new calculated column. The dimension attribute will have its IsAggregatable value set to false, to indicate that the value needs to be calculated at each hierarchy level and can't just be summed up.

- In addition, a static default member will be created to be the default member for use when slicing the cube (remember that when you slice a cube, every dimension must have a member selected—making the default member here a static member aids scalability).

After those changes are made, you just have to deploy the cube and process it. The calculations will show up as an attribute hierarchy under the time dimension you selected in the wizard. You can then use them like any hierarchy and members.

Account Intelligence

Another enhancement available in the Business Intelligence Wizard is Define Account Intelligence. To understand account intelligence, consider an account book—columns of assets and liabilities. We add assets together and subtract liabilities; you'll look at how to handle this later in the "Unary Operators" section. Then there's the issue of income vs. assets: income is a recurring value that should be added together, but assets are fixed, and shouldn't. For example, retail receipts are recorded every day, and need to be added together to give the value for a month, a quarter, or a year. On the other hand, assets are recorded each month—15 motor pool cars on hand on January 1, February 1, March 1, and so forth. We can't total these numbers, or we'd end the year with 180 autos. So to report on our assets, we need to take the last number reported.

Account intelligence enables us to annotate a dimension to indicate how to handle various values in a measure so they are aggregated appropriately in a cube. You select a dimension in the wizard to apply the changes to (for example, the Account dimension), and then map the attributes of the dimension to the attributes that Analysis Services is expecting, as shown in Figure 10-3.

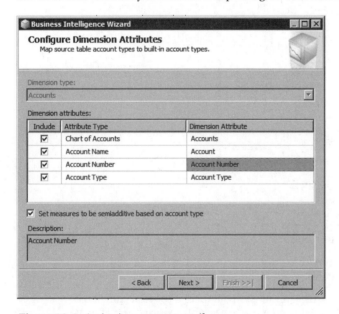

Figure 10-3. Assigning account attributes

The next step is to map the Account Type in the dimension to the built-in account types, which SSAS understands how to handle. The options here are as follows:

Asset or Liability: These are fixed values that should not be aggregated over time. Instead Analysis Services takes the last nonempty value and then converts it based on the currency conversion rate.

Expense or Income: These values will be summed over time, and will also be converted based on the currency rate.

Balance: A recorded bookkeeping value, SSAS will take the last nonempty value, and convert it to local currency.

Statistical: A nonmonetary value, this does not aggregate over time and is not converted. If there is no value in place, none will be reported (no roll-up).

Flow: A numerical value that aggregates over time (such as income or expense), but is not converted with a currency exchange rate.

The comments about exchange rates are another factor in account business intelligence. This wizard establishes the metadata on accounts for use by the currency conversion put in place by the Currency Conversion Business Intelligence Wizard.

After the account attributes are defined, the final wizard page summarizes the changes to be made. Clicking the Finish button makes the changes. You will have to deploy and process the cube for the changes to take effect.

Dimension Intelligence

Dimension intelligence is the more abstract version of account intelligence. Essentially, both these functions assign well-known types to dimensions to normalize them and make them more recognizable. Then client applications can take advantage of the dimension in a standard way. For example, assigning the Product dimension type to the Product dimension, and the well-known attribute types to the attributes, could enable a client application to use the dimensions to fill catalog pages. Whereas account business intelligence triggers specific behaviors from Analysis Services itself, dimension intelligence tags only the dimension and attributes for use by other applications.

Operators, Functions, and More

Part of working with a cube is to have the right data. But usually you'll want to do something with that data, manipulate it in different ways so that you can analyze it. You have the ability to manipulate data via predefined unary operators, and through custom expressions that you write yourself by using the Business Intelligence Wizard. You can order attributes in a dimension. You can even deal with the seemingly trivial, but really quite important issue of currency conversion.

Unary Operators

Specifying a unary operator goes hand in hand with adding account intelligence to a cube. As I pointed out in the section on account intelligence, an accounts ledger will have income and expenses, assets and liabilities. In account intelligence, we specified that the first pair should be added over time, and the second pair should not.

What do we do in cases where we have income and expenses in a list of accounts? Income is added, and expenses subtracted. Assets have positive value, and liabilities are negative. How do we ensure that the proper signs are accounted for?

If you look at the accounts table that the dimension is based on (open the Adventure Works data source view, find the Accounts table, right-click, and select Explore Data), you'll see a column named Operator, as shown in Figure 10-4. Running the Unary Operator Enhancement enables you to map this attribute to a specific operator attribute that Analysis Services understands to apply when summing members of the hierarchy.

AccountKey	ParentAccountKey	AccountCodeAlternateKey	ParentAccountCodeAlternateKey	AccountDescription	AccountType	Operator	CustomMembers	ValueType
1		1		Balance Sheet		~		Currency
2	1	10	1	Assets	Assets	+		Currency
3	2	110	10	Current Assets	Assets	+		Currency
4	3	1110	110	Cash	Assets	+		Currency
5	3	1120	110	Receivables	Assets	+		Currency
6	5	1130	1120	Trade Receivables	Assets	+		Currency
7	5	1140	1120	Other Receivables	Assets	+		Currency
8	3	1150	110	Allowance for Bad Debt	Assets	+		Currency
9	3	1160	110	Inventory	Assets	+		Currency
10	9	1162	1160	Raw Materials	Assets	+		Currency
11	9	1164	1160	Work in Process	Assets	+		Currency

Figure 10-4. *The Operator column in the underlying data source*

Figure 10-5 shows the Account dimension in the browser in BIDS. The column to the right shows the Unary Operator attribute, but more importantly, if you look at the icon to the left of each member, you'll see a small + or – sign, indicating how that member will operate when summed.

⊟ 🌑 Net Income	+
⊟ 🌑 Operating Profit	+
⊞ 🌑 Operating Expenses	-
⊞ 🌑 Gross Margin	+
⊞ 🌑 Other Income and Expense	+
🌑 Taxes	-

Figure 10-5. *Members of the Account hierarchy, showing unary operators*

If you go to the cube browser in BIDS and pull the Accounts hierarchy to rows, and then pull Financial Reporting ➤ Amount member to the values, you can browse through the various accounts to see how the account mapping and unary operator affect the various totals.

Custom Member Formulas

The Create a Custom Member Formula enhancement, when chosen from the Business Intelligence Wizard, enables you to indicate a specific formula that will be used to define the members of a dimension. The custom member formula will be stored as an MDX statement in a column in the data source for the dimension. The wizard here is very straightforward; select a dimension, and then select the attribute that will use the custom formula, and the column where the formula is located.

Attribute Ordering

The Attribute Ordering Enhancement is a quick way to clean up how you want all the attributes ordered in a dimension. You may not always want attributes ordered alphabetically or by the index. For example, you may want the abbreviations for items ordered in accordance with the full names of the members of the dimension.

The wizard will (as usual) have you select a dimension, and then will present a list of attributes and allow you to select how to order them, as shown in Figure 10-6. For each attribute, you can select the attribute to order it by, and which aspect of that attribute (the key or the name) to order it by.

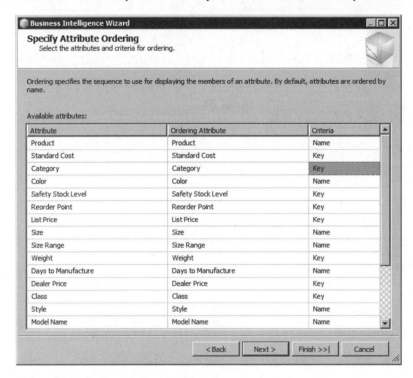

Figure 10-6. *Setting attribute ordering*

After making any changes, the Next button will take you to the standard summary page, and clicking Finish will affect the changes.

Currency Conversion

I mentioned currencies in the section on account intelligence. SQL Server Analysis Services has the ability to automatically convert currency in order to unify reporting. In order to do so, the currency conversion feature must be enabled by using this option in the BI Wizard.

After selecting the Currency Conversion option, the next page enables you to select the measure group that contains exchange rates to be used in the conversions, as shown in Figure 10-7. After you've selected an exchange rate measure, the wizard will list the values in the attribute group marked as destination currency to select a pivot currency.

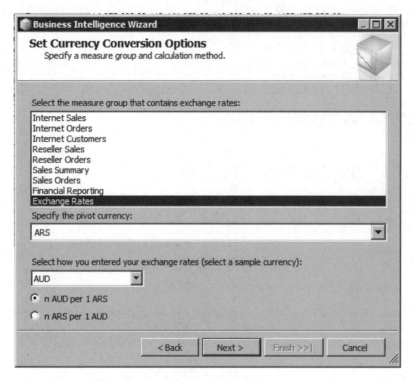

Figure 10-7. Setting up currency conversions

The *pivot currency* is the currency for which all the other exchange rates are defined. For example, if you choose US dollars (USD), then for USD in the exchange rate table, 1USD = 1USD. The next entry enables you to define in which direction the exchange rate is defined.

The next page of the wizard enables you to select where to apply the exchange rate metadata changes; you can select one or more measures, the Account hierarchy, or members of the Account hierarchy. After that, you select whether the cube is defined with one currency that needs to be converted to multiple local currencies, multiple local currencies that must be converted to a single central currency, or if conversions will happen between multiple currencies. Finally, you select where the destination currency code is defined in the measure, and what currencies it will be necessary to report on. Finish the wizard and reprocess the cube.

Calculations Tab

Next we're going to take a look at calculated measures and named sets. These are accessible on the toolbar of the Calculations tab. A typical view of the Calculations tab is shown in Figure 10-8, which displays the Script Organizer pane, the Calculation Tools pane, and the MDX Form view.

New Calculated Member   New Named Set

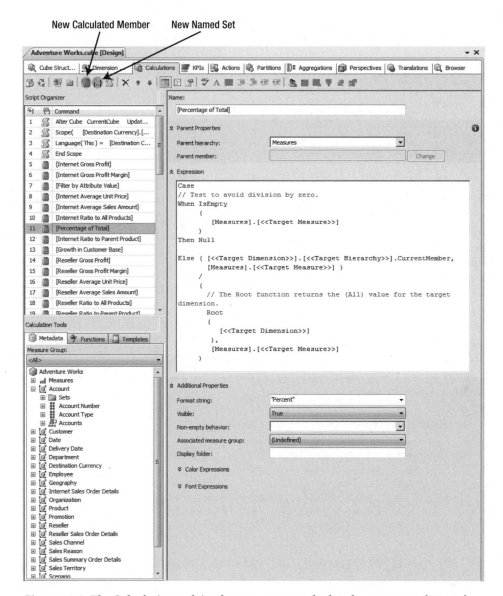

Figure 10-8. The Calculations tab is where you create calculated measures and named sets.

Calculated Measures

When there's no other way to make the numbers do what you need them to, you can always write MDX. Calculated measures enable you to create extended totals, represent taxes, perform averages, calculate net and gross values, and so on. After you've created a calculated measure, it's available just as any other measure.

Open the AdventureWorks cube in BIDS, and then select the Calculations tab. This is where you'll find script commands, calculated members, and named sets (which are all defined with bits of MDX script). In the top left is the Script Organizer, which lists all the existing scripts, with an icon to indicate script command, calculated measure, or named set, as shown in Figure 10-9.

Script Organizer			
⚵	🗒	Command	▲
24	{..}	[Negative Margin Products]	
25	{..}	[New Product Models FY 2006]	
26	{..}	[New Product Models FY 2007]	
27	{..}	[New Product Models FY 2008]	
28	{..}	[Long Lead Products]	
29	{..}	[Core Product Group]	
30	{..}	[Large Resellers]	
31	{..}	[High Discount Promotions]	
32	🖩	[Gross Profit]	
33	🖩	[Gross Profit Margin]	
34	🖩	[Expense to Revenue Ratio]	
35	🖩	[Ratio to All Products]	
36	🖩	[Ratio to Parent Product]	
37	📜	Scope ([Date].[Fiscal Year].&[2009], [Date].[Fiscal].[Fiscal Quarter].Member...	
38	📜	This = ParallelPeriod ([Date].[Fiscal].[Fiscal Year], ...	
39	📜	Scope ([Date].[Fiscal Year].&[2006], [Date].[Fiscal].[Month].Members)	
40	📜	This = [Date].[Fiscal].CurrentMember.Parent / 3	
41	📜	End Scope	
42	📜	Freeze ([Date].[Fiscal].[Fiscal Quarter].Members, [Measures].[Sales Amou...	

Figure 10-9. The Script Organizer pane

If you select any script with a calculator icon, you'll see the calculated measure window come up in the center. Shown in Figure 10-10 is the definition for [Internet Ratio to All Products]. You can see that you define the name and select a parent hierarchy, as well as a parent member in a multilevel hierarchy. At the bottom are formatting options: the format string, whether or not the member is visible, a folder to use, font, and color.

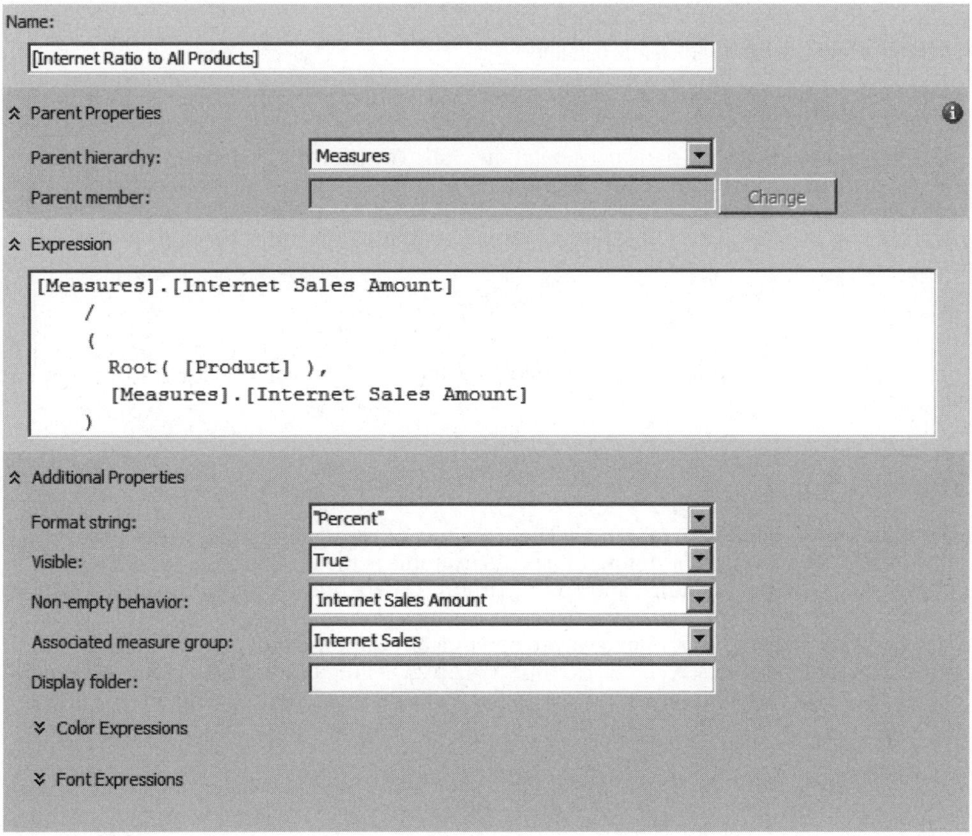

Figure 10-10. *Calculated measure for products ratio*

The Non-Empty Behavior setting and Associated Measure Group are used to point to a measure that SSAS can use as a guide if the calculated measure should be evaluated. If the cell in the associated measure group is empty, the calculated measure won't be evaluated.

Last but not least is the Expression in the center window. The expression is MDX. In this case, we have a pretty straightforward expression:

```
[Measures].[Internet Sales Amount]
    /
    (
      Root( [Product] ),
      [Measures].[Internet Sales Amount]
    )
```

This is saying that for any cell (sliced by any dimensions), the ratio of sales to all products is the Internet Sales Amount divided by value of the Internet Sales Amount at the root of the Product hierarchy. In other words, the ratio is the value of the current product, category, or subcategory divided by the total of products sold.

■ **Tip** The Expression window autoexpands for its contents.

In the lower-left corner of BIDS is a pane of calculation tools. The Metadata tab lists all the objects in the cube; you can drag and drop to get an idea of proper syntax for defining measures, dimensions, and other objects. The Functions tab lists all the MDX functions available and is good as a quick reference. When you drag a function over, it will have a small bit of hint text indicating how it should be used. For example:

```
«Hierarchy».ALLMEMBERS
```

Finally, the Templates tab lists some useful complex templates that you can use as a starting point if you need to go that route.

Named Sets

Named sets are essentially a collection of members. There are two types of named sets: static and dynamic. *Static named sets* have a fixed set of members. An example is the set Southeastern States, which consists of Florida, Georgia, Alabama, and South Carolina. There may be a specific set of products you always want to keep an eye on, or certain machines in the shop you want a report on.

The other type, *dynamic named sets*, offer a way to dynamically filter the members of a dimension based on some aspect of the measures. A classic example is Top 25 Selling Products. To create a named set, we simply enter the MDX expression that defines the set. For example, Top 25 Selling Products is as follows:

```
TopCount
    (
        [Product].[Product].[Product].Members,
        25,
        [Measures].[Sales Amount]
    )
```

This is pretty easy to figure out—this will return a set of the top 25 products based on the total sales. But notice that we're filtering based on the total sales amount; this will return our all-time top 25 selling products. If you want the top 25 products for a given year, you'll have to write a query and make that part of the SELECT statement.

Other Cube Features

In this final section, you will look at four other areas of the Cube Designer. Key performance indicators offer you several ways to measure and trend your business's performance, while actions can pass XMLA data from your cube to your OLAP client. You will finish this chapter with perspectives and translations.

Key Performance Indicators

Key performance indicators (KPIs) are metrics that enable you to define a target value for a specific measure value. A standard example is revenue. Every company has specific revenue targets, and they want to understand how they are performing relative to that target. One problem that many organizations have faced is differences in how a metric is defined. When you consider revenue, should returns be subtracted back out? What about recalls?

An important consideration here is that the actual definition isn't as critical as consistency. When you are trying to identify root-cause problems, the ability to compare "apples to apples" across an organization makes the difference between easy analysis and simply not being able to do any analysis.

If we embed our metrics into our data mart, they are standardized across the data structure. If *revenue* is defined as product sales, not including tax, minus returns, but not deducting recalls, then by defining the metric in the cube, everyone will be on the same page. Also, by defining the KPI in the cube, we can get a uniform representation of the indicator in various user interfaces. In Figure 10-11, you can see as Analysis Services KPI in Excel.

	A	B	C	D	E
1		Column Labels ▼			
2		FY 2008			
3	Row Labels ▼	Internet Revenue	Goal	Ind.	Trend
4	⊞ Accessories	$667,015.32	$667,015.32 △		⬆
5	⊟ Bikes	$15,483,926.11	$6,338,347.73 ●		⬆
6	⊞ Mountain Bikes	$6,397,228.51	$2,435,851.02 ●		⬆
7	⊞ Road Bikes	$5,241,896.55	$3,902,496.71 ●		⬆
8	⊞ Touring Bikes	$3,844,801.05	$3,844,801.05 △		⬆
9	⊞ Clothing	$322,676.62	$322,676.62 △		⬆
10	Grand Total	$16,473,618.05	$6,338,347.73 ●		⬆

Figure 10-11. Analysis Services KPIs in Excel

KPIs are set up on the KPIs tab of the Cube Designer in BIDS. The lower-left pane is the Calculation Tools collection you learned about from the Calculations tab. In the top left is the KPI organizer, which lists all the KPIs in the cube. In the center is the KPI Designer. In the designer, you select the measure group for the KPI to be associated with (although because all the KPI metrics are MDX, you're not restricted to that measure group for defining value, goal, and so forth).

The Value Expression is very straightforward; this is generally going to be a measure value, although you could enter a calculation as well. The Goal Expression (target) may be a lookup from another table, a set of fixed values, or a calculation based on other measures (for example, the Internet Revenue KPI uses the Internet Sales Amount for the previous year plus ten percent).

The Status, which drives the status indicator (shown in Figure 10-12), enables you to select an indicator type and then enter a status expression. Note the expression is a CASE statement that evaluates the value as compared to the target, and returns –1, 0, or 1. Each of those values will be translated by the display interface to match up to a state of the status indicator (in this case, –1 is Red, 0 is Yellow, and 1 is Green).

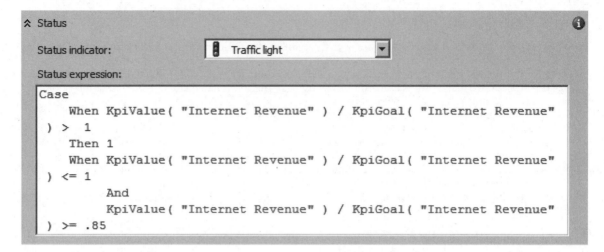

Figure 10-12. The KPI status indicator in BIDS

■ **Note** An indicator is sent to the client via XMLA only as a text value (for example, Traffic light/Red). So there is likely to be some degree of visual difference in the user interface. But the essential part should be consistent as described.

The next section is a trend indicator. This works exactly like the status indicator—a graphical image and an MDX expression that returns values to set the appearance of the indicator. In this case, the indicator is an arrow that points up, across, or down to indicate the trend of the KPI. The MDX here, as you can see, compares the current value to the previous value.

The trend indicator uses the `ParallelPeriod` function to dynamically select the previous fiscal year. The member for the previous fiscal year, combined with the member for the KPI value, makes a tuple for the KPI value for the previous year. The equation then boils down to this:

`(Current value - Previous Value)/Previous Value`

This results in a normalized value for the trend—positive is up, and negative is down. The statement compares the normalized value to +/− .02 to allow for some rounding error on the "no change" arrow.

The additional properties allow for definition of a folder to organize the KPIs, assignment of a parent KPI to create a hierarchy, current time member to create a KPI that changes with the calendar, and weighting. When you have a collection of KPIs, you may want some considered more strongly than others when rolling them together. (Or give the KPI a weight of zero and it won't get rolled in.)

■ **Note** Although it's great to have KPIs centralized in the Analysis Server, there are a lot of capabilities in a scorecard that aren't provided—custom indicators, multiple targets, associated reports, and so on. If you're interested in a more powerful scorecard platform, check out one of Philo's books: *Pro PerformancePoint Server 2007* (Apress, 2008) if you're on MOSS 2007, or *Building Integrated Business Intelligence Solutions with SQL Server 2008 R2 and Office 2010* (McGraw-Hill, 2010) if you're on SharePoint Server 2010.

Actions

Very often in a business intelligence solution, you want to provide your end users with some type of amplifying detail— detailed chart for a measure value, a map to show the location of a store that's a member of a dimension, or a form with some of the attribute values for a member.

OLAP solutions have a standard metadata construct called an *action* that can be attached to a measure value. The data is passed to the client through XMLA; it's up to the client to implement the action in some form. Clients have various levels of compliance with SSAS actions. Your best bet is always to experiment directly with a client to see whether an action you are contemplating is implemented. Actions are defined on the Actions tab for the cube. There are three types of actions, defined when you create the action: general action, drillthrough action, and reporting action.

A *drill-through action*, shown in Figure 10-13, defines a record set to be returned to the client. When the user chooses to execute an action on a cell containing a measure value, SSAS will return a defined recordset of the underlying data values, with the columns defined in the Drillthrough Columns section. Also note the Condition section, which enables you to limit the scope of the recordset returned. For example, if there are 20 years of detailed manufacturing data in the data mart, you may not want end users trying to drill down on all 20 years worth of data. In that case, you could put a filter that would restrict drill-down data to the last two years. You can also limit the size of returned recordsets with the Maximum rows value.

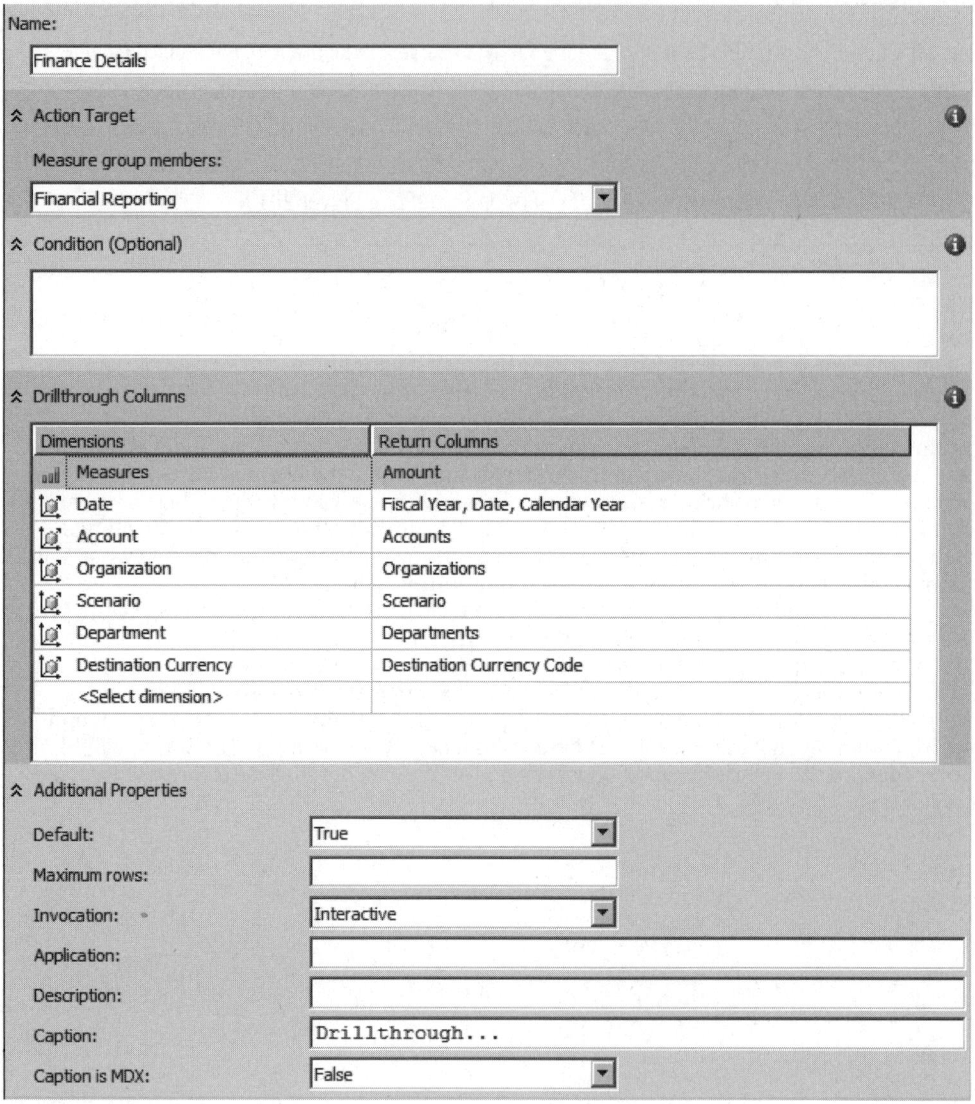

Figure 10-13. *A drill-through action*

To execute the action, open the context menu for a value in Excel. Select the Additional Actions option, and you should see any available actions in the flyout, as shown in Figure 10-14. For the drill-through action, Excel will create a new sheet in the current workbook and place the data table there. You can have more than one action available for a given measure. All actions will be listed on the context menu.

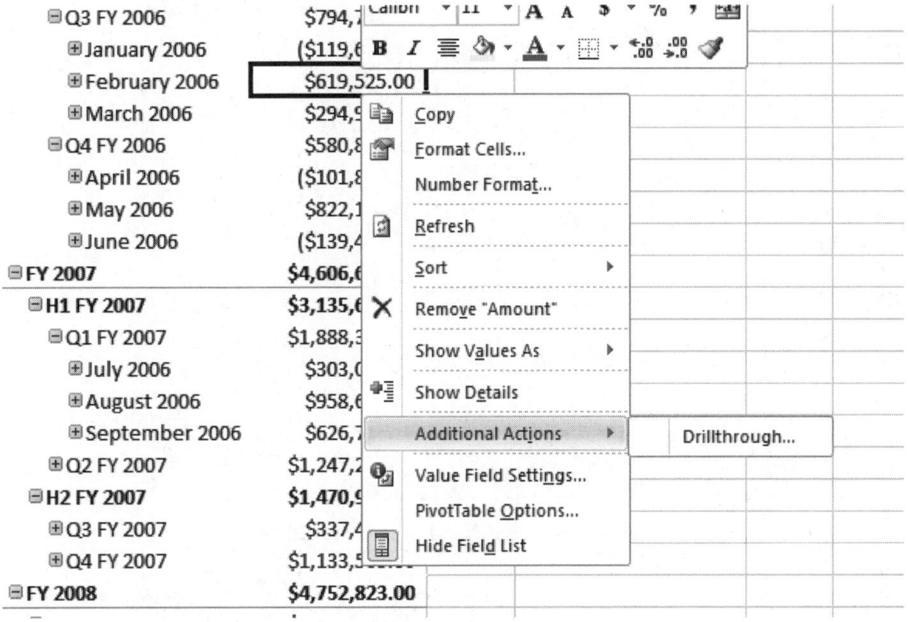

Figure 10-14. *Invoking an action via Excel*

A *reporting action* will open a SQL Server Reporting Services report when invoked, as shown in Figure 10-15. Note that the Target type allows selecting from dimensions, members, and attributes. So you can create a report that shows, for example, the product data from a specific product if selected. You select the target type, and then the target object will populate with the appropriate objects to select from. Then you configure the report server and report, and add any parameters you need to add. (Most important—if you want a contextual report, you'll need to pass some kind of identifying information to the report!)

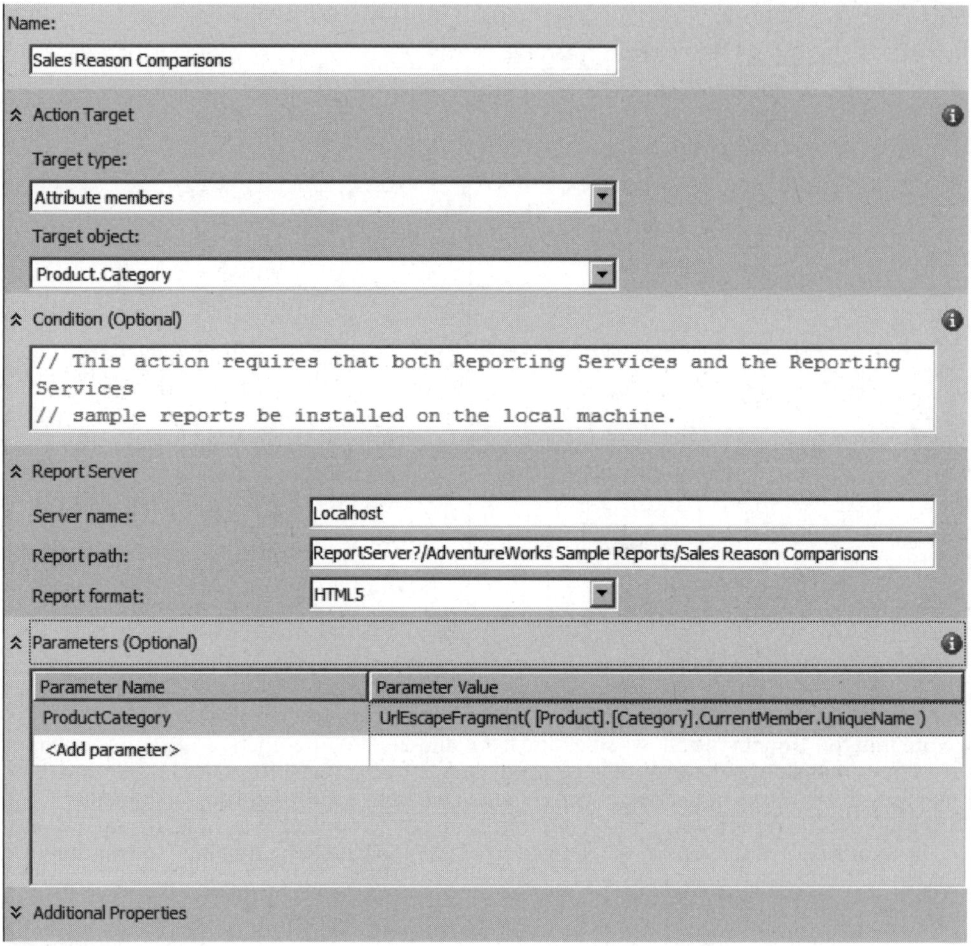

Name:

Sales Reason Comparisons

⌃ Action Target ⓘ

Target type:

Attribute members ▾

Target object:

Product.Category ▾

⌃ Condition (Optional) ⓘ

```
// This action requires that both Reporting Services and the Reporting
Services
// sample reports be installed on the local machine.
```

⌃ Report Server

Server name: | Localhost

Report path: | ReportServer?/AdventureWorks Sample Reports/Sales Reason Comparisons

Report format: | HTML5 ▾

⌃ Parameters (Optional) ⓘ

Parameter Name	Parameter Value
ProductCategory	UrlEscapeFragment([Product].[Category].CurrentMember.UniqueName)
<Add parameter>	

⌄ Additional Properties

Figure 10-15. Setup for a reporting action

Finally, let's look at general actions; the designer is shown in Figure 10-16. With a general action, you can again select which type of object to apply it to (attribute members, cells, cube, dimension members, hierarchy, and hierarchy members, level and level members). After you select a target type, you can select the specific object from the Target Object drop-down.

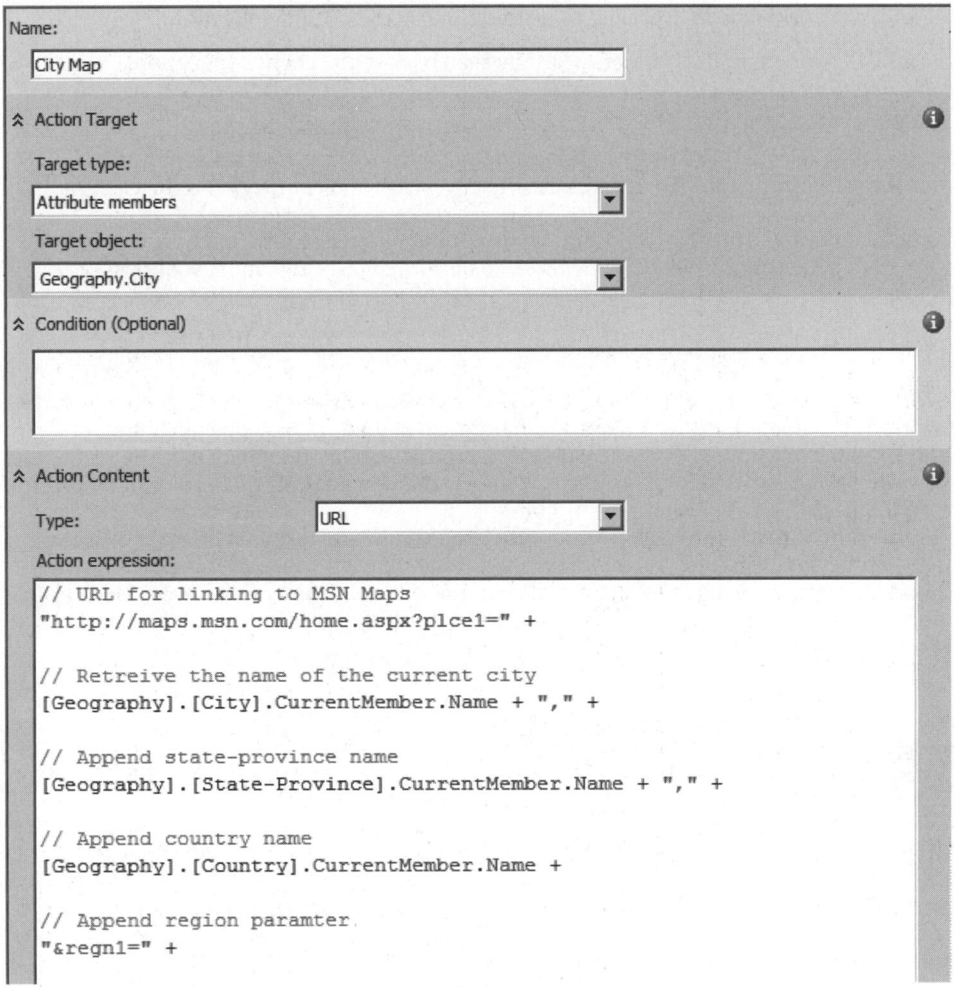

Name:

City Map

∧ Action Target   ⓘ

Target type:

Attribute members ▼

Target object:

Geography.City ▼

∧ Condition (Optional)   ⓘ

∧ Action Content   ⓘ

Type: URL ▼

Action expression:

```
// URL for linking to MSN Maps
"http://maps.msn.com/home.aspx?plce1=" +

// Retreive the name of the current city
[Geography].[City].CurrentMember.Name + "," +

// Append state-province name
[Geography].[State-Province].CurrentMember.Name + "," +

// Append country name
[Geography].[Country].CurrentMember.Name +

// Append region paramter
"&regn1=" +
```

Figure 10-16. The Action Designer for general actions

The action Type determines what will be returned to the client—data set, proprietary, row set, statement, or URL. The client will implement the action based on the action type. For example, the URL type will be launched into the default browser. Here are the specifics:

Data set: An MDX statement that parses to a multidimensional data set. The data set is passed to the client, which will act on it (perhaps opening another instance with the data set).

Proprietary: A string value (which can be dynamically built with MDX) that will be interpreted by the client application defined by the application string in the additional properties below the action expression.

Rowset: Similar to the data set, except in this case it returns a tabular data set.

Statement: A string that is a command to be executed by the client application.

URL: An MDX expression that renders a URL passed back to the client application (which can render it with a browser.)

And as I mentioned, the action expression is an MDX expression that renders the appropriate content for the action type selected. The City Map action in AdventureWorks is a URL type, and the expression assembles a URL string for MSN maps, with the city, state, and country from the Geography hierarchy.

That sums it up for actions. The most important thing to remember, again, is that how various actions are supported depends on the client application, so be sure to test early and test often!

Perspectives

The AdventureWorks database has dozens of measures, dozens of dimensions (each of which may have one or more attributes and hierarchies), 12 KPIs, and a handful of actions. If you consider a finance analyst who is concerned with a few account measures and product-oriented dimensions, the rest of the cubes will just be taking up space in his or her UI. Worst case is that the analyst will have to scroll up and down to get the right objects for any given report or query.

Also, as you've worked with the AdventureWorks cube, you've probably tripped over a few times where selecting the wrong dimension for a given measure doesn't do what you expect it to, as shown in Figure 10-17. This is a pivot table using Reseller Sales Amount and the Customer Geography. Because there is no relationship between the Customer dimension and the measure group, it doesn't break down the values, and you get the repeating top-level values.

		Fiscal Year ▼	
		⊞ FY 2006	Grand Total
Country ▼	Product ▼	Reseller Sales Amount	Reseller Sales Amount
⊞ Australia		$3,192,232.07	$3,192,232.07
⊞ Canada		$3,192,232.07	$3,192,232.07
⊞ France		$3,192,232.07	$3,192,232.07
⊞ Germany		$3,192,232.07	$3,192,232.07
⊞ United Kingdom		$3,192,232.07	$3,192,232.07
⊞ United States		$3,192,232.07	$3,192,232.07
Grand Total		$3,192,232.07	$3,192,232.07

Figure 10-17. *When a dimension doesn't slice properly*

Perspectives can solve both of these problems by offering a user a more targeted selection of objects to work with. The Perspective Designer is shown in Figure 10-18. Creating a perspective is pretty straightforward; after you create it and name it, it's just a matter of deselecting the objects you don't want displayed (when you create a new perspective, all objects are selected by default).

Cube Objects	Object Type	Perspective Name	Perspective Name	Perspective Name	Perspective Name	Perspective Name
Adventure Works	Name	Direct Sales	Channel Sales	Sales Summary	Finance	Sales Targets
Reseller Sales Amount	DefaultMeasure	Internet Sales Amount	Reseller Sales Amount	Sales Amount	Amount	Sales Amount Quota
⊟ Measure Groups						
⊟ Internet Sales	MeasureGroup	☑	☐	☐	☐	☐
Internet Sales Amount	Measure	☑	☐	☐	☐	☐
Internet Order Quantity	Measure	☑	☐	☐	☐	☐
Internet Extended Amount	Measure	☑	☐	☐	☐	☐
Internet Tax Amount	Measure	☑	☐	☐	☐	☐
Internet Freight Cost	Measure	☑	☐	☐	☐	☐
Internet Unit Price	Measure	☐	☐	☐	☐	☐
Internet Total Product Cost	Measure	☑	☐	☐	☐	☐
Internet Standard Product Cost	Measure	☑	☐	☐	☐	☐
Internet Transaction Count	Measure	☐	☐	☐	☐	☐
⊟ Internet Orders	MeasureGroup	☑	☐	☐	☐	☐
Internet Order Count	Measure	☑	☐	☐	☐	☐
⊟ Internet Customers	MeasureGroup	☑	☐	☐	☐	☐
Customer Count	Measure	☑	☐	☐	☐	☐
⊟ Sales Reasons	MeasureGroup	☑	☐	☐	☐	☐
Sales Reason Count	Measure	☑	☐	☐	☐	☐
⊞ Reseller Sales	MeasureGroup	☐	☑	☐	☐	☑
⊟ Reseller Orders	MeasureGroup	☐	☑	☐	☐	☑
Reseller Order Count	Measure	☐	☑	☐	☐	☑
⊞ Sales Summary	MeasureGroup	☐	☐	☑	☐	☐
⊟ Sales Orders	MeasureGroup	☐	☐	☑	☐	☐
Order Count	Measure	☐	☐	☑	☐	☐
⊟ Sales Targets	MeasureGroup	☐	☑	☐	☐	☑
Sales Amount Quota	Measure	☐	☑	☐	☐	☑
⊟ Financial Reporting	MeasureGroup	☐	☐	☐	☑	☐
Amount	Measure	☐	☐	☐	☑	☐
⊞ Exchange Rates	MeasureGroup	☑	☑	☑	☑	☐
⊟ Dimensions						
⊞ Date	CubeDimension	☑	☑	☑	☑	☑
⊞ Ship Date	CubeDimension	☑	☑	☑	☐	☐
⊞ Delivery Date	CubeDimension	☑	☑	☑	☐	☐
⊞ Customer	CubeDimension	☑	☐	☐	☐	☐
⊞ Reseller	CubeDimension	☐	☑	☐	☐	☐
⊞ Geography	CubeDimension	☐	☑	☐	☐	☐

Figure 10-18. *The BIDS Perspective Designer*

After you've created the perspective, deployed it, and processed it, it's available to end users. The perspective looks just like a cube, as seen in the Data Connection Wizard from Excel in Figure 10-19. Now our Finance user has a Finance perspective. By selecting that, the user will have a more focused view of the AdventureWorks cube for their purposes.

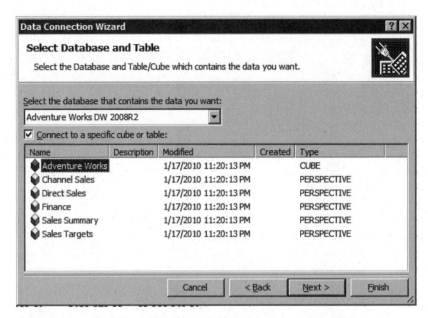

Figure 10-19. *Note the perspectives in the Database Selection Wizard.*

■ **Note** Perspectives are for convenience only; they are not intended for security.

Translations

The final cube feature we're going to look at is the *translation*. This is very straightforward. The Translation Designer is shown in Figure 10-20. You can add as many languages as you choose, and entry for each language is optional. This will provide your non-English-speaking users a cube with labels in their native language.

Default Language	Object Type	Spanish (Spain)	French (France)
Adventure Works	Caption	Adventure Works	Adventure Works
⊟ Measure Groups			
⊟ Internet Sales	Caption	Ventas por Internet	Ventes par Internet
Internet Sales Amount	Caption	Cantidad de Ventas por I...	Montant des Ventes (Inte...
Internet Order Quantity	Caption	Cantidad de Ordenes por ...	Nombre de Commandes (I...
Internet Extended Am...	Caption	Cantidad Extendida por I...	Montant Étendu (Internet)
Internet Tax Amount	Caption	Cantidad de Impuesto por...	Taxes (Internet)
Internet Freight Cost	Caption	Costo de Flete por Internet	Coût de Transport (Inter...
Internet Unit Price	Caption	Precio Unitario por Internet	Prix Unitaire (Internet)
Internet Total Product ...	Caption	Costo Total del Producto ...	Coût Total Produit (Intern...
Internet Standard Prod...	Caption	Costo Estándar del Produ...	Coût Standard Produit (In...
Internet Transaction C...	Caption	Cuenta de Transacción po...	Nombre de Transactions (...
⊟ Internet Orders	Caption	Ordenes por Internet	Commandes par Internet
Internet Order Count	Caption	Conteo de Ordenes por I...	Nombre de Commandes (I...

Figure 10-20. *Translations for the AdventureWorks cube*

In addition to the caption (display title) for each object, many objects have other aspects that can be translated—descriptions, display folder names, and a few other labels. Client applications can then request data in a specific language. If the client does not request a specific language or the computer doesn't have the language requested, then the default translation is used.

Summary

As I said, this was a whirlwind tour around many of the ancillary objects in Analysis Services. Although they're not all directly in service of the cube, they really show the power of this platform. In the next chapter, you're going to look at figuring out answers when you don't even know what question to ask—by data mining.

■ ■ ■

CHAPTER 11

Data Mining

Data-mining technologies provide business users with the means to analyze and sift through gargantuan amounts of information stored in corporate data warehouses. Data-mining tools implement a variety of algorithms that, when applied to the enterprise data warehouse (EDW), greatly aid the decision-making process. In today's hypercompetitive climate, adding mining to your analytics bag of tricks can help you find that "needle in the haystack" prediction that puts your ability to act ahead of your competitors'.

In this chapter, you will create and apply mining models to the AdventureWorks data warehouse. These models will be used to predict buyers of AdventureWorks products. While predicting buyers, you will discover not only potential customers, but the products they are most interested in. Finally, you will analyze the possible sales growth to AdventureWorks from these customers. You'll see how to create models that help you find a niche or an entirely new way to grow your business.

Why Mine Data?

Data mining provides a powerful decision support mechanism. Data mining is used in virtually every industry by marketing departments, customer relationship management groups, network (e-mail) security, finance/credit departments, human resources (employee performance), and others.

The following examples offer a few ways that I have seen data mining used:

Loss mitigation: A finance company mined data related to customer payment, auction sale recovery, auction sale expenses, customer income, and demographics to predict future troubled accounts. This data was used to predict repossession and auction probabilities. Based on the discoveries made, collection activities were focused on the customers most likely to perform in the future. For loans that were predicted not to perform, new loan modification products and processes were created and actively presented to the at-risk borrowers.

Cost reduction: A remanufacturing company mined manufacturer repair data, bill of materials (BOM) usage, parts cost trending, and parts availability trending to predict future costs of goods for each product refurbished. One outcome of this effort resulted in the creation of new business processes at the receiving dock, which included holding back product from technician workstations that did not have parts in stock. This change alone improved service-level agreement (SLA) performance to manufacturers by reducing turn-around time for products with stocked parts. In addition, these process updates improved costs by allowing technicians to focus on their part of the remanufacturing process.

Worldwide strategic marketing: A telecom equipment manufacturer mined government data for cell phones owned, computers owned, and gross income at the regional, national, and per-capita level. Competitive data was also gathered/purchased that provided service provider penetration and relationships; larger providers own chunks of smaller ones. These efforts resulted in an extremely focused marketing plan, targeting second-tier service providers in emerging economies, and resulted in several successes in the Chinese and Brazilian marketplaces.

Education: An independent school district mined statewide standardized testing scores and other demographic data to predict students at risk of underperforming or failing these tests at the grade and subject levels. These efforts resulted in a new initiative by the district to group at-risk students together, not only in smaller classes, but in smaller classes focused on the particular subjects these students were struggling in. Along with this, teacher assignments were reevaluated to enable teachers who performed better with at-risk students to head these classes. Not only did test scores show improvement, but discipline-related issues decreased as well.

Using Data-Mining Algorithms

The mining algorithms implemented in SSAS are your core toolset for creating predictive models. In this section, I describe a few of the more essential data-mining algorithms and their general purpose. For each algorithm, I introduce the algorithm type and define three data requirements: input data, the key column, and the prediction column.

Microsoft Naïve Bayes

The Microsoft Naïve Bayes algorithm creates classification predictions. The Naïve Bayes algorithm is simply a true-or-false mechanism, and does not account for any types of dependencies that may exist in the data you are analyzing. In order to create a Naïve Bayes model, you need to prepare your data and create a training data set. This need for a training data set marks Naïve Bayes as one of the *supervised* mining algorithms.

A Microsoft Naïve Bayes model requires the following training data:

Input: Input data columns must be discretized, because Naïve Bayes is not designed to be used with continuous data, and your inputs should be fairly independent of each other. One method of discretization is known as *banding*, or *binning*, data. Two common examples of banding data include age and income. Age is often discretized as Child, Teenager, Adult, and Senior Citizen. It may also be banded as < 11, 12–17, 18–54, and 55+. Income is often banded in brackets, such as < 20,000, 20,000–50,000, and 50,000+.

Key column: A unique identifier is required for each record in the training set. Two examples of unique identifiers are a surrogate key and a globally unique identifier (GUID). A surrogate key in the EDW is usually a SQL Server Identity column, which is basically a numeric counter, guaranteed to be unique by the database engine.

Prediction column: This is the the prediction attribute. Will the customer buy more of our products? Is the applicant a good credit risk?

Microsoft Clustering

Microsoft Clustering is a segmentation algorithm. It creates *segments*, or *buckets*, of groups that have similar characteristics, based on the attributes that you supply to the algorithm. Microsoft Clustering uses equal weighting across attributes when creating segments. Many times, these segments may not be readily apparent but are important nonetheless.

A Microsoft Clustering model requires the following:

Input: Input data columns for this algorithm can be continuous in nature. Continuous input is represented by the Date, Double, and Long data types in SQL Server. These data types can be divided into fractional values.

Key column: A unique identifier is required for each record in the training set. Two examples of unique identifiers are a surrogate key and a GUID. A surrogate key in the EDW is usually a SQL Server Identity column, which is basically a numeric counter, guaranteed to be unique by the database engine.

Prediction column: This is the the prediction attribute. Use of a prediction column is optional with this algorithm. Segments will be created automatically by the algorithm when a prediction column is not supplied. I suggest trying this with every Clustering model you make. It may teach you something you did not know about your data!

Microsoft Decision Trees

The Microsoft Decision Trees algorithm creates predictions for classification and association. A decision tree resembles the structure of a B-tree. As part of the data-mining business process, decision trees are often run against individual segments from a Clustering model.

A Microsoft Decision Tree model requires the following:

Input: Input data can be either discrete or continuous.

Key column: Like Naïve Bayes and Clustering, a unique identifier is required for each record in the training set.

Prediction column: This is the prediction attribute. It can be either discrete or continuous for this algorithm. If you use a discrete attribute, the tree can be well represented by a histogram.

■ **Note** *Prediction columns* have two possible settings: Predict and PredictOnly. PredictOnly columns, as the name suggests, attempt to create predictions on the column chosen. A Predict column, on the other hand, uses the chosen data point to train your model as well.

Creating the Accessory Buyers Marketing Campaign

The AdventureWorks marketing group has come to you for their next campaign. The Bike Buyers campaign has completed successfully, and your management would like to build on this success.

For the remainder of this chapter, you will create data-mining models using the AdventureWorks data warehouse. You will create three models to support the launch of the Accessory Buyers campaign, which is being created to supplement the recently concluded Bike Buyers campaign.

You will use a different method to create each of the needed models. First, you will use the Data Mining Wizard to create a mining model. Second, you will use Data Mining Extensions (DMX) to generate a mining model, and last, you will create a mining model by using an existing AdventureWorks cube.

Preparing the Data Warehouse

Your first step is to discuss this campaign in more detail with your marketing group. Based on these discussions and further review of the Bike Buyers campaign, you've decided that bike buyers always buy helmets with a new bicycle. The marketing group states that the reasons for this are twofold: first, bike buyers want a helmet to match the new bicycle, and second, local laws in many areas prohibit riding a bicycle without a helmet. In addition, marketing asks that you include the Clothing product category in this campaign.

Creating the Accessory Buyers Views in AdventureWorks

With the requirements in mind, you decide on two views in the AdventureWorks DW2008 database: vDMPrepAccessories and vAccessoryBuyers. These two new views are based on the original vDMPrep and vTargetMail views that already exist.

The first view, vDMPrepAccessories, is listed here:

```
--      Purpose:        Create vDMPrepAccessories
Use AdventureWorksDW2008
GO

If      Exists
(
   Select * From sys.views Where object_id = OBJECT_ID(N'dbo.vDMPrepAccessories')
)
   Drop View dbo.vDMPrepAccessories
GO

Create View dbo.vDMPrepAccessories
As

Select
   PC.EnglishProductCategoryName,
   PSC.EnglishProductSubcategoryName,
   Coalesce(P.ModelName, P.EnglishProductName) As Model,
   C.CustomerKey,
   S.SalesTerritoryGroup As Region,

   Case
      When Month(GetDate()) < Month(C.BirthDate)
         Then DateDiff(yy, C.BirthDate,GetDate()) - 1
      When Month(GetDate()) = Month(C.BirthDate) And Day(GetDate()) < Day(C.BirthDate)
         Then DateDiff(yy, C.BirthDate,GetDate()) - 1
      Else DateDiff(yy, C.BirthDate,GetDate())
   End As Age,

   Case
      When C.YearlyIncome < 40000
         Then 'Low'
      When C.YearlyIncome > 60000
         Then 'High'
```

```
        Else 'Moderate'
    End As IncomeGroup,

    D.CalendarYear,
    D.FiscalYear,
    D.MonthNumberOfYear As Month,
    F.SalesOrderNumber As OrderNumber,
    F.SalesOrderLineNumber As LineNumber,
    F.OrderQuantity As Quantity,
    F.ExtendedAmount As Amount
From
    dbo.FactInternetSales F Inner Join
    dbo.DimDate D On
    F.OrderDateKey = D.DateKey Inner Join
    dbo.DimProduct P On
    F.ProductKey = P.ProductKey Inner Join
    dbo.DimProductSubcategory PSC On
    P.ProductSubcategoryKey = PSC.ProductSubcategoryKey Inner Join
    dbo.DimProductCategory PC On
    PSC.ProductCategoryKey = PC.ProductCategoryKey Inner Join
    dbo.DimCustomer C On
    F.CustomerKey = C.CustomerKey Inner Join
    dbo.DimGeography G On
    C.GeographyKey = G.GeographyKey Inner Join
    dbo.DimSalesTerritory S On
    G.SalesTerritoryKey = S.SalesTerritoryKey
Where
    PC.EnglishProductCategoryName Not In ('Bikes','Components') And
    PSC.EnglishProductSubcategoryName Not In ('Helmets')
Go
```

This code creates the vDMPrepAccessories view that you will be using throughout the rest of this chapter. This view joins several tables in the AdventureWorks DW2008 database, and creates three bands for YearlyIncome. Finally, the view will return only accessories that are not Helmets.

The second view, vAccessoryBuyers, is listed here:

```
--      Purpose:        Create vAccessoryBuyers
Use AdventureWorksDW2008
GO

If      Exists( Select * From sys.views Where object_id↵
 = OBJECT_ID(N'dbo.vAccessoryBuyers'))
    Drop View dbo.vAccessoryBuyers
GO

Create View dbo.vAccessoryBuyers
As

Select
    C.CustomerKey,
    C.GeographyKey,
    C.CustomerAlternateKey,
```

```
        C.Title,
        C.FirstName,
        C.MiddleName,
        C.LastName,
        C.NameStyle,
        C.BirthDate,
        C.MaritalStatus,
        C.Suffix,
        C.Gender,
        C.EmailAddress,
        C.YearlyIncome,
        C.TotalChildren,
        C.NumberChildrenAtHome,
        C.EnglishEducation,
        C.SpanishEducation,
        C.FrenchEducation,
        C.EnglishOccupation,
        C.SpanishOccupation,
        C.FrenchOccupation,
        C.HouseOwnerFlag,
        C.NumberCarsOwned,
        C.AddressLine1,
        C.AddressLine2,
        C.Phone,
        C.DateFirstPurchase,
        C.CommuteDistance,
        D.Region,
        D.Age,

        Case D.Clothing
            When  0  Then  0
            Else  1
        End As ClothingBuyer,

        Case D.Accessories
            When  0  Then  0
            Else  1
        End As AccessoryBuyer
From
    dbo.DimCustomer As C Inner Join
    (
        Select
            CustomerKey,
            Region,
            Age,

            Sum(Case EnglishProductCategoryName
                    When  'Clothing'      Then  1
                    Else  0
                End)  As Clothing,

            Sum(Case EnglishProductCategoryName
```

```
           When  'Accessories'  Then  1
           Else  0
        End)  As Accessories
   From
      dbo.vDMPrepAccessories
   Group By
      CustomerKey,
      Region,
      Age
 ) As D  On
 C.CustomerKey  =  D.CustomerKey
GO
```

This code creates the vAccessoryBuyers view that we will be using throughout the rest of this chapter. This view joins DimCustomer to a derived table, D, which is based on the vDMPrepAccessories view you created earlier. You now have your ClothingBuyer and AccessoryBuyer data points.

Creating the Accessory Campaign Data Source View

In addition to the views defined in the preceding section, you will need a new data source view (DSV) that references the vAccessoryBuyers view and the ProspectiveBuyer table. The ProspectiveBuyer table is populated with your campaign targets. In Exercise 11-1, you will create the AccessoryCampaign DSV. Because this DSV is virtually identical to the one you created in Chapter 5, I will list the instructions only, without the dialog box figures.

Exercise 11-1. Create the AccessoryCampaign DSV

Following are the steps to create a DSV for the AccessoryCampaign:

1. Open the AdventureWorks solution that you've been using for these exercises.

2. Right-click the Data Source Views folder in the Solution Explorer pane, and click New Data Source View. The Data Source View Wizard introduction dialog box appears; click Next.

3. In the Select a Data Source dialog box, choose Adventure Works DW from your Relational Data Sources, and click Next.

4. For the Select Tables and Views dialog box, choose vAccessoryBuyers and ProspectiveBuyer from the Available Objects and move them to the Included Objects area. Click Next.

5. Name this DSV **AccessoryCampaign** and click Finish.

6. AccessoryCampaign.dsv should now appear in your Data Source Views folder.

Finding Accessory Buyers by Using the AdventureWorks EDW

Now that you have your views implemented, your next exercise is to create a new data-mining model. Do that by following the steps in Exercise 11-2. You will use the Microsoft Decision Trees algorithm to mine the AdventureWorks data warehouse. Your goal is to find a target population of potential accessory buyers.

Exercise 11-2. Use the Data Mining Wizard

Follow these steps to create a data-mining model:

1. Open the AdventureWorks solution that you've been using for these exercises.

2. Right-click the Mining Structures folder in the Solution Explorer pane, and click New Mining Structure. The Data Mining Wizard introduction dialog box appears; click Next.

3. The Select the Definition Method dialog box appears, as shown in Figure 11-1; choose From Existing Relational Database or Data Warehouse, and click Next.

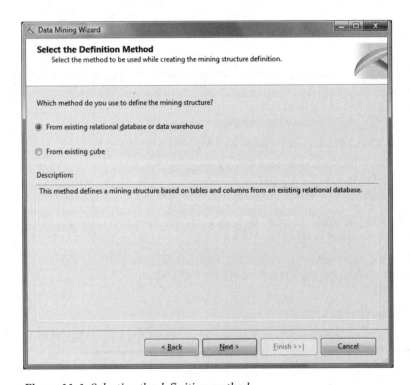

Figure 11-1. Selecting the definition method

4. In the Create the Data Mining Structure dialog box, shown in Figure 11-2, leave Create Mining Structure with a Mining Model selected, and choose Microsoft Decision Trees from the Which Data Mining Technique Do You Want to Use drop-down. Click Next.

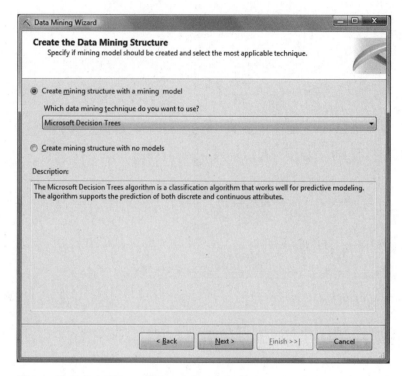

Figure 11-2. Selecting a data-mining technique

5. In the Select Data Source View dialog box, shown in Figure 11-3, choose AccessoryCampaign from your Available Data Source Views, and click Next.

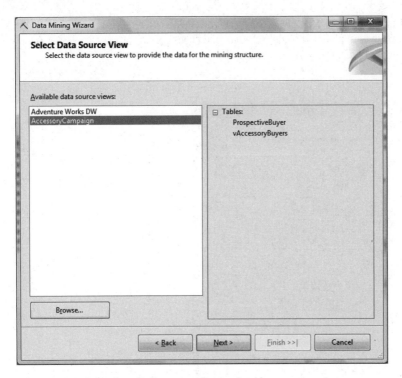

Figure 11-3. Selecting the AccessoryCampaign data source view

6. You will now see the Specify Table Types dialog box. As shown in Figure 11-4, select the Case check box for vAccessoryBuyers. Click Next.

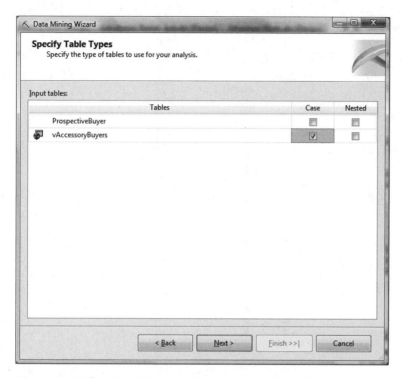

Figure 11-4. Choosing vAccessoryBuyers as your Case table

7. Now you get to think about the actual analysis you'll be doing. In the Key column, leave CustomerKey selected as your Key. For your Input columns, choose the following thirteen fields: Age, CommuteDistance, EnglishEducation, EnglishOccupation, Gender, GeographyKey, HouseOwnerFlag, MaritalStatus, NumberCarsOwned, NumberChildrenAtHome, Region, TotalChildren, and finally YearlyIncome. This model requires two Predictable columns; choose AccessoryBuyer and ClothingBuyer. Your selections should mimic Figure 11-5. When you are finished, click Next.

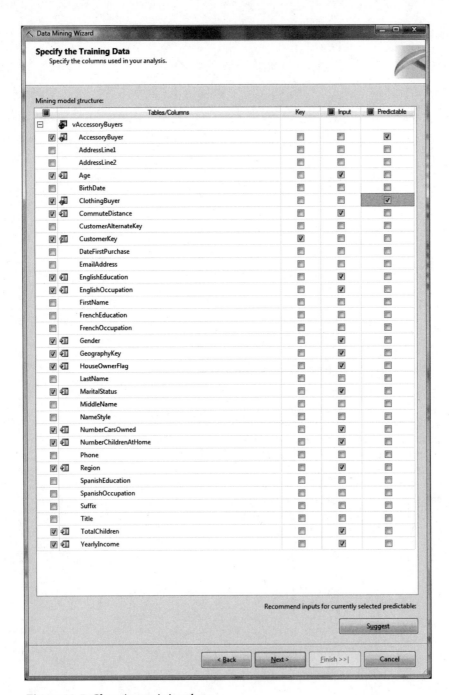

Figure 11-5. Choosing training data

8. In the Specify Columns' Content and Data Type dialog box, shown in Figure 11-6, there are two changes that you need to make. Change Accessory Buyer and Clothing Buyer to Discrete in the Content Type column. Click Next.

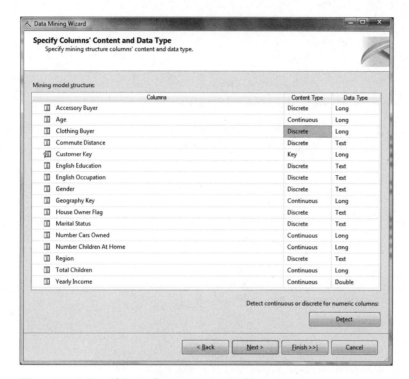

Figure 11-6. Specifying column content and type

9. The Create Testing Set dialog box, shown in Figure 11-7, is where you will specify some inputs regarding how you would like your model to be trained. For this exercise, leave the Percentage of Data for Testing option set to 30 percent, and Maximum Number of Cases in Testing Data blank. Click Next.

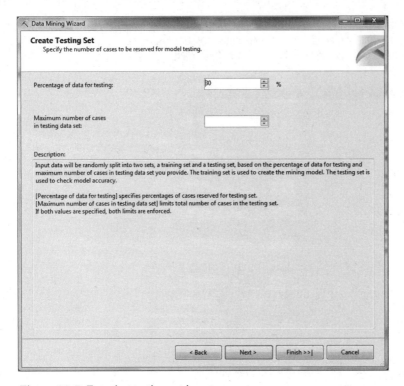

Figure 11-7. Entering testing set inputs

10. Now you can finalize your wizard entries by entering a name for your structure and model. Enter **AccessoryBuyersCampaign** as your Mining Structure Name and **AB_DecisionTree** for your Mining Model Name. Finally, as shown in Figure 11-8, be sure to select the Allow Drill Through check box. Click Finish.

Figure 11-8. *Completing the Data Mining Wizard*

After the wizard completes, the Data Mining Model Designer will fill your workspace. Next, you will explore the functionality of each of the five tabs within the designer.

Using the Data Mining Model Designer

The Data Mining Model Designer will be your main work area, now that you have finished defining your model with the Data Mining Wizard. The Data Mining Model Designer consists of the following tabs:

Mining Structure: This is where you modify and process your mining structure.

Mining Models: Here you create or modify models from your mining structure.

Mining Model Viewer: This view enables you to explore the models you have created.

Mining Accuracy Chart: Here you can view various mining charts. Later in this chapter, you will use this tab to look at and review a lift chart.

Mining Model Prediction: Using this view, you will create and review the predictions your mining model asserts.

The Mining Structure View

The Mining Structure view is separated into two panes, as shown in Figure 11-9. The leftmost pane displays your mining structure columns, and your data source view is shown on the right. You will also process your mining model here, using the Process the Mining Structure button on the toolbar. Click the Process the Mining Structure button (leftmost button in view toolbar) now to begin processing.

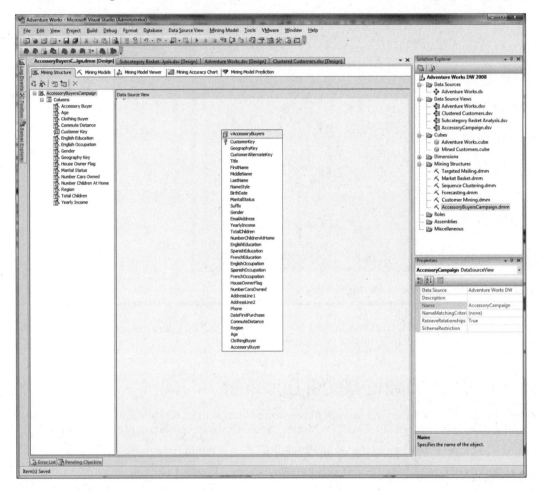

Figure 11-9. The Mining Structure tab

After completing some preprocessing tasks, the Process Mining Structure dialog box appears. For this model, as shown in Figure 11-10, simply click the Run button at the bottom of the dialog.

Figure 11-10. The Process Mining Structure, ready to process our campaign

When the Process Progress dialog box appears and the Status area displays Process Succeeded, click Close. This returns you to the Process Mining Structure dialog box. Click Close again.

The Mining Models View

With our processing complete, you can now explore the other tabs. Click the Mining Models tab, as shown in Figure 11-11. In this view, you can review your Structure Mining columns and Mining Model columns. Also notice that your mining model name is shown at the top of the Mining Model columns. Update both the Accessory Buyer and Clothing Buyer to Predict from PredictOnly.

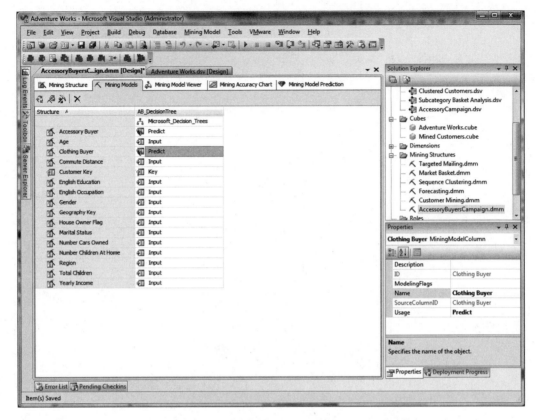

Figure 11-11. The Mining Models view, with both Buyer columns set to Predict

Next, process the model to reflect our Buyer column changes. When this completes, click the Mining Model Viewer tab.

The Mining Model Viewer View

The Mining Model Viewer is a container that supports several viewers. Choosing Microsoft Tree Viewer from the Viewer drop-down will load your Accessory Buyers campaign into the Tree Viewer. After the Tree Viewer has loaded, ensure that Accessory Buyer is selected in the Tree drop-down, which is just below the Viewer drop-down list. In this section, I will show you the Microsoft Tree Viewer, the Mining Legend, and this model's drill-through capability.

Exploring the Decision Tree

The decision tree is built as a series of splits, from left to right, as shown in Figure 11-12. The leftmost node is All, which represents the entire campaign. The background color of each node is an indication of that node's population; the darker the color, the greater the population.

In your model, All is the darkest. To the right of All, the Clothing Buyer = 0 node is larger than Clothing Buyer = 1. If you hover over Clothing Buyer = 0, an infotip will show the size of this node as 6,338. Doing this on the Clothing Buyer = 1 node will display 4,767.

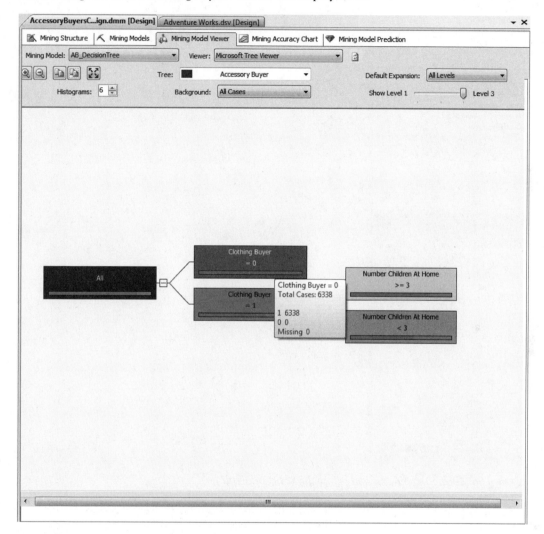

Figure 11-12. Clothing Buyer = 0 node, with an infotip displayed

Each node in the bar under the node condition is a histogram that represents our buyers. The blue portion of the bar is our True state, and the pink our False state. Click the All node and select Show Legend. After the legend appears, dock it in the lower-right corner, as shown in Figure 11-13. Doing this makes it easier to watch the values as you navigate the tree.

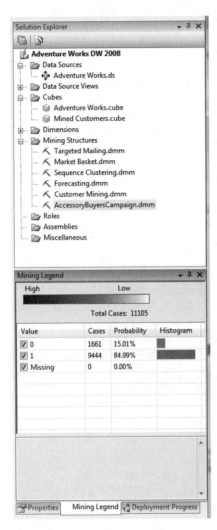

Figure 11-13. The Mining Legend, docked below the Solution Explorer

The Mining Legend has four columns of information about your model. The first column, Value, shows 0, 1, and Missing. 0 is our false, or nonbuyer case, while 1 is our true, or buyer case. The Missing value of 0 is good to see, as it indicates that our data is clean and fully populated. Our Cases column displays the population of each case, and the Probability calculates this distribution as a percentage. Finally, the Histogram column mimics the node's histogram.

Reading the decision tree is done in a left-to-right manner, as shown in Figure 11-14. In our Accessory Buyer model, the most significant factor that determines our accessory buyers is whether they are also clothing buyers. This factor alone can enable the creation of a focused campaign. But in our case, we see another valuable factor is at work here. The Number Children At Home node contains some interesting values. The Number Children At Home >= 3 has a higher probability of purchase per its histogram than the Number Children At Home < 3. On the other hand, the Number Children At Home < 3 is a darker node, meaning it has a greater population. Let's take a closer look.

Click the Number Children At Home >= 3 node and review the Mining Legend. Our accessory buyers equal 722, with a probability of 78.52 percent. Reviewing the Number Children At Home < 3 reveals 2,384 cases, with a probability of 61.95 percent. Based on these figures, your marketing group may decide to target the group with fewer than three children at home first, because it has more buyers.

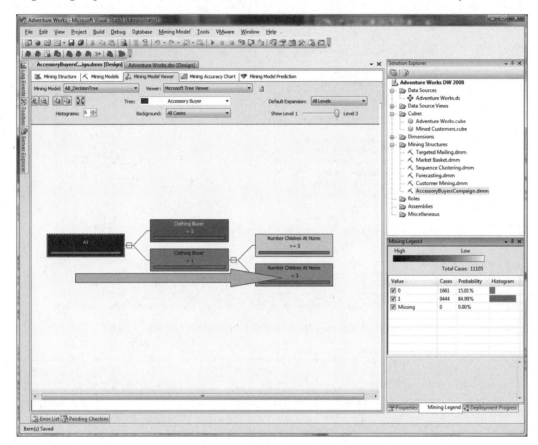

Figure 11-14. *The decision tree is read in a left-to-right manner.*

Using Drill-Through to View a Model's Training Cases

Using drill-through, which you enabled when creating the AccessoryBuyers mining model, enables you to see the underlying data that belongs to a particular node. Right-click the Number Children At Home < 3 node and select Drill Through, followed by Model Columns Only.

The Drill Through grid, shown in Figure 11-15, displays your training cases classified to Clothing Buyer = 1 and Number Children At Home < 3. The model's data points are displayed in alphabetical order (I've hidden a few to narrow the view). Take note of the Number Children At Home column. It contains values of 0, 1, and 2. This Number Children At Home bucket was created by the Decision Tree algorithm during model processing.

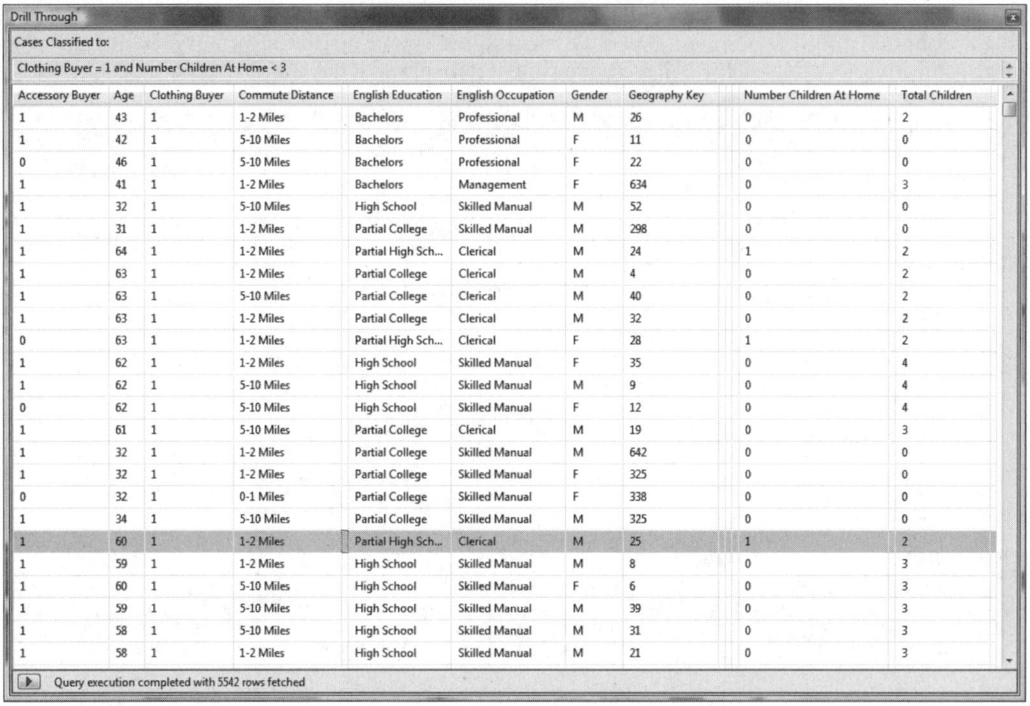
Drill Through

Cases Classified to:

Clothing Buyer = 1 and Number Children At Home < 3

Accessory Buyer	Age	Clothing Buyer	Commute Distance	English Education	English Occupation	Gender	Geography Key	Number Children At Home	Total Children
1	43	1	1-2 Miles	Bachelors	Professional	M	26	0	2
1	42	1	5-10 Miles	Bachelors	Professional	F	11	0	0
0	46	1	5-10 Miles	Bachelors	Professional	F	22	0	0
1	41	1	1-2 Miles	Bachelors	Management	F	634	0	3
1	32	1	5-10 Miles	High School	Skilled Manual	M	52	0	0
1	31	1	1-2 Miles	Partial College	Skilled Manual	M	298	0	0
1	64	1	1-2 Miles	Partial High Sch...	Clerical	M	24	1	2
1	63	1	1-2 Miles	Partial College	Clerical	M	4	0	2
1	63	1	5-10 Miles	Partial College	Clerical	M	40	0	2
1	63	1	1-2 Miles	Partial College	Clerical	M	32	0	2
0	63	1	1-2 Miles	Partial High Sch...	Clerical	F	28	1	2
1	62	1	1-2 Miles	High School	Skilled Manual	F	35	0	4
1	62	1	5-10 Miles	High School	Skilled Manual	M	9	0	4
0	62	1	5-10 Miles	High School	Skilled Manual	F	12	0	4
1	61	1	5-10 Miles	Partial College	Clerical	M	19	0	3
1	32	1	1-2 Miles	Partial College	Skilled Manual	M	642	0	0
1	32	1	1-2 Miles	Partial College	Skilled Manual	F	325	0	0
0	32	1	0-1 Miles	Partial College	Skilled Manual	F	338	0	0
1	34	1	5-10 Miles	Partial College	Skilled Manual	M	325	0	0
1	60	1	1-2 Miles	Partial High Sch...	Clerical	M	25	1	2
1	59	1	1-2 Miles	High School	Skilled Manual	M	8	0	3
1	60	1	5-10 Miles	High School	Skilled Manual	F	6	0	3
1	59	1	5-10 Miles	High School	Skilled Manual	M	39	0	3
1	58	1	5-10 Miles	High School	Skilled Manual	M	31	0	3
1	58	1	1-2 Miles	High School	Skilled Manual	M	21	0	3

Query execution completed with 5542 rows fetched

***Figure 11-15.** The Drill Through grid, displaying the Number Children At Home < 3 node*

■ **Tip** If you right-click on the grid contents and select Copy All, a copy of your cases will be placed into the copy buffer, complete with column headings. This data can then be pasted into an Excel worksheet for further analysis.

Using the Dependency Network

You use the Dependency Network to view the input/predictable attribute dependencies in your model. Clicking the Number Children At Home node will change the node display in the following ways:

Selected node: The background color will change to turquoise.

Predicts this node: If the selected node predicts this node, this node's background color will change to a slate-blue.

Next, click the Clothing Buyer node. Choosing Clothing Buyer will change the Dependency Network view in the following ways:

Selected node: The background color will change to turquoise.

Predicts both ways: The selected node and this node predict each other. This node's background color is changed to purple.

Predicts the selected node: This node predicts the node you selected, and its background color is changed to rust.

Finally, select the Accessory Buyer node, right-click, and select Copy Graph View. This copies your Dependency Network, shown in Figure 11-16, to the Windows Clipboard.

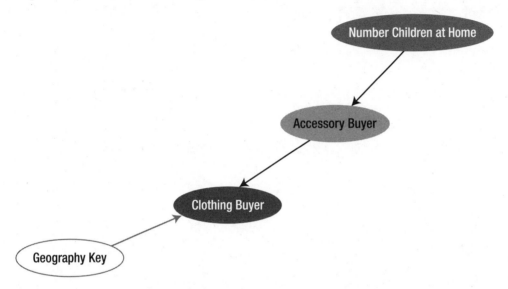

Figure 11-16. The Dependency Network, with the Accessory Buyer node selected

The Mining Accuracy Chart View

The Mining Accuracy Chart tab, shown in Figure 11-17, is where you will validate your mining models. For your Decision Tree model, you will be using the Input Selection and Lift Chart tabs. In the Input Selection tab's Predictable Column Name column, select Accessory Buyer. After this, select 1 for Predict Value. From the Select Data Set to be Used for Accuracy Chart section, confirm that Use Mining Model Test Cases is selected.

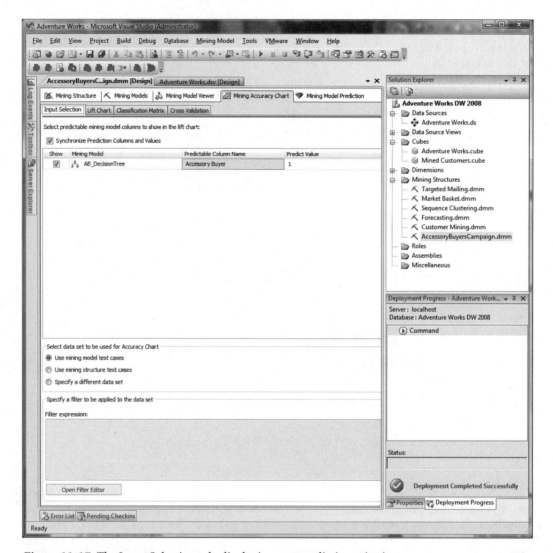

Figure 11-17. The Input Selection tab, displaying our prediction criteria

With the Input Selection tab completed, choose the Lift Chart tab. After a moment of processing, the Data Mining Lift Chart for Mining Structure: AccessoryBuyersCampaign, and the Mining Legend that you docked earlier will appear.

The lift chart, shown in Figure 11-18, will help you determine the value of your model. The lift chart evaluates and compares your targets by using an *ideal line* and a *random line*. The percentage amount difference between your *lift line* and your random line is referred to as *lift* in the marketing world. The lift chart lines are defined as follows:

Random line: The random line is drawn as a straight diagonal line, from (0, 0) to (100, 100). For a mining model to be considered a productive model, it should show some lift above this baseline.

Ideal line: Shown in green for this model this line reaches 100 percent Overall Population at 85 percent Target Population. The ideal line for the Accessory Buyers campaign suggests that a perfectly constructed model would reach 100 percent of your targets by using only 85 percent of the target population.

Lift line: This coral-colored line represents your model. The Accessory Buyers campaign follows the ideal line for quite a ways, meaning that our model is performing quite well. Deciding to contact more than 60 percent of the overall population will be a discussion point with marketing, because this is where your model begins to trend back toward the random line.

Measured location line: This is the vertical gray line in the middle of the chart. This line can be moved simply by clicking inside the chart. Moving this line will automatically update the Mining Legend with the appropriate values.

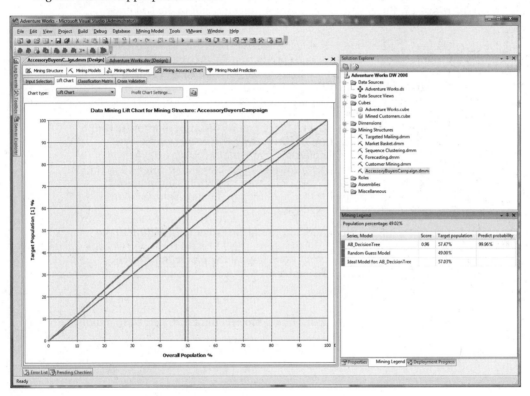

Figure 11-18. The Accessory Buyers campaign lift chart

The Mining Model Prediction View

Now you are ready to create a prediction. Selecting the Mining Model Prediction tab displays the Prediction Query Builder in Design view. Initially, there will be no selections in the Select Input Tables(s) area, or in the Source/Field grid at the bottom of the designer.

To create your prediction query, complete the following steps:

1. Select a mining model: if AB_DecisionTree is not loaded in the Mining Model box, click the Select Model button to navigate your data models. Expand the AccessoryBuyersCampaign, and choose the AB_DecisionTree model.

2. Select your input table: in the Select Input Table(s) box, click the Select Case Table button. In the Select Table dialog box, pick AccessoryCampaign from the Data Source drop-down. After you have done this, choose the ProspectiveBuyer table that marketing purchased, and click OK. Notice that seven fields are automatically mapped between the mining model and the ProspectiveBuyer table.

3. Map columns: in addition to the preceding columns that were mapped automatically, we need to add our predict column. To do this, simply drag and drop the Accessory Buyer column in your mining model onto the Unknown column in the input table.

4. Design the query: In the Source/Field grid, you will select the specific data points and data types to output for your prediction. Begin by choosing Prediction Function from the Source drop-down list. In the Field drop-down list, select PredictProbability. The last thing needed for our prediction function is to drag and drop the Accessory Buyer column from the mining model to the Criteria/Argument cell. Doing this will replace the cell's text with `[AB_DecisionTree].[Accessory Buyer]`. Next, you will add another row to the grid. Choose AB_DecisionTree from the drop-down list in the row below Prediction Function. In the Field column, ensure that Accessory Buyer is chosen. You want to predict who your future buyers may be, so enter = **1** in the Criteria/Argument column. Finally, you will want to identify your possible customers. Do this by selecting ProspectiveBuyer in the next Source column, and ProspectiveBuyerKey in the Field column.

5. Add other prospect information: Add additional customer information to the grid by creating six new rows. For the first row, select ProspectiveBuyer as your source, followed by FirstName as your field. Repeat the preceding steps five more times, adding LastName, AddressLine1, City, StateProvinceCode, and PostalCode.

When you have completed these steps, your Mining Model Prediction view should look like Figure 11-19.

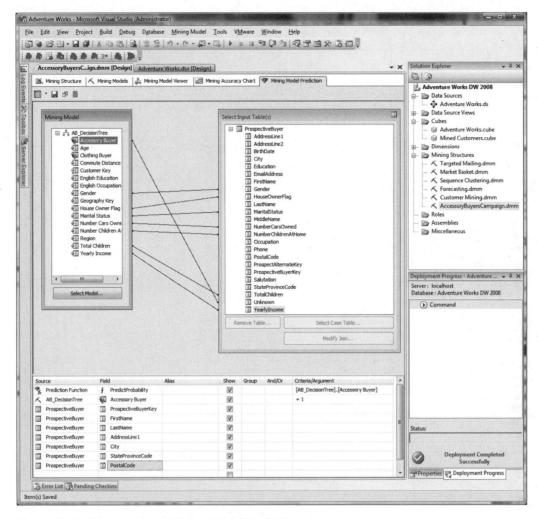

Figure 11-19. The completed Mining Model Prediction view

It is now time to run the prediction and view our results. Click the Switch to Query Result View (leftmost) button in this tab's toolbar. Your prediction query will process, and you will be presented with a grid, shown in Figure 11-20, containing your accessory buying targets.

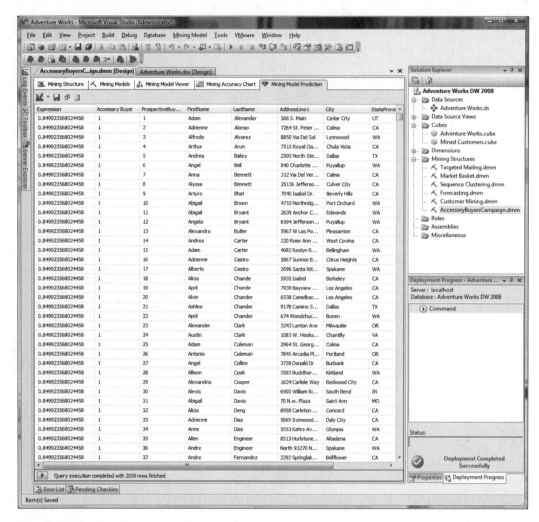

Figure 11-20. *Here they are, our future accessory buyers!*

These targets can be saved to your database by clicking the Save Query Result button (second from the left), which is also in this tab's toolbar. In the Save Data Mining Query Result dialog box, Select Adventure Works DW from the Data Source drop-down. Enter **AccessoryBuyerTargets** as your table name, as shown in Figure 11-21, and click Save.

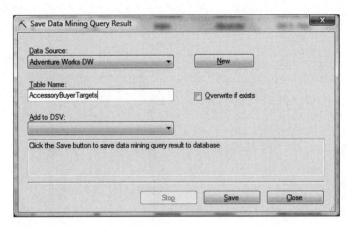

Figure 11-21. Use the Save Data Mining Query Result dialog box to save your targets to a table.

Finding Accessory Buyers by Using Data Mining Extensions (DMX)

It's time to turn our attention to Microsoft Data Mining Extensions (DMX). DMX, as a query language, has a syntax that is quite similar to Transact-SQL (T-SQL). In this section of the chapter, you will use DMX to create, train, and explore the Accessory Buyers campaign with the Microsoft Naïve Bayes algorithm.

Use the DMX Development Environment

To ready your development environment to create the DMX queries, launch SQL Server Management Studio (SSMS). In the Connect to Server dialog box, select Analysis Services from the Server Type drop-down list. Choose or enter your Server Name and Authentication information, and click Connect. Right-click the Analysis Services (SSAS) instance you just connected to, point to New Query, and click DMX. Figure 11-22 displays a new DMX query window.

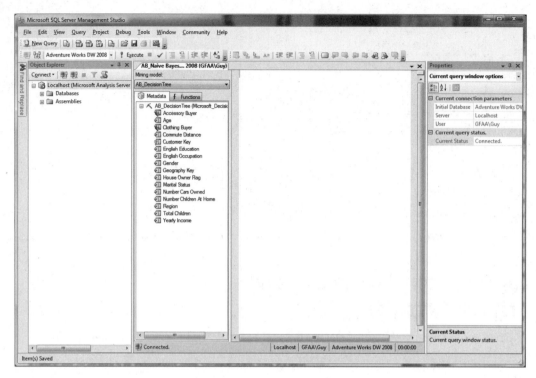

Figure 11-22. SSMS, with a new DMX query window at the ready

Create the Accessory Buyers Mining Structure

Your first task will be to drop the existing AccessoryBuyersCampaign mining structure. To do this, simply enter the following code into the query window, and click Execute in the toolbar:

```
--Drop structure, if needed
Drop Mining Structure AccessoryBuyersCampaign;
```

The next step is to create the AccessoryBuyersCampaign mining structure using DMX. Enter the following code into the query window, highlight it, and click Execute:

```
--Create AccessoryBuyersCampaign mining structure
Create Mining Structure [AccessoryBuyersCampaign]
(
    [Customer Key]          Long      Key,
    [Accessory Buyer]       Long      Discrete,
    [Age]                   Long      Discretized(Automatic, 10),
    [Commute Distance]      Text      Discrete,
    [English Education]     Text      Discrete,
    [English Occupation]    Text      Discrete,
    [Gender]                Text      Discrete,
```

```
[House Owner Flag]            Text     Discrete,
[Marital Status]             Text     Discrete,
[Number Cars Owned]          Long     Discrete,
[Number Children At Home]    Long     Discrete,
[Region]                     Text     Discrete,
[Yearly Income]              Double   Discretized(Automatic, 10)
)
With Holdout (30 Percent);
```

In the preceding code, you can see the similarity of syntax in DMX and T-SQL. The holdout of 30 percent is what you entered in the Create Testing Set earlier in this chapter. Note the Discretized(Automatic, 10) content type. In this statement, you are stating that you want the algorithm to automatically create ten buckets. It is common to start with ten buckets when creating new structures.

Add a Naïve Bayes Mining Model to the Accessory Buyers Campaign

Now that your mining structure is in place, you can use DMX to create mining models. Use the following code to add a Naïve Bayes model to the mining structure you have created:

```
--Add a Naive Bayes model for the Accessory Buyers Campaign
Alter Mining    Structure    AccessoryBuyersCampaign
Add   Mining    Model        AB_NaiveBayes
(
    [Customer Key],
    [Accessory Buyer]        Predict,

    [Age],
    [Commute Distance],
    [English Education],
    [English Occupation],
    [Gender],
    [House Owner Flag],
    [Marital Status],
    [Number Cars Owned],
    [Number Children At Home],
    [Region],
    [Yearly Income]
)
Using Microsoft_Naive_Bayes;
```

In the preceding code, you added the AB_NaiveBayes model to the AccessoryBuyersCampaign mining structure. Note the explicit assignment of Accessory Buyer for prediction.

Process the Accessory Buyers Campaign

Now that you have successfully created a mining structure and added a model to it, it is time to process the Accessory Buyers campaign. To process this structure, enter the following code and execute it:

```
--Process the Accessory Buyer mining structure
Insert Into Mining Structure AccessoryBuyersCampaign
(
    [Customer Key],
    [Accessory Buyer],

    [Age],
    [Commute Distance],
    [English Education],
    [English Occupation],
    [Gender],
    [House Owner Flag],
    [Marital Status],
    [Number Cars Owned],
    [Number Children At Home],
    [Region],
    [Yearly Income])
OpenQuery
(
    [Adventure Works DW],
    'SELECT
       CustomerKey,
       AccessoryBuyer,
       Age,
       CommuteDistance,
       EnglishEducation,
       EnglishOccupation,
       Gender,
       HouseOwnerFlag,
       MaritalStatus,
       NumberCarsOwned,
       NumberChildrenAtHome,
       Region,
       YearlyIncome
    FROM
       dbo.vAccessoryBuyers'
);
```

■ **Tip** The first parameter for OpenQuery is the object name of your data source. In the preceding example, the parameter is [Adventure Works DW].

View the Accessory Buyers Mining Model

To view the Accessory Buyers mining model, you use a variant of the standard T-SQL Select statement. The main difference here is that instead of viewing data from a table, you view the content of a mining model. Enter the following into a query window and execute it:

```
Select   Distinct   [Age]
From     AB_NaiveBayes
```

This simple query will return the distinct ages in your model. This is useful for discretized columns, as the returned data set represents the buckets chosen by the algorithm.

Age
32
38
44
48
53
59
64
68
74
81

This next DMX query will look at the decisions made by the algorithm. Note that the mining columns listed are not defined in your mining structure. These columns are added during model processing to hold values calculated by the model. The Node_Caption column contains the name of each node, while the Node_Support column displays the number of cases belonging to the node. The Node_Distribution column is expandable, and contains the same data as the mining legend.

Enter the next query and execute it:

```
--Review node data in this model
Select
    Node_Caption,
    Node_Support,
    Node_Distribution
From
    AB_NaiveBayes.Content
Where
    Node_Support > 0
Order By
    Node_Support Desc
```

After executing the preceding query, the Results grid shown in Figure 11-23 will display your data.

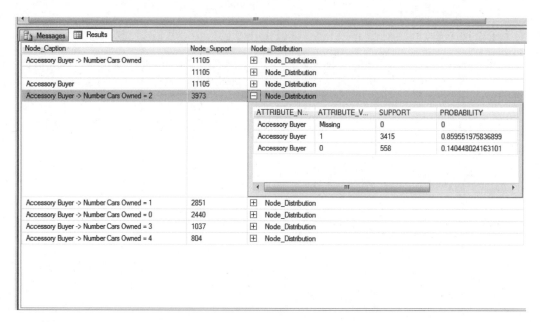

Figure 11-23. *The results of executing the DMX, with a Node_Distribution column expanded*

Predict Our Accessory Buyers

The final product is at hand! It's now time to predict your accessory buyers. To do this, you will create and execute a DMX query that joins your mining model to your prospective buyers. This type of join is called a `Prediction Join`, and uses the same `OpenQuery` syntax you used in the processing section. Note the `PredictProbability` call to `Accessory Buyer`, which returns the probability of this target being an accessory buyer. The following query returns our predicted buyers:

```
--Predict our Accessory Buyers
Select
    PB.FirstName,
    PB.LastName,
    PB.City,
    PB.StateProvinceCode,
    PB.PostalCode,
    PB.Phone,
    PB.NumberChildrenAtHome,
    PB.Occupation,
    PB.ProspectiveBuyerKey,
    AB_NaiveBayes.[Accessory Buyer],
    PredictProbability([Accessory Buyer]) As PredictedProbability
From
    AB_NaiveBayes
Prediction Join
    OpenQuery
    (
```

```
        [Adventure Works DW],
        'SELECT
            ProspectiveBuyerKey,
            FirstName,
            LastName,
            City,
            StateProvinceCode,
            PostalCode,
            Education,
            Occupation,
            Phone,
            HouseOwnerFlag,
            NumberCarsOwned,
            NumberChildrenAtHome,
            MaritalStatus,
            Gender,
            YearlyIncome
        FROM
            dbo.ProspectiveBuyer'
    ) As PB
On
    AB_NaiveBayes.[Marital Status] = PB.[MaritalStatus] And
    AB_NaiveBayes.[Gender] = PB.[Gender] And
    AB_NaiveBayes.[Yearly Income] = PB.[YearlyIncome] And
    AB_NaiveBayes.[Number Children At Home] = PB.[NumberChildrenAtHome] And
    AB_NaiveBayes.[English Education] = PB.[Education] And
    AB_NaiveBayes.[English Occupation] = PB.[Occupation] And
    AB_NaiveBayes.[House Owner Flag] = PB.[HouseOwnerFlag] And
    AB_NaiveBayes.[Number Cars Owned] = PB.[NumberCarsOwned]
Where
    AB_NaiveBayes.[Accessory Buyer] = 1
Order By
    PredictProbability([Accessory Buyer]) Desc;
```

Enter this query into your query window and execute it. A portion of the predicted accessory buyers, per Naïve Bayes, appears in Figure 11-24.

FirstName	LastName	City	StateProvinceCode	PostalCode	Phone	NumberChildrenAtHome	Occupation	ProspectiveBuyerKey	Accessory Buyer	PredictedProbability
Nelson	Alvarez	Tacoma	WA	98403	1 (11) 500 555-0167	4	Management	1527	1	0.985964889917269
Terry	Anand	Lake Oswego	OR	97034	1 (11) 500 555-0187	4	Professional	1913	1	0.985740384932596
Levi	Kapoor	Tacoma	WA	98403	1 (11) 500 555-0153	5	Management	1283	1	0.985694163744909
Manuel	Gonzalez	Olympia	WA	98501	1 (11) 500 555-0134	5	Management	1396	1	0.985694163744909
Aimee	Yang	Columbus	GA	31901	1 (11) 500 555-0130	4	Professional	187	1	0.985627396742234
Carmen	Sai	Lynnwood	WA	98036	1 (11) 500 555-0164	5	Management	448	1	0.985580809989085
Bethany	Sharma	Colma	CA	94014	1 (11) 500 555-0192	5	Management	279	1	0.985580809989085
Josue	Navarro	Bellevue	WA	98004	1 (11) 500 555-0142	4	Management	1029	1	0.985073502323777
Devon	Nath	Colma	CA	94014	1 (11) 500 555-0193	4	Management	580	1	0.985073502323777
Jon	Xie	Colma	CA	94014	1 (11) 500 555-0191	4	Management	1108	1	0.984985711967043
Kaitlin	Fernandez	Bellflower	CA	90706	1 (11) 500 555-0152	4	Management	1154	1	0.98495524312532
Kaitlyn	Davis	El Cajon	CA	92020	1 (11) 500 555-0151	4	Management	1146	1	0.98495524312532
Alfredo	Alvarez	Lynnwood	WA	98036	1 (11) 500 555-0143	4	Professional	3	1	0.984834772994438
Teresa	Hernandez	National City	CA	91950	1 (11) 500 555-0112	5	Management	1933	1	0.984665091842426
Henry	Rodriguez	City Of Commerce	CA	90040	1 (11) 500 555-0197	4	Professional	853	1	0.984370564690409
Arthur	Martinez	Imperial Beach	CA	91932	1 (11) 500 555-0116	4	Professional	90	1	0.984370564690409
Richard	Morgan	Rio Rancho	NM	87124	1 (11) 500 555-0154	4	Professional	1724	1	0.984370564690409
Hector	Alvarez	Lynnwood	WA	98036	1 (11) 500 555-0111	4	Professional	823	1	0.984370564690409
Darrell	Nara	Colma	CA	94014	1 (11) 500 555-0110	4	Professional	578	1	0.984370564690409
Bruce	Verma	San Francisco	CA	94109	1 (11) 500 555-0166	4	Professional	293	1	0.984370564690409
Nathaniel	Stewart	Burien	WA	98168	1 (11) 500 555-0125	5	Management	1577	1	0.984319917296913
Jared	Cooper	Burlingame	CA	94010	1 (11) 500 555-0177	5	Management	934	1	0.984319917296913
Lydia	Kapoor	Odessa	MO	64076	1 (11) 500 555-0159	4	Professional	1282	1	0.984246757408286
Paula	Suarez	Burbank	CA	91502	1 (11) 500 555-0137	4	Professional	1639	1	0.984246757408286
Mayra	Rodriguez	Ballard	WA	98107	1 (11) 500 555-0132	5	Management	1469	1	0.984195710582744
Mayra	Kapoor	W. Linn	OR	97068	1 (11) 500 555-0113	5	Management	1416	1	0.984195710582744
Harold	Verma	Downey	CA	90241	1 (11) 500 555-0184	4	Professional	862	1	0.9837770260321279
Todd	Ye	Tacoma	WA	98403	1 (11) 500 555-0121	5	Management	1991	1	0.983724464671915
Micheal	Jimenez	San Gabriel	CA	91776	1 (11) 500 555-0141	5	Management	1413	1	0.983724464671915

Figure 11-24. Our predicted buyers, complete with contact information and sale probability

Summary

This chapter began by making a case for data mining as an integral part of any business intelligence effort. After reviewing a few of the more common algorithms, we delved into the Accessory Buyer marketing campaign. Using Analysis Services and the Business Intelligence Development Studio, you created a mining model, validated your model, and predicted future customers by using Microsoft Decision Trees.

In the next section, you used DMX to create, validate, and predict future accessory buyers—only this time, you did so in SQL Server Management Studio, using a query-only approach. Now you will move on to PowerPivot, where you will use Excel to work with multidimensional data from your desktop.

■ ■ ■

CHAPTER 12

PowerPivot

PowerPivot is an exciting combination of improvements to Microsoft Excel that enable you to do your own data analysis and mining from your desktop. You can connect to a database, pull data down into Excel, perform analysis on that data, and push the results back up to your server for others in your organization to view. PowerPivot provides extraordinary functionality, giving you full control over your data analysis and mining efforts.

PowerPivot Support in SQL Server 2008 R2

PowerPivot is closely linked to SQL Server. It depends on some key features that are new in SQL Server 2008 Release 2 (R2). Before looking at PowerPivot, it's worth getting familiar with the key features that form its foundation.

About two years after the launch of SQL Server 2008, Microsoft released SQL Server 2008 R2. This launch was in conjunction with Office 2010 and SharePoint Server 2010 as part of a "business intelligence wave." The biggest changes we're interested in are as follows:

Master Data Services: A centralized repository for business data, such as product lists, office locations, customers, and so forth.

VertiPaq mode: A new way of installing an Analysis Services instance that is oriented toward in-memory storage so that queries can be answered in real time. Also referred to as *PowerPivot for SharePoint*, this provides a new layer in the business intelligence structure, for self-service BI.

Excel writeback: We discussed writeback previously, but one problem we had was that there was no real client that could write back to SSAS cubes. Excel 2010 has writeback capabilities, opening some interesting new doors for what-if analysis.

We're going to provide a brief overview of Master Data Services. Because the technology is so new, it's still a little "wait and see," especially from the perspective of Analysis Services. Then we'll take a deep dive into PowerPivot, and finally look at Excel Writeback and what it can do for us.

Master Data Services

Much of our discussion about SQL Server Analysis Services has been around the idea of "one version of the truth." However, on occasion we may create our own stovepipe in the data warehouse. We want the dimensions and their members to be representative of the business, but we're either pulling them from a random business system or setting them up in the EDW ourselves. Although this is great for our BI system, it can be problematic for other business systems that may want to reflect the same structure.

Master Data Services (MDS), a new feature in SQL Server 2008 R2, provides a central repository for the canonical data of an organization. It's an answer to part of the problem with stovepipes, in which each stovepipe refers to some aspect of the business in a different way. It also provides the data of record for new systems being designed (as opposed to spending an entire four-hour meeting on "we have three customer lists—which one is the right one?").

Figure 12-1 shows the data management interface for Master Data Services.

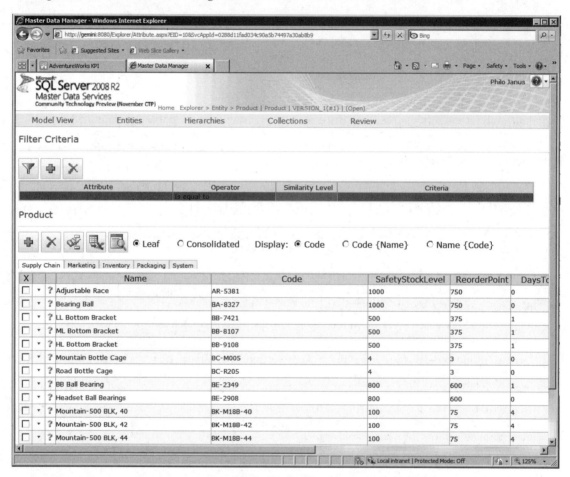

Figure 12-1. Master Data Services web-based data management interface

---

■ **Note** MDS is a separate installation from SQL Server 2008 R2. You can find the installer, `MasterDataServices.msi`, in the installation media for SQL Server 2008 R2.

So the question is—why do you, the Analysis Services DBA, care? If MDS is being implemented as a master data repository, you should plan to take advantage of it for any and all dimensions that use master data. It's possible to use the MDS database as a data source for a dimension, but the problem is that changes to MDS may happen without warning, which would cause serious problems with the relationships between the dimensions and the fact data. There is version data in the MDS schema, so it would be possible to filter by a specific version; then subsequent changes wouldn't affect the dimension until the DBA deals with it.

There is an alternative approach: MDS has a robust business rule engine, which can activate and pass data to a SharePoint workflow. As a result, you could design a workflow to be kicked off when the master data you are tracking for a dimension changes. Perhaps if the change doesn't affect fact data, you could allow the workflow to make the change in place; otherwise, send a notification to the DBAs. Alternatively, you could have the workflow follow an approval chain in which responsible DBAs indicate to either process the change, or to log and hold it until they can evaluate the changes.

Master data management is a field unto itself, and the handling of Master Data Services will evolve as the technology in SQL Server matures. For more information on master data management, especially the business aspects of MDM, I recommend *Master Data Management* by David Loshin (Morgan Kaufmann, 2008). For technical information, the best starting point is always the Books Online: http://msdn.microsoft.com/en-us/library/ee633763(SQL.105).aspx.

Excel Writeback

We've discussed writeback to some degree, regarding the ability to change dimensions and edit fact data. In Excel 2010 we have the ability to edit cube fact data directly from Excel. Well, we don't actually edit the fact data; what happens is that when we enable writeback on a cube partition, Analysis Services creates a table in the underlying data source, and a writeback partition in the cube. Then when queries are performed against the fact table that has a parallel writeback partition, the values are combined on the fly.

Writeback is enabled in the Partitions tab of the Cube Designer; right-click on a partition and select Writeback Settings to enable writeback. This will create a writeback partition related to the main partition. That writeback "partition" will actually map to a table in one of the data sources (if you want the writeback tables kept in a unique location, obviously you can create a data source for writeback tables). Analysis Services stores the *difference* value in the table, along with the user that entered it and a timestamp.

■ **Note** You can enable writeback only on a partition where all the measures are aggregated with the SUM function. The big gotcha here is that by default, measure groups usually have a COUNT member.

After changes have been made via writeback, they will be seen by any other user querying the cube. After you've written data back to the database, there are three follow-up actions that you can take:

Convert to Partition: This makes the writeback changes a permanent part of the cube. This is accomplished by right-clicking on the writeback partition and selecting Convert to Partition.

Clear Writeback Data: Deleting the contents of the writeback table effectively "resets" the writeback data. Just remember that the same table is used by all users!

Disable Writeback: Right-clicking on the writeback partition in the Partitions tab and selecting Writeback Settings brings up the Disable Writeback dialog box. Clicking OK disables writeback for that partition.

Because every user sees the writeback data, your having these three choices can be problematic if a user is changing values for what-if scenarios. The best solution is to have a Scenario dimension. For those users who need to perform writeback analysis, you can give them control over the Scenario dimension. (That dimension can even have a user-based hierarchy.) Any other user can have a default member of Actuals, with no writeback abilities to the dimension.

WHAT IS LEAF LEVEL, AND WHY DO I CARE?

When you work with writeback, you can quickly find yourself getting "out of memory" errors. The reason for this is that when you are writing back to a cube, you can write back to either leaf-level values or nonleaf values. *Leaf-level* values are those values that have a lowest-level member selected for every dimension associated with the measure group.

For example, on the Products dimension, the default member is All Products. If a measure was related to Region, Product, and Employee, and I selected a specific region and specific employee, I would have a value representing sales in all products, which is not a leaf-level value. I have to choose a specific product to get to a leaf-level value. You can see which measures you have to worry about by looking at the Dimension Usage tab in the Cube Designer. For many measures in the AdventureWorks cube, you can see that there are a large number of associated dimensions. To get to a leaf-level value in those measures, you would have to select a member from each dimension (or the dimension would have to have a default member at the lowest level of the hierarchy defined).

Look at the dimension usage for Sales Targets. You'll see that there are only three dimensions associated. Also note how the Date dimension is associated—the *granularity attribute* is the Calendar Quarter. This means that you can drill down to only the quarter level, but more important, you have to select only a quarter to get to a leaf-level value.

Why does all this matter? Because if you write back to a *non*-leaf-level value, the value you are writing back must be allocated among the leaves of what you're writing back to. For example, if you wrote a value back to a measure that had All Products selected, that value would have to be allocated among four categories, and then 37 subcategories, and then 395 products. And the allocation works geometrically; if you didn't select a region, you would be dividing by 6 countries, 71 provinces, and 587 cities. But the value has to be allocated by geography *and* product (395 x 587 = 231,865). And that's just two dimensions.

So the important thing to realize is that when you are designing a cube for what-if analysis, you will most likely want dedicated measure groups for the analysis, and to be brutal on minimizing the dimensions associated with those measure groups.

Let's take Excel's new writeback capability for a spin. Follow the instructions in Exercise 12-1 to experience writeback for yourself.

Exercise 12-1. Use Excel Writeback

1. Ensure that Writeback is enabled on the Sales Targets partition in the AdventureWorks cube, as shown in Figure 12-2.

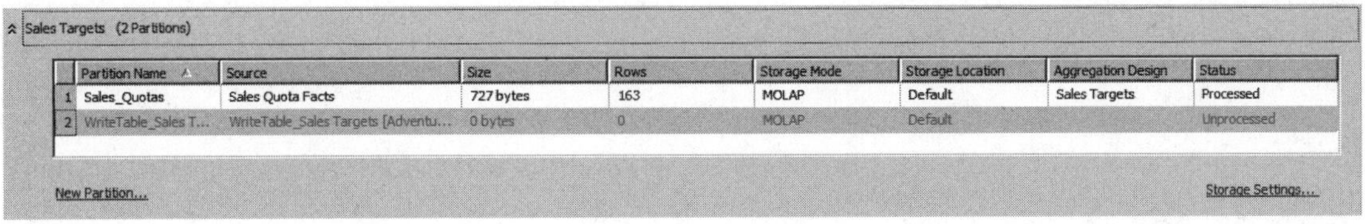

	Partition Name	Source	Size	Rows	Storage Mode	Storage Location	Aggregation Design	Status
1	Sales_Quotas	Sales Quota Facts	727 bytes	163	MOLAP	Default	Sales Targets	Processed
2	WriteTable_Sales T...	WriteTable_Sales Targets [Adventu...	0 bytes	0	MOLAP	Default		Unprocessed

New Partition...   Storage Settings...

Figure 12-2. Ensuring the Sales_Quotas partition has a writeback partition

2. Open Excel 2010.

3. Click the Data tab in the Ribbon.

4. In the Get External Data panel, click From Other Sources and then select From Analysis Services.

5. Enter the Server name and then click Next.

6. Select the AdventureWorks cube and then click Next.

7. Click Finish to close the Data Connection dialog box. When the Import Data dialog box opens, leave the default selection of PivotTable Report and click OK.

8. After the pivot table is created, in the Pivot Table Field List select Sales Amount Quota.

9. Under the Date dimension, open the Calendar folder, and drag Date.Calendar Year to the Column Labels area.

10. Under the Employee dimension, drag Employees to Row Labels. The setup should look like Figure 12-3, and the resulting pivot table should look like Figure 12-4.

Figure 12-3. *Pivot table setup*

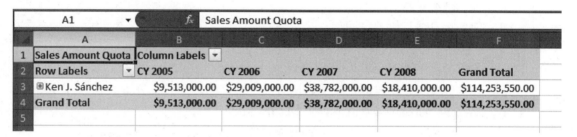

A1		f_x	Sales Amount Quota		
	A	B	C	D	E	F

	A	B	C	D	E	F
1	Sales Amount Quota	Column Labels ▾				
2	Row Labels ▾	CY 2005	CY 2006	CY 2007	CY 2008	Grand Total
3	⊞ Ken J. Sánchez	$9,513,000.00	$29,009,000.00	$38,782,000.00	$18,410,000.00	$114,253,550.00
4	Grand Total	$9,513,000.00	$29,009,000.00	$38,782,000.00	$18,410,000.00	$114,253,550.00
5						

Figure 12-4. *The resulting pivot table*

11. Right-click on the cell labeled Column Labels and select PivotTable Options to open the PivotTable Options dialog box.

12. Click the Display tab, as shown in Figure 12-5, and select the Show Items with No Data on Columns check box.

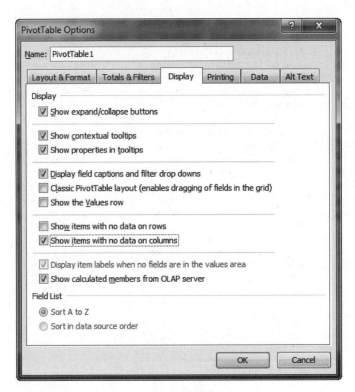

Figure 12-5. Setting pivot table options

13. Click OK. This shows the CY 2010 member of the Date dimension.

14. Open the Employees tree to show Stephen Jiang's subordinates. The pivot table should now look like Figure 12-6.

	A	B	C	D	E	F	G
1	Sales Amount Quota	Column Labels ▼					
2	Row Labels ▼	CY 2005	CY 2006	CY 2007	CY 2008	CY 2010	Grand Total
3	⊟Ken J. Sánchez	$9,513,000.00	$29,009,000.00	$38,782,000.00	$18,410,000.00		$114,253,550.00
4	⊟Brian S. Welcker	$9,513,000.00	$29,009,000.00	$38,782,000.00	$18,410,000.00		$114,253,550.00
5	⊞Amy E. Alberts		$4,270,000.00	$10,027,000.00	$4,937,000.00		$24,202,000.00
6	⊟Stephen Y. Jiang	$9,513,000.00	$24,739,000.00	$27,716,000.00	$12,620,000.00		$87,336,050.00
7	David R. Campbell	$579,000.00	$1,371,000.00	$1,438,000.00	$637,000.00		$4,876,850.00
8	Garrett R. Vargas	$600,000.00	$1,478,000.00	$1,617,000.00	$670,000.00		$5,049,450.00
9	Jillian Carson	$1,437,000.00	$4,750,000.00	$4,350,000.00	$1,661,000.00		$13,778,850.00
10	José Edvaldo. Saraiva	$1,292,000.00	$2,114,000.00	$2,293,000.00	$1,399,000.00		$8,516,850.00
11	Linda C. Mitchell	$1,418,000.00	$3,668,000.00	$4,682,000.00	$2,018,000.00		$13,844,750.00
12	Michael G. Blythe	$923,000.00	$3,805,000.00	$4,716,000.00	$1,718,000.00		$13,288,250.00
13	Pamela O. Ansman-Wolfe	$634,000.00	$1,001,000.00	$1,183,000.00	$733,000.00		$3,981,650.00
14	Shu K. Ito	$1,009,000.00	$2,813,000.00	$2,644,000.00	$1,338,000.00		$9,066,250.00
15	Tete A. Mensa-Annan		$321,000.00	$1,481,000.00	$951,000.00		$3,449,600.00
16	Tsvi Michael. Reiter	$1,586,000.00	$2,963,000.00	$2,768,000.00	$1,224,000.00		$9,823,500.00
17	⊞Syed E. Abbas			$1,039,000.00	$853,000.00		$2,715,500.00
18	Grand Total	$9,513,000.00	$29,009,000.00	$38,782,000.00	$18,410,000.00		$114,253,550.00
19							

Figure 12-6. The pivot table ready for use

15. Now in the Ribbon, under PivotTable Tools/Options, click the What-If Analysis option and then select Enable What-If Analysis.

16. In the CY 2010 column for Jillian Carson, enter **$5,000,000**.

17. Note the little tag that appears in the lower right of the cell. If you click on the cell, you'll get a context button. Click it to see the context menu shown in Figure 12-7.

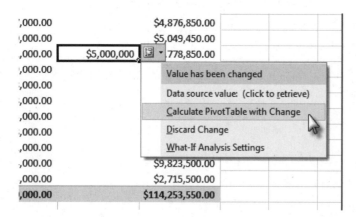

Figure 12-7. The what-if context dialog box

18. Select Calculate PivotTable with Change. Note that the $5,000,000 value has been propagated up the hierarchy, totaled for CY 2010, and added to the Grand Total for Jillian.

19. At this point, the changes exist only in Excel. If anyone else queried the cube, that person would still see the underlying values. If we didn't want to keep this result, we could select Discard Change, or use the Discard Changes option on the What-If option in the Ribbon to discard all the session values.

20. Now let's select Publish Changes from the What-If Analysis button. This commits our writeback changes to the cube, and they will show up for anyone else running a cube report.

21. Let's look at our "written back" data. Open SQL Server Management Studio.

22. Connect to the AdventureWorksDW2008R2 database. This should be the data source for your cube. Look for the table named dbo.WriteTable_Sales Targets.

23. Right-click on the table and select Select Top 1000 Rows to view the contents of the table. You should see something like Figure 12-8.

	SalesAmountQuota_0	EmployeeKey_1	SalesTerritoryKey_2	CalendarYear_3	CalendarQuarter_4	MS_AUDIT_TIME_5	MS_AUDIT_USER_6
1	454545.454545455	283	7	2010	4	2010-03-01 13:21:33.000	SAINTCHAD\philo
2	454545.454545455	283	8	2010	4	2010-03-01 13:21:33.000	SAINTCHAD\philo
3	454545.454545455	283	10	2010	4	2010-03-01 13:21:33.000	SAINTCHAD\philo
4	454545.454545455	283	11	2010	4	2010-03-01 13:21:33.000	SAINTCHAD\philo
5	454545.454545455	283	6	2010	4	2010-03-01 13:21:33.000	SAINTCHAD\philo
6	454545.454545455	283	3	2010	4	2010-03-01 13:21:33.000	SAINTCHAD\philo
7	454545.454545455	283	2	2010	4	2010-03-01 13:21:33.000	SAINTCHAD\philo
8	454545.454545455	283	1	2010	4	2010-03-01 13:21:33.000	SAINTCHAD\philo
9	454545.454545455	283	5	2010	4	2010-03-01 13:21:33.000	SAINTCHAD\philo
10	454545.454545455	283	4	2010	4	2010-03-01 13:21:33.000	SAINTCHAD\philo
11	454545.454545455	283	9	2010	4	2010-03-01 13:21:33.000	SAINTCHAD\philo

Figure 12-8. Writeback data

The EmployeeKey is the Primary Key (PK) for Jillian. CalendarYear looks like we expect, as does Quarter. But what are those values, and why are the Sales Territory keys all over the place? Well, remember that we didn't select a territory, so Excel indicated to Analysis Services to spread the values between the territories. There are 11 territories, and 5,000,000 / 11 = 454,545.4545.

24. As I mentioned previously, Analysis Services will now seamlessly combine these values with the cube values for anyone who browses the cube. If you want to drop the writeback data, the simplest way is to simply delete the records from this table. (Yes, this is nontrivial, but there isn't an easy client tool for this at this time.)

Although this was a very simplistic solution, you can imagine more-complex cubes, and adding formulas on the Excel side to perform what-if analysis on stock levels, consultant hourly rates, quotas, and so on.

That covers the essentials of Excel writeback capabilities. As you can see, it's a fairly straightforward feature, but an incredibly powerful tool in the analyst's toolbox. Writeback is simply part of Excel 2010 out of the box, and requires SQL Server Enterprise Edition on the server side, but those are the only special requirements. For real analytic power on the desktop, let's talk about PowerPivot.

PowerPivot from Excel

Of course the biggest news in SQL Server 2008 R2 (in fact, the whole reason for the release) is the new PowerPivot functionality. PowerPivot is the next logical step in eliminating *spreadmarts*, or the use of Excel as a database/data mart. PowerPivot is a combination of the following:

- Improvements to Excel

- New architecture in SharePoint

- A new way of installing Analysis Services known as *VertiPaq mode*

What PowerPivot enables the user to do is to design a cube with data sources in Excel, and work with that cube in pivot tables and pivot charts just as if it were an Analysis Services cube. In addition, an analyst can publish an Excel workbook with PowerPivot cube(s) to a SharePoint Server that has been provisioned with Analysis Services, Excel Services, and PowerPivot. After an Excel workbook with a PowerPivot cube has been published, it can be used as a data source by any client that can connect to Analysis Services.

■ **Note** In VertiPaq mode, SQL Server Analysis Services runs as an in-memory service, and keeps all its cubes in memory for rapid query response. Furthermore, you cannot publish solutions from BIDS. A VertiPaq SSAS instance is intended solely to service PowerPivot databases.

The Excel add-in for PowerPivot is simply a free download from Microsoft.com. The best way to find it is to start at `www.powerpivot.com/download.aspx`, which has links to both the 32-bit and 64-bit versions of the Excel add-in.

■ **Note** The 32-bit or 64-bit versions are with respect to what bitness of *Excel* you installed, not the operating system. If you have 32-bit (x86) Excel running on 64-bit (x64) Windows 7, then use the 32-bit add-in.

The Excel client adds a tab to the Ribbon for PowerPivot, as shown in Figure 12-9. This is used for managing the PowerPivot client. Think of PowerPivot as a cube running under Excel—the PowerPivot tab gives you access to the designer, and enables you to create calculated measures from the underlying data, add pivot tables and charts based on the PowerPivot cube to the workbook, or link data from tables in the workbook.

Figure 12-9. *PowerPivot tab in Excel 2010*

■ **Note** The PowerPivot help file comes with a great tutorial for creating a PowerPivot workbook. The data files necessary are located on CodePlex. In lieu of creating another exercise that duplicates the tasks in that tutorial, I recommend walking through the tutorial; that model is what we'll be looking at as we move forward.

Clicking the PowerPivot Window button opens the PowerPivot Designer, which is like a stripped-down version of Excel, focused on data. Figure 12-10 shows the PowerPivot Designer. If you note the tabs along the bottom, you will see multiple tables of data. You can import data from any number of sources. Out of the box PowerPivot supports connections to the following:

- Microsoft SQL Server
- Microsoft SQL Azure
- Microsoft Office Access
- Oracle Database
- Teradata Database
- Sybase
- IBM Informix
- IBM DB2
- Microsoft Excel
- Text files
- Reporting Services as a data feed
- Other web services
- Others via OLE DB/ODBC

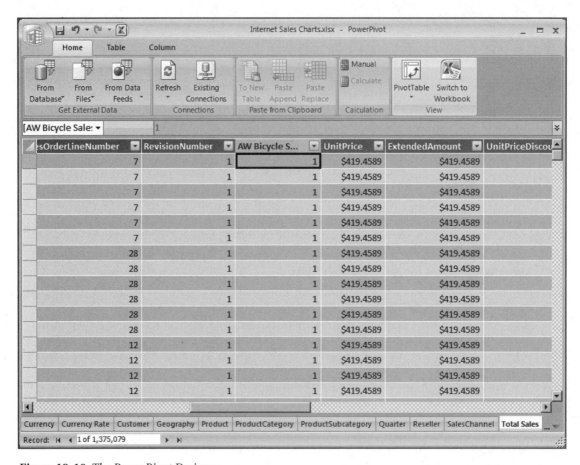

Figure 12-10. The PowerPivot Designer

Note once again that the data in Figure 12-10 is *imported* from a database, so connecting to large data sources may take some time and memory. However, PowerPivot has been optimized to enable data tables with millions of rows to be usable. When you import data from a relational source, you can import multiple tables at once, as shown in Figure 12-11. When you import multiple tables, PowerPivot will recognize and maintain relationships between the tables. In addition, for each table you can select which columns to import, and you can filter the data (for example, if you want to bring in sales from only a specific year). If you are working with large data sets, try to minimize what you are importing in every way possible.

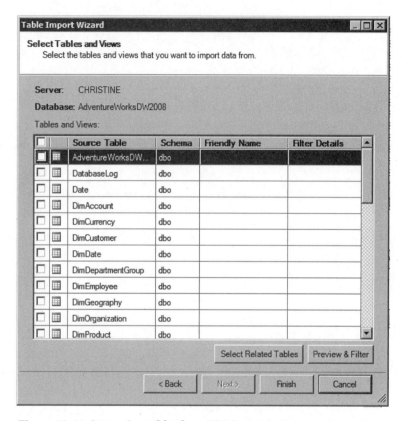

Figure 12-11. *Importing tables from SQL Server*

■ **Note** If you're an experienced DBA, you may look for some form of entity relationship designer to graphically set up relationships. There isn't one. The reason is twofold: time constraints in getting the product finished, and some philosophical considerations over how to best present the concept of entity relationship design to a nontechnical analyst audience.

You can also create relationships in the PowerPivot Designer between tables from unrelated sources. When you create the relationship, PowerPivot will verify the suitability of the data in the columns for a relationship—that the data types match, and that one column has unique values.

After you have imported the necessary data and created the relationships you need, you can create a pivot table and chart(s) from the PowerPivot Designer, or from the PowerPivot tab in Excel. When you insert a pivot table, it should look pretty familiar, as shown in Figure 12-12. You have the task pane with panels for values, columns, rows, and filters. You also have two new panels, for horizontal and vertical *slicers*. These are new features of Excel 2010 to provide attractive, dynamic filtering for pivot tables.

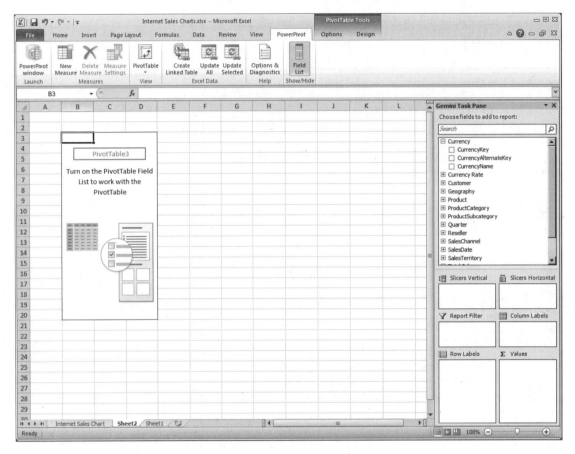

Figure 12-12. A PowerPivot pivot table in Excel

You can design a pivot table here just as we did in the earlier exercise. If you get a string of the same values, as shown in Figure 12-13, you know that there isn't a relationship between the dimension in the labels and the values in the table.

Sum of Industry Bicycle Sales	Column Labels			
Row Labels	1	2	3	4
Australia	$ 7,173,355.00	$ 7,332,882.00	$ 7,582,891.00	$ 10,172,834.00
Canada	$ 7,173,355.00	$ 7,332,882.00	$ 7,582,891.00	$ 10,172,834.00
France	$ 7,173,355.00	$ 7,332,882.00	$ 7,582,891.00	$ 10,172,834.00
Germany	$ 7,173,355.00	$ 7,332,882.00	$ 7,582,891.00	$ 10,172,834.00
United Kingdom	$ 7,173,355.00	$ 7,332,882.00	$ 7,582,891.00	$ 10,172,834.00
United States	$ 7,173,355.00	$ 7,332,882.00	$ 7,582,891.00	$ 10,172,834.00
Grand Total	$ 7,173,355.00	$ 7,332,882.00	$ 7,582,891.00	$ 10,172,834.00

Figure 12-13. Measure not related to the Geography dimension used for row labels

You can add slicers by dragging a field from a table (what we're used to thinking of as a dimension attribute) down to the slicers area. Excel will then create a filter area with the values from the filter field, as shown in Figure 12-14. You can have multiple slicers, and selecting a slicer will enable the Slicer Tools tab in the Ribbon, with which you can customize the slicer, as shown in Figure 12-15. (Remember to click in the pivot table or chart to get the task pane back if you need it.)

Year				Sum of AW Bicycle Sales	Column Labels				
				Row Labels	1	2	3	4	Grand Total
2001	2002	2003	2004	Australia	$ 43,182.00	$ 49,979.00	$ 51,942.00	$ 52,450.00	$ 197,553.00
2005	2006	2007	2008	Canada	$ 65,806.00	$ 62,476.00	$ 65,576.00	$ 58,845.00	$ 252,703.00
2009				France	$ 33,388.00	$ 30,386.00	$ 30,292.00	$ 37,643.00	$ 131,709.00
				Germany	$ 24,677.00	$ 29,013.00	$ 22,894.00	$ 25,137.00	$ 101,721.00
				United Kingdom	$ 32,041.00	$ 28,865.00	$ 37,607.00	$ 36,972.00	$ 135,485.00
				United States	$ 201,022.00	$216,038.00	$198,145.00	$218,572.00	$ 833,777.00
				Grand Total	$ 400,116.00	$416,757.00	$406,456.00	$429,619.00	$1,652,948.00

Figure 12-14. Adding a slicer to our table

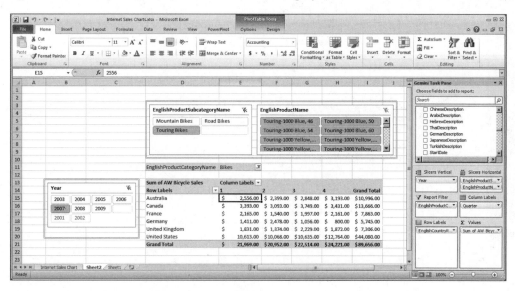

Figure 12-15. The Slicer Tools tab on the Ribbon

You can add multiple filters and slicers, and the slicers will cascade as necessary. Figure 12-16 shows a pivot table with a filter, horizontal slicers on Subcategory and Product, and a vertical slicer on the Year.

Figure 12-16. Robust pivot table with multiple slicers

If you've been through the tutorial and you've tried some of these reports, you should notice that building the reports is pretty quick—selecting a slicer filters the report in subsecond time. This is even though the Total Sales table we're building it from has almost 1.4 million rows of data! Remember, this isn't an Excel pivot table; we actually have SQL Server Analysis Services running on the desktop and performing these aggregations for us!

Building and manipulating these models on the desktop is great, but the next great thing is collaborative analysis. We want to be able to share what we've created. In SharePoint 2007, we had the ability to publish a spreadsheet to Excel Services, and it was somewhat interactive for our users. PowerPivot opens a whole new world.

PowerPivot with SharePoint Server 2010

SharePoint Server 2010 integrates with SQL Server Analysis Services, leveraging the VertiPaq mode to run PowerPivot models on the server. The net result of this is that after you build a model in Excel, you can publish the workbook to Excel Services on SharePoint, and that model will be available just like an Analysis Services cube for any client that can connect to Analysis Services.

The interface point is a site named the PowerPivot Gallery on the SharePoint Server. When you're ready to publish a PowerPivot workbook, click the File tab above the Ribbon, and then select Share and then Publish to Excel Services, as shown in Figure 12-17. This will publish the Excel workbook to Excel Services, where it can be rendered in a browser, and publish the PowerPivot model to the integrated SQL Server Analysis Services instance (and create the necessary connections between them).

Figure 12-17. Publishing a PowerPivot workbook to SharePoint Server

After your workbook is published to SharePoint, it will appear in the PowerPivot Gallery (see Figure 12-18). From the gallery you can manage your PowerPivot workbooks, with a screen preview of the reports and charts contained in each workbook. Of course, you can still switch to the standard SharePoint document library views if you need to.

■ **Caution** When you publish a PowerPivot Excel workbook to SharePoint, you are saving the entire data store, which can get pretty large pretty fast. PowerPivot uses a lot of data compression to get the file size smaller, but be aware that with a PowerPivot Gallery, you are going to be seeing file sizes you're not used to.

Figure 12-18. PowerPivot Gallery in SharePoint 2010

You can click on a PowerPivot document to open it in Excel Services, as shown in Figure 12-19. This is a lightweight interactive spreadsheet view of your document, with parsing and processing being performed on the server. You can also click the Edit link at the top left, and if the document is available for editing, it will open in the new Excel Web App—an interactive version of Excel Services, or a browser-based spreadsheet application.

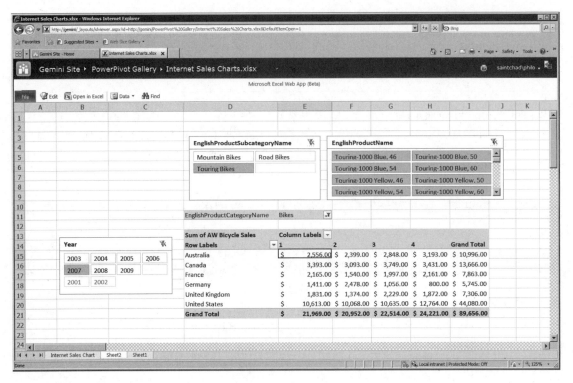

Figure 12-19. A PowerPivot workbook open in Excel Services

Now is where it really gets cool. Open Report Builder (or Excel, or Reporting Services, or any application that connects to SQL Server Analysis Services), create a connection to Analysis Services, and enter the document URL for the server name. You should then get one option for the database—a Sandbox, as shown in Figure 12-20.

Figure 12-20. *Connecting to a PowerPivot worksheet*

After you've connected, you will have the standard view of a dimensional data source. Figure 12-21 shows our PowerPivot demo workbook loaded in the Matrix Designer for Report Builder. This is part of the power of the new designer: all the client tools are exactly the same. Analysts can build models by using Excel, publish them to SharePoint (where they can be viewed online), and then the data sources can be consumed in any client you choose to use.

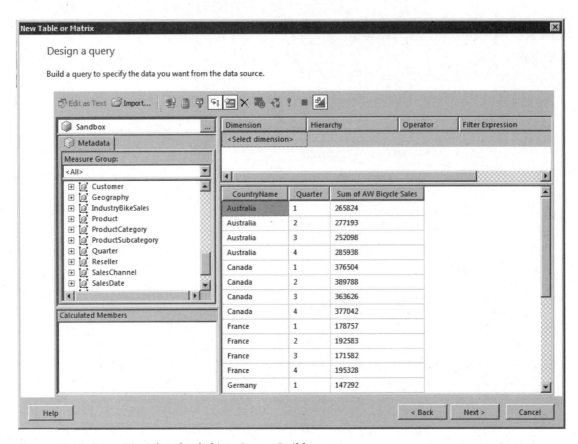

Figure 12-21. PowerPivot data loaded into Report Builder

Remember that the data stored in a PowerPivot workbook was a copy? Part of the PowerPivot Services on SharePoint runs on a timer service to automatically refresh data in the workbook (for that data that has live connections). Of course, you'll have to be sure that the appropriate security is in place; the query will run under the credentials of the timer service. Incidentally, to verify those credentials, you can even connect to a PowerPivot workbook with SQL Server Management Studio, shown in Figure 12-22. Again, the server address is the URL of the workbook.

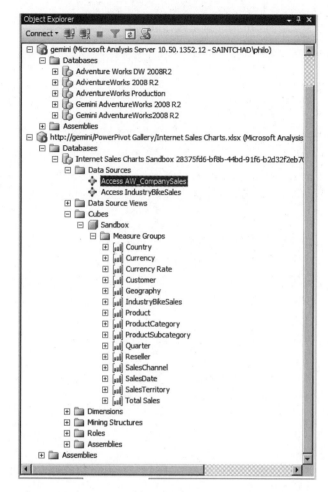

Figure 12-22. *Connecting to PowerPivot with SSMS*

Summary

Those are the fundamental areas of change in SQL Server Analysis Services 2008 R2. Of course, three little terms—Master Data Services, Excel writeback, and PowerPivot—each create whole new areas of discovery in their own rights. This interim release has given us the tools to manage our corporate data, bringing us one step closer to "one version of the truth." It's given end users the ability to perform what-if scenarios with the multidimensional models that we deliver via Analysis Services, and we can empower end users to build their own cubes and publish them for reuse around the enterprise.

In Chapter 13, let's take a look at how we're going to keep all of this stuff running. We'll look at Analysis Services administration, and we'll talk about how to manage an Analysis Services environment.

Administration

This chapter covers the DBA side of Analysis Services. We'll start with processing cubes, and what happens when you process an SSAS cube. We'll look at Analysis Services security, and finally end with some coverage of aspects of SSAS performance management, how to design for performance, and considerations for scalability.

DBA Tasks

After you build a cube—create the dimensions, map them to measures, create your attributes and hierarchies—none of it actually *does* anything until you deploy and process the cube. Deploying is effectively "saving the files to the server," but let's take an in-depth look at what happens when we process objects in an Analysis Services solution.

Processing a Cube

You can process various objects in SSAS: databases, cubes, dimensions, measure groups, partitions, data mining models, and data mining structures. When you process an object, all the child objects are reviewed for processing. If they are in an unprocessed state, they are then processed as well.

To process the Adventure Works cube, select *Process…* from the *Adventure Works.cube* context menu, or click *Process* in the cube designer toolbar. This will display the Process Cube dialog, as shown in Figure 13-1.

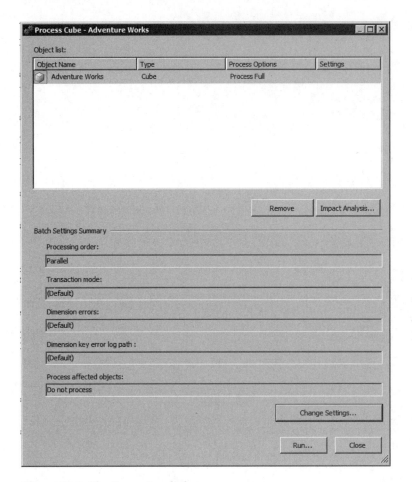

Figure 13-1. The Processing dialog

How the cube will be processed depends on the processing option selected. You have the following options available via the Process Options drop-down list:

Process Full: Processes the object and all its children. If an object has already been processed, then SSAS drops the data and processes it again. This is necessary if there's been a structural change that invalidates the existing aggregations.

Process Default: Detects the status of the object and its children, and processes them if necessary.

Process Incremental: (Measure groups and partitions only) Adds new fact data.

Process Update: (Dimensions only) Forces a re-read of data to update dimension attributes. Good for when you add members to a dimension.

Process Index: (Cubes, dimensions, measure groups, and partitions only) Creates and rebuilds aggregations and indexes for affected objects. Can be run only on processed objects; otherwise throws an error.

Process Data: (Cubes, dimensions, measure groups, and partitions only) Drops all data in the cube and reloads it. Will not rebuild aggregations or indexes.

Unprocess: Drops the data in the objects.

Process Structure: (Cubes and Mining structures only) Processes dimensions and cube definitions. Won't load data or create aggregations.

Process Clear Structure: (Mining structures only) Removes all training data.

Processing Options

Clicking the *Change Settings* button will open the Change Settings dialog, which has two tabs, Processing options and Dimension key errors. In this section I'll focus on the Processing options tab, which is shown in Figure 13-2.

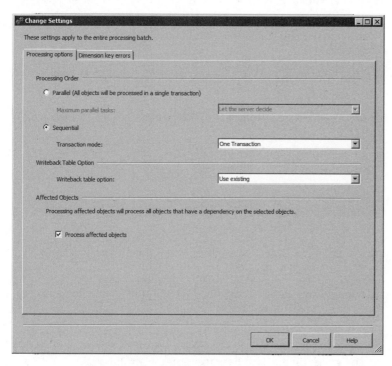

Figure 13-2. The processing settings dialog

Under Processing options you can select whether to process the objects in parallel or sequentially. Parallel processing splits a task and allows jobs to run in parallel. However, during parallel processing, objects will be taken offline as they're processed. The result is that the cubes will be unavailable during most of the processing process. Parallel processing is wrapped in a single transaction, so if the processing fails, all the changes are rolled back.

Sequential processing has two options: `One Transaction` or `Separate Transactions`. When processing an SSAS job as a single transaction, the entire task is wrapped in a transaction—if there's an error at any time during processing, the whole thing is rolled back. The major benefit is that the cube stays available to users until the transaction is committed, at which point the whole update is slammed in place at once.

If you process a task as separate transactions, each job is wrapped in its own transaction, and each is committed as they're completed. The benefit here is that if you have a large, complex processing task, it can be painful to have the whole thing roll back because of one bad piece of data. Processing a series of jobs sequentially as separate transactions means everything up to the error will be kept—only the current job will roll back.

Writeback tables are also something you can configure from the Processing options tab. Writeback tables are relational tables created when you enable writeback on a partition. Certain changes to cube structure can invalidate the writeback tables if they no longer map to the cube structure. The Writeback Table Option indicates what processing should do with existing writeback tables.

Use Existing: Will use an existing writeback table, but will have no effect if there isn't one.

Create: Creates a writeback table if there isn't one. If there is one, this option will cause the process to fail.

Create Always: This option indicates that the processing task should create a new writeback table, and overwrite the table if it already exists.

Finally, `Process Affected Objects` will process objects that have a dependency on objects being processed. For example, if you process a dimension and select this option, the cubes that depend on that dimension will also be processed.

So let's take a look at the mechanics of processing an object in Analysis Services, which I feel helps reinforce the understanding of what's going on behind the curtain.

Processing Architecture

When Analysis Services processes a cube, the end result is a number of binary hash files on disk for the dimensions, attributes, and aggregates generated to optimize response time for dimensional queries. How do we get from a collection of `.xml` files that define the cube to the processed hashed binary files that are the cube?

When you issue a process command, the Analysis Server analyzes the command, and evaluates the objects affected and their dependencies. From that the processing engine will build a list of jobs to be executed and their dependencies. The jobs are then processed by a job execution engine, which performs the tasks necessary to build the file structures for the cube. Jobs without dependencies on each other can be executed in parallel. The server property `CoordinatorExecutionMode` controls how many jobs can run in parallel at once.

A dimension-processing job will create jobs to build the attribute stores, creating the relationship store, name, and key stores. The Attribute job will then build the hierarchy stores, and, at the same time, the decoding stores, followed by bitmap indexes on the decoding stores. As the attribute-store processing job is the first step, it's the bottleneck in processing; this is why having a well-structured attribute relationship is so important.

The decoding stores are how the storage engine retrieves data for the attributes in a hierarchy. For example, to find the part number, price, and picture of a specific bicycle, the storage engine fetches them from the decoding stores. The bitmap indexes are used to find the attribute data in the relationship store. The processing engine can spend a lot of time building the bitmap indexes while processing, especially in dimensions with a large number of members. If a collection of attributes is unique, the processing engine may spend more time building bitmap indexes than they're worth.

A cube-processing job will create child jobs for measure groups, and measure-group jobs will create child jobs for partitions. Finally, partition jobs will create child jobs to process fact data and build the aggregations and bitmap indexes. To process the fact data, the job-execution engine queries the data sources from the data-source view to retrieve the relational data. Then it determines the appropriate keys from the dimension stores for each relational record and populates them in the processing buffer. Finally, the processing buffer is written out to disk.

Once the fact data is in place, then jobs are launched to build the aggregations and bitmap indexes. They scan the fact data in the cube matrix, then create the bitmap indexes in memory, writing them out to disk segment by segment. Finally, using the bitmap indexes and the fact data, the job engine will build the aggregations and aggregation indexes in memory, writing them out to disk when they're complete.

Profiler

Your first tool in evaluating the processing of a cube is the SQL Server Profiler. Profiler is an application that allows you to log all events being performed on or by a SQL Server service. In the case of Analysis Services, we want to run a trace on SSAS while performing an action of interest (in this case, processing a cube.) Figure 13-3 shows part of a trace from a cube-processing job. From the Profiler trace, you can identify the start time and completion time of various tasks, and from there figure out how long various tasks are taking, and, more importantly, where most of the time is being spent processing.

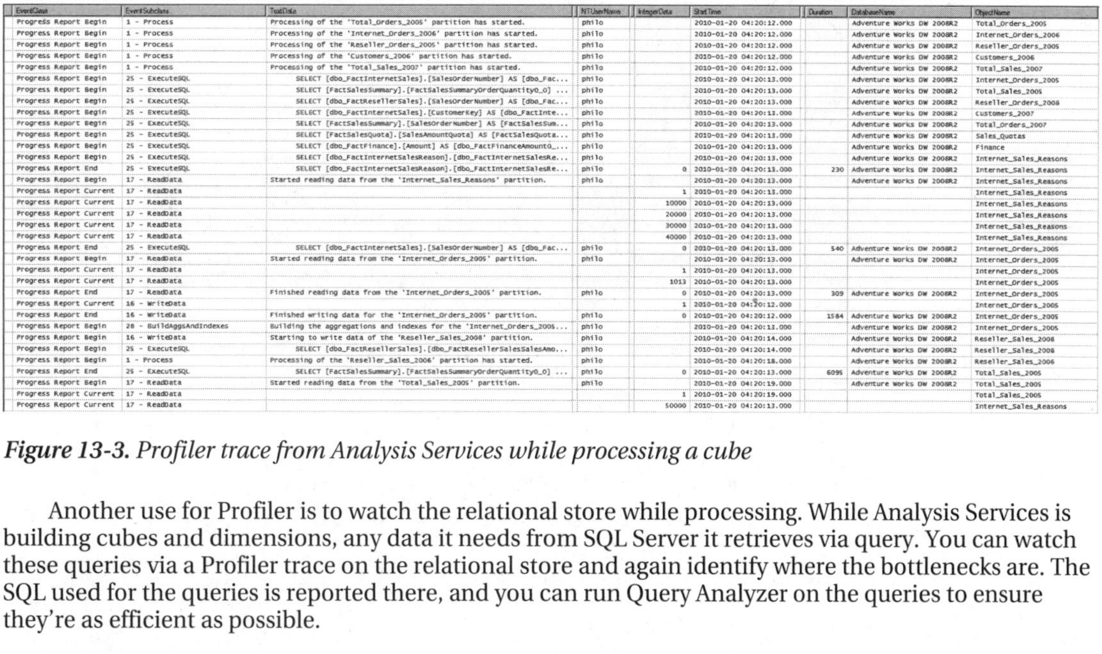

EventClass	EventSubclass	TextData	NTUserName	IntegerData	StartTime	Duration	DatabaseName	ObjectName
Progress Report Begin	1 - Process	Processing of the 'Total_Orders_2005' partition has started.	philo		2010-01-20 04:20:12.000		Adventure Works DW 2008R2	Total_Orders_2005
Progress Report Begin	1 - Process	Processing of the 'Internet_Orders_2006' partition has started.	philo		2010-01-20 04:20:12.000		Adventure Works DW 2008R2	Internet_Orders_2006
Progress Report Begin	1 - Process	Processing of the 'Reseller_Orders_2005' partition has started.	philo		2010-01-20 04:20:12.000		Adventure Works DW 2008R2	Reseller_Orders_2005
Progress Report Begin	1 - Process	Processing of the 'Customers_2006' partition has started.	philo		2010-01-20 04:20:12.000		Adventure Works DW 2008R2	Customers_2006
Progress Report Begin	1 - Process	Processing of the 'Total_Sales_2007' partition has started.	philo		2010-01-20 04:20:13.000		Adventure Works DW 2008R2	Total_Sales_2007
Progress Report Begin	25 - ExecuteSQL	SELECT [dbo_FactInternetSales].[SalesOrderNumber] AS [dbo_Fac...	philo		2010-01-20 04:20:13.000		Adventure Works DW 2008R2	Internet_Orders_2005
Progress Report Begin	25 - ExecuteSQL	SELECT [FactSalesSummary].[FactSalesSummaryOrderQuantity0_0] ...	philo		2010-01-20 04:20:13.000		Adventure Works DW 2008R2	Total_Sales_2006
Progress Report Begin	25 - ExecuteSQL	SELECT [dbo_FactResellerSales].[SalesOrderNumber] AS [dbo_Fac...	philo		2010-01-20 04:20:13.000		Adventure Works DW 2008R2	Reseller_Orders_2005
Progress Report Begin	25 - ExecuteSQL	SELECT [dbo_FactInternetSales].[CustomerKey] AS [dbo_FactInte...	philo		2010-01-20 04:20:11.000		Adventure Works DW 2008R2	Customers_2007
Progress Report Begin	25 - ExecuteSQL	SELECT [FactSalesSummary].[SalesOrderNumber] AS [FactSalesSum...	philo		2010-01-20 04:20:13.000		Adventure Works DW 2008R2	Total_Orders_2007
Progress Report Begin	25 - ExecuteSQL	SELECT [FactSalesQuota].[SalesAmountQuota] AS [FactSalesQuota...	philo		2010-01-20 04:20:13.000		Adventure Works DW 2008R2	Sales_Quotas
Progress Report Begin	25 - ExecuteSQL	SELECT [dbo_FactFinance].[Amount] AS [dbo_FactFinanceAmount0...	philo		2010-01-20 04:20:13.000		Adventure Works DW 2008R2	Finance
Progress Report End	25 - ExecuteSQL	SELECT [dbo_FactInternetSalesReason].[dbo_FactInternetSalesRe...	philo	0	2010-01-20 04:20:13.000	230	Adventure Works DW 2008R2	Internet_Sales_Reasons
Progress Report Begin	17 - ReadData	Started reading data from the 'Internet_Sales_Reasons' partition.	philo		2010-01-20 04:20:13.000		Adventure Works DW 2008R2	Internet_Sales_Reasons
Progress Report Current	17 - ReadData			1	2010-01-20 04:20:13.000			Internet_Sales_Reasons
Progress Report Current	17 - ReadData			10000	2010-01-20 04:20:13.000			Internet_Sales_Reasons
Progress Report Current	17 - ReadData			20000	2010-01-20 04:20:13.000			Internet_Sales_Reasons
Progress Report Current	17 - ReadData			30000	2010-01-20 04:20:13.000			Internet_Sales_Reasons
Progress Report Current	17 - ReadData			40000	2010-01-20 04:20:13.000			Internet_Sales_Reasons
Progress Report End	25 - ExecuteSQL	SELECT [dbo_FactInternetSales].[SalesOrderNumber] AS [dbo_Fac...	philo	0	2010-01-20 04:20:13.000	540	Adventure Works DW 2008R2	Internet_Orders_2005
Progress Report Begin	17 - ReadData	Started reading data from the 'Internet_Orders_2005' partition.	philo		2010-01-20 04:20:13.000		Adventure Works DW 2008R2	Internet_Orders_2005
Progress Report Current	17 - ReadData			1	2010-01-20 04:20:13.000			Internet_Orders_2005
Progress Report End	17 - ReadData	Finished reading data from the 'Internet_Orders_2005' partition.	philo	1013	2010-01-20 04:20:13.000	309	Adventure Works DW 2008R2	Internet_Orders_2005
Progress Report Current	16 - WriteData			1	2010-01-20 04:20:13.000			Internet_Orders_2005
Progress Report End	16 - WriteData	Finished writing data for the 'Internet_Orders_2005' partition.	philo	0	2010-01-20 04:20:13.000	1584	Adventure Works DW 2008R2	Internet_Orders_2005
Progress Report Begin	28 - BuildAggsAndIndexes	Building the aggregations and indexes for the 'Internet_Orders_200S...	philo		2010-01-20 04:20:13.000		Adventure Works DW 2008R2	Internet_Orders_2005
Progress Report Begin	16 - WriteData	Starting to write data of the 'Reseller_Sales_2008' partition.	philo		2010-01-20 04:20:14.000		Adventure Works DW 2008R2	Reseller_Sales_2008
Progress Report Begin	25 - ExecuteSQL	SELECT [dbo_FactResellerSales].[dbo_FactResellerSalesSalesAmo...	philo		2010-01-20 04:21:04.000		Adventure Works DW 2008R2	Reseller_Sales_2006
Progress Report Begin	1 - Process	Processing of the 'Reseller_Sales_2006' partition has started.	philo		2010-01-20 04:20:18.000		Adventure Works DW 2008R2	Reseller_Sales_2006
Progress Report End	25 - ExecuteSQL	SELECT [FactSalesSummary].[FactSalesSummaryOrderQuantity0_0] ...	philo	0	2010-01-20 04:20:13.000	6095	Adventure Works DW 2008R2	Total_Sales_2006
Progress Report Begin	17 - ReadData	Started reading data from the 'Total_Sales_2005' partition.	philo		2010-01-20 04:20:19.000		Adventure Works DW 2008R2	Total_Sales_2005
Progress Report Current	17 - ReadData			1	2010-01-20 04:20:19.000			Total_Sales_2005
Progress Report Current	17 - ReadData			50000	2010-01-20 04:20:13.000			Internet_Sales_Reasons

Figure 13-3. Profiler trace from Analysis Services while processing a cube

Another use for Profiler is to watch the relational store while processing. While Analysis Services is building cubes and dimensions, any data it needs from SQL Server it retrieves via query. You can watch these queries via a Profiler trace on the relational store and again identify where the bottlenecks are. The SQL used for the queries is reported there, and you can run Query Analyzer on the queries to ensure they're as efficient as possible.

Performance Monitor

Windows Performance Monitor (found in Administrative Tools on the Start Menu) is a tool that provides resources for analyzing system performance. Figure 13-4 shows an analysis in progress. You can analyze operating systems, server software, and hardware performance in real time, collect data in logs, create reports, set alerts, and view past data. The metrics tracked by Performance Monitor are performance counters. Windows comes with hundreds of counters built in (for disk performance, CPU, memory, caching, and so on). Third-party software, as well as other software from Microsoft, can install additional performance counters.

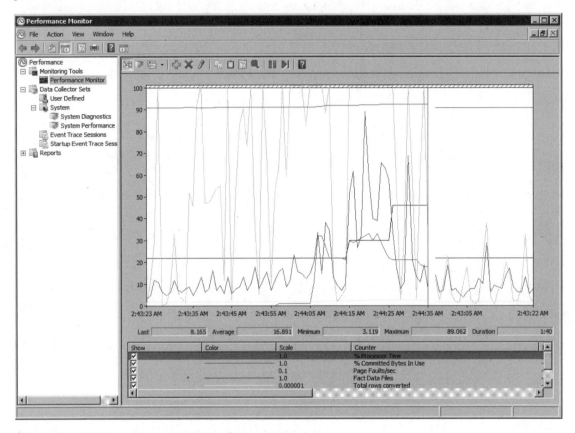

Figure 13-4. *Windows Server 2008 R2 Performance Monitor*

When you install Analysis Services, it adds a number of new counter groups:

- MSAS 2008: Connection

- MSAS 2008: Data Mining Model Processing

- MSAS 2008: Data Mining Prediction

- MSAS 2008: Locks

- MSAS 2008: MDX

- MSAS 2008: Memory

- MSAS 2008: Proactive Caching

- MSAS 2008: Proc Aggregations

- MSAS 2008: Proc Indexes

- MSAS 2008: Processing

- MSAS 2008: Storage Engine Query

- MSAS 2008: Threads

■ **Note** As of this writing (Nov CTP), the counters for SQL Server 2008 R2 are still labeled "MSAS 2008."

Each of the counter groups will have a number of counters subordinate to them. For example, MSAS 2008:Processing has six counters for reading, writing, and converting rows of data. By tracking certain performance counters you can evaluate the performance of your Analysis Services server. If you're getting complaints of bogged-down performance that you can't pin down, you can run the performance counters to a log file for review later. For more information about the Windows Performance Monitor, check out the TechNet article at `http://technet.microsoft.com/en-us/library/cc749249.aspx`.

Automation

Once you have an Analysis Services database in production, you will need to process the cube periodically to refresh it. Remember that only ROLAP partitions and dimensions will dynamically show changes in the underlying data, so if we expect the data in the data sources to change, the cube will have to be processed to reflect the changes.

For active partitions, you will have to decide on a processing schedule depending on the business requirements. For example, cubes that are used for strategic analysis may need to be processed only once a week, as the focus is on the previous year at the quarterly level. On the other hand, a project manager may want analytic data to be no more than 24 hours old, as it provides data for her project dashboard, which she relies on heavily on a daily basis. Finally, archive partitions for previous years may be able to sit untouched for long periods of time, being proactively reprocessed should the underlying data change.

In any event, all these requirements indicate a need for automated processing. The good news is that there is a lot of flexibility in the processing of SSAS objects. Essentially, there are two ways to initiate the process: either via an XMLA query, or by calling the AMO object model. However, we have a number of ways to accomplish those two tasks.

■ **Note** The user context that attempts to process an SSAS object must have the permission to do so. I strongly recommend using Windows Security Groups to manage user permissions—see the section on Security later in this chapter.

XML for Analysis

You can process Analysis Services objects with XML for Analysis (XMLA) queries. This means that essentially any tool that can connect to SSAS and issue a query can process objects, if that connection has the appropriate privileges. One great benefit is that if you want a cube or database processed a certain way, you can create the script and store it separately, then use the same script to process the object(s) manually, through code, via tools, and so on.

A basic example is shown here, but the processing options in XMLA are full-featured enough to duplicate anything you can do through the BIDS UI.

```
<Process xmlns="http://schemas.microsoft.com/analysisservices/2003/engine">
    <Object>
        <DatabaseID>Adventure Works DW</DatabaseID>
        <CubeID>Adventure Works</CubeID>
    </Object>
    <Type>ProcessFull</Type>
    <WriteBackTableCreation>UseExisting</WriteBackTableCreation>
</Process>
```

You can fire off multiple tasks in parallel by using the `<Batch>` XMLA command. SSAS will attempt to execute as many multiple statements within a `<Batch>` command in parallel as possible, and, of course, execution will hold at the end of the `<Batch>` command until all subordinate commands are completed.

For more information about processing SSAS objects with XMLA queries, check the TechNet article at `http://technet.microsoft.com/en-us/library/ms187199.aspx`.

Analysis Management Objects

Analysis Management Objects (AMO) are the members of the `Microsoft.AnalysisServices` class library that provide for the automation of Analysis Services. Working with Analysis Services objects via AMO is very straightforward. (I'll cut to the chase—the code to process a cube is `Cube.Process(ProcessType.ProcessFull)`—startling, isn't it?) The upside to using AMO is that it's very intuitive, and the .NET code can make it easy to perform some pretty arcane tasks.

For example, consider a complex Analysis Services database that has various dependencies, and you want to enable your users to request a cube to be reprocessed. For a given cube, there are various dimensions that need to be reprocessed, depending on when it was last processed and the age of the data. Now if the structure of what needs to be processed in which order is static, similar to what's shown in Figure 13-5, then creating the process job in XMLA makes sense—structure the query once, and store it.

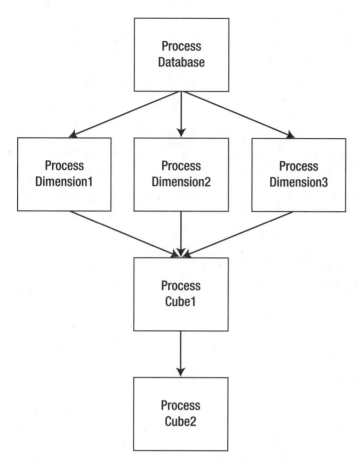

Figure 13-5. *Simple execution of a batch process*

When you have a complex cube structure, you may want to create specific dependencies (if you process x dimension, then process y cube). You may also want to verify partition structures, data freshness, or even compare the last processing data against the age of the data to determine if you need to reprocess an object. So you may end up with a process similar to that in the flow chart shown in Figure 13-6. In this case, it may make more sense to craft the reprocessing logic in code. It's going to be easier to trace the flow through code, and it will be easier to instrument and debug.

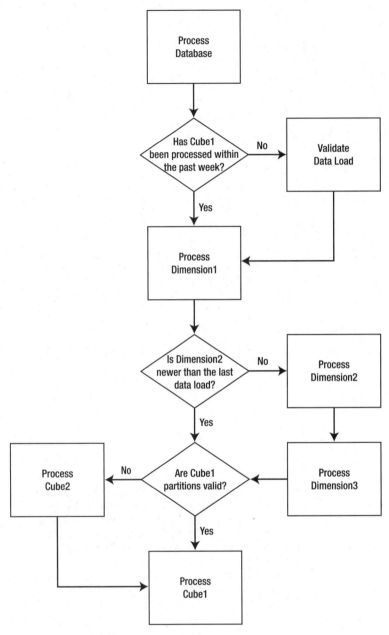

Figure 13-6. *A notional complex logic flow for processing a database*

The code is very straightforward:

```
Server server;
Database db=new Database();

//connect to analysis server
server = new Server();
server.Connect(@"data source=<server name>");
db = server.Databases["<database name>"];

foreach(Cube cube in db.Cubes)
{
    //Processes cube and child objects
    cube.Process(ProcessType.ProcessFull);
}
```

Remember that cubes own measure groups; dimensions are owned by the database.

PowerShell

A common problem when dealing with automation is the need for a framework to run the code in. Very often you may find the easiest way to tackle an administrative task is with code, but you end up taking the time to create a small application to run the code. And next time you need to run the code, you don't have the application handy, so you have to do it again.

In the Unix world, this is rarely a problem—since everything is command-line interface, then any administrative task can usually be performed by writing a script. Windows administrators didn't really have this option for a long time—there wasn't anything between 1980s-era batch files and modern fourth generation code. That is, until Microsoft created PowerShell.

PowerShell is a Windows component introduced a few years ago to enable scripting for enterprise applications. It works very similar to scripting shells in Unix, with the exception that instead of piping text from one process to another, PowerShell enables passing actual objects from one to the next. It's available as a download for Windows XP, Vista, and Server 2003, and is an installable component with Windows 7 and Server 2008 and 2008R2.

Since PowerShell can leverage .NET assemblies, the code in PowerShell will be similar to the .NET code in the previous section. The benefit here is that you can write and execute that code against an Analysis Services server without having to build a harness or application for it. You can learn more about PowerShell at `http://technet.microsoft.com/en-us/scriptcenter/dd742419.aspx`.

Scheduling

Now that we have a solid collection of ways to automate processing of Analysis Services objects, we need a way to kick it off. Perhaps we need to trigger the processing on a regular schedule, or we want to call it when we load data into the data mart. Alternatively, it's possible that we don't have control over the data-loading process, and need to trigger processing when data in a warehouse changes.

We're going to take a look at SQL Server Agent, a process that runs in SQL Server and can be used to schedule cube processing. We'll also take a look at SQL Server Integration Services—the primary use of SSIS is to extract, transform, and load (ETL) data; it's often used to load data warehouses from production systems. Since SSIS is doing the loading, it can also trigger or schedule the processing of the affected objects in Analysis Services.

SQL Server Agent

SQL Server Agent is a component of SQL Server, and runs as a Windows Service. You can configure SQL Server Agent either through T-SQL or via SQL Server Management Studio, as shown in Figure 13-7.

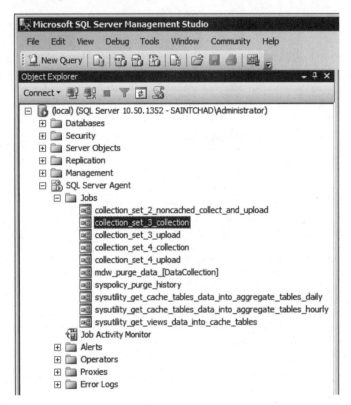

Figure 13-7. *Managing SQL Server Agent through SSMS*

Each job can contain a number of steps, as shown in Figure 13-8, and for each step you can indicate what action to take depending on whether the job step exits reporting success or failure (exit the job, reporting success or failure, or continue on to the next step). Each step can also write to an output file, log results to a table, or place its results in the job history. You can also set the user context for each specific job step.

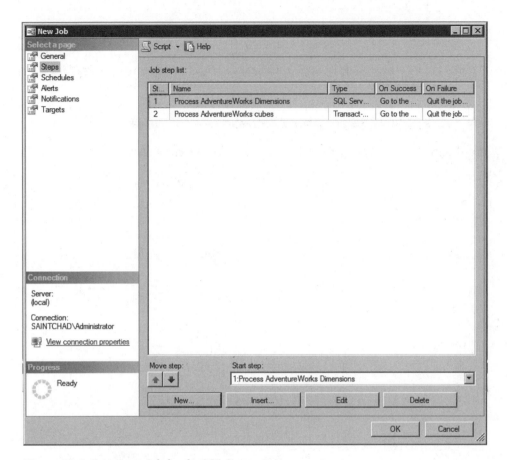

Figure 13-8. *Creating a job for the SQL Server Agent*

Job steps can be a number of things, each option having its own options for execution. While there are twelve different types of job steps, the ones we're interested in are PowerShell, SQL Server Analysis Services Command, and SQL Server Analysis Services Query. As you can see from the methods we've covered previously, these options allow us to schedule Analysis Services processing jobs by calling a PowerShell script, running an XMLA query, or using the Analysis Services command directly.

Once we have the processing job set up the way we want, then we can select or build a schedule, and set up the alerts and notifications we want to execute when the job finishes.

SQL Server Integration Services

While the SQL Server Agent is the best way to schedule a recurring action you want to perform on Analysis Services, if you need actions executed as part of a process, your best bet is SQL Server Integration Services (SSIS). There's a good chance that in an Analysis Services BI solution, you'll be using SSIS to move data into your data mart or data warehouse. The nice thing is that while using SSIS to move data (which means you'll need to reprocess the cubes and/or dimensions at some point), you can also trigger the reprocessing directly.

SSIS uses a drag-and-drop designer for data-oriented workflows, as shown in Figure 13-9. For an ETL type job, the central point of the workflow will be the Data Flow Task, which contains logic for moving and transforming data. However, in the Control Flow, there is a task for processing Analysis Services objects.

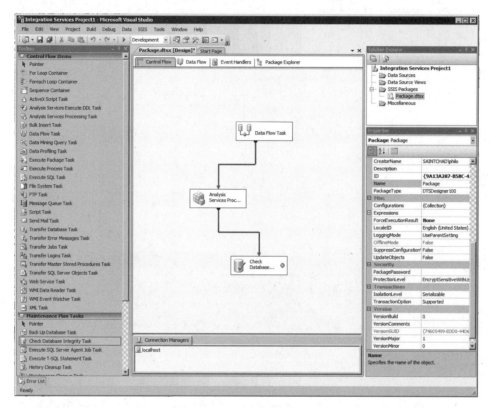

Figure 13-9. *SQL Server Integration Services designer in BIDS*

You can make paths conditional, so as we've discussed, you may want to process cubes only when the underlying data has changed. You can also have different exit paths based on success or failure of the previous component. This is beneficial if a data load job fails—you don't want to waste downtime either processing a cube on unchanged data, or, worse, processing a cube on bad data.

Configuration of the Analysis Services task is very straightforward. First you need to create a connection to the Analysis Server, either by right-clicking in the connection manager pane and selecting New Analysis Services Connection or by opening the properties pane of the Analysis Services Processing Task and clicking New next to the connection manager selection drop-down list, as shown in Figure 13-10.

Figure 13-10. *Selecting a connection to Analysis Services*

The processing settings themselves reflect the settings dialog we're familiar with, as shown in Figure 13-11. Once you select the connection manager, then you will be able to add objects from the database selected in the connection manager. You can select cubes, measure groups, partitions, dimensions, or mining models. Once the selections are listed in the dialog, you can change the processing settings for each object individually.

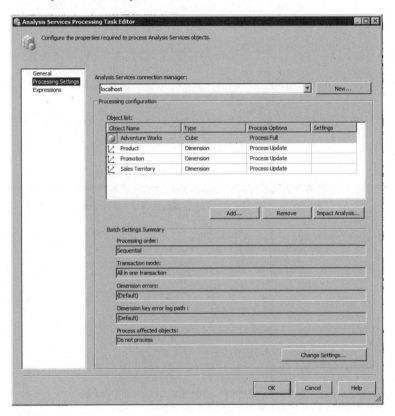

Figure 13-11. *Processing settings in SSIS*

The green arrow from the bottom of the task leads to the next task you want completed. You can have multiple arrows leading to multiple tasks, and the arrows can be configured to be conditional based on the success or failure of the task, or more intricately based on an expression, which may depend on variables defined in the Integration Services package.

If you're interested in learning more about Integration Services, I highly recommend the Integration Services tutorials online at `http://msdn.microsoft.com/en-us/library/ms167031.aspx`. You can also read more about SSIS in *SQL Server 2008 Integration Services Unleashed* (Sams, February 2009) by Kirk Haselden, one of the SSIS architects at Microsoft.

■ **Note** There are no significant changes in Integration Services for 2008R2.

A few times I've referred to security and permissions with respect to working with Analysis Services. Due to the sensitive nature of both the information contained in SSAS cubes and the data that underlies them, implementing security properly is of paramount importance. Let's take a look at some aspects of security in SSIS.

Security

The first, most important idea (and this shouldn't be news) is that security is neither a feature you just add on, nor something you worry about only with Analysis Services. Security should be a system-wide philosophy. However, having said that, I'll cover a couple of key security concepts regarding Analysis Services.

We'll look at authentication vs. authorization—the difference between "who gets in?" and "what they are allowed to do." We'll look at SSAS roles and how to use them to permit access to SSAS objects and actions. Finally, we'll look at the impacts of using security to restrict access to specific parts of the cube, down to the dimension member and measure cell level.

AUTHENTICATION VS. AUTHORIZATION

Security in Analysis Services, like any enterprise application, is a two-step process, consisting of authentication, then authorization. *Authentication* is about verifying that someone is allowed into the server—something of a "who goes there?" *Authorization* happens once someone is authenticated into the server—"now that you're here, what are you allowed to do?"

Authentication

Analysis Services recognizes only Windows Integrated Authentication, also known as Active Directory security. As of SQL Server 2008 R2, there is no other authentication method available. With respect to connecting from the Internet, this can be problematic, as generally your firewalls won't have the necessary ports open for NT authentication. In this case you can configure SSAS for HTTP access, which uses an ISAPI filter to connect to SSAS from IIS on the web server.

■ **Note** If you use HTTP access, and want to leverage the user credentials for authorization, you could run into the double-hop problem, since users will authenticate against the IIS server, and the IIS server must forward the user credentials to the SSAS server. For more information about the double-hop problem, see
`http://blogs.msdn.com/knowledgecast/archive/2007/01/31/the-double-hop-problem.aspx`.

HTTP access can be performed with any authentication mode on IIS—NTLM, Basic (username/password), or even anonymously (using the IIS application pool account to authenticate into SSAS). The general architecture and authentication flow will look like Figure 13-12. A great article regarding setting up HTTP access for SQL Server 2008 Analysis Services on Windows Server 2008 can be found at `http://bloggingabout.net/blogs/mglaser/archive/2008/08/15/configuring-http-access-to-sql-server-2008-analysis-services-on-microsoft-windows-server-2008.aspx`.

Figure 13-12. User authenticating against Web application

However, don't be in a rush to expose SSAS to the Internet in this manner—remember that you are exposing a bare XMLA query interface (so the Internet user can use Excel or another Analysis tool). If, on the other hand, you will be delivering only analytic data to Internet users via an application (Reporting Services, Excel Services, PerformancePoint, etc.), then you need to provide only for that application's authentication of Internet users, then handle authentication of the application against Analysis Services. (Again, if you choose to pass through user credentials, be mindful of the double-hop problem.)

Authorization

Once we've authenticated against an Analysis Services server, then we have to do something. What we're allowed to do depends on what *Authorization* our user is granted. Within SSAS, authorization is handled via role-based security—roles are objects in Analysis Services that are containers for the permissions that users authenticated into the server will receive.

Roles are defined near the bottom of the tree in the Solution Explorer pane—if you right-click on the Roles folder, you can create a new role. New roles default to not having any permissions. When you create a new role, you'll see the General tab for the role, as shown in Figure 13-13.

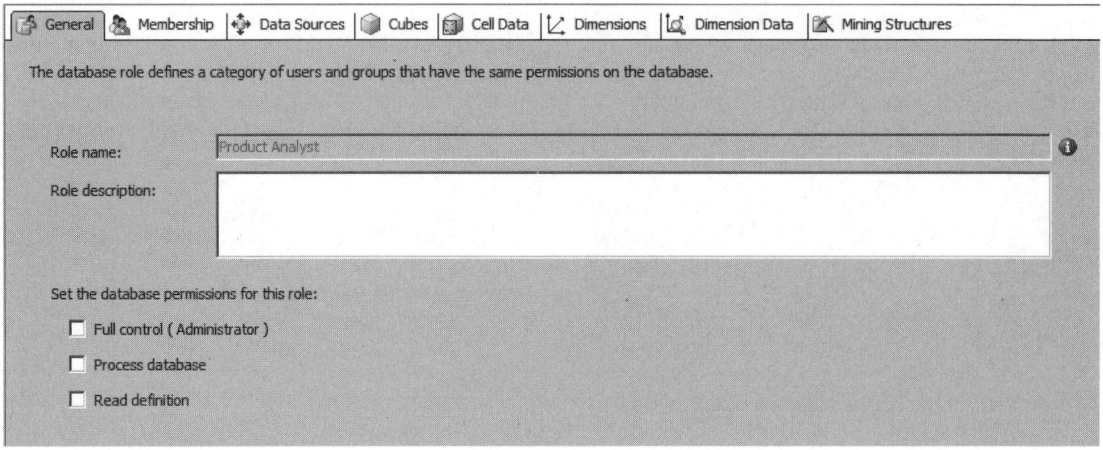

Figure 13-13. The General tab for a newly created role

The *Role name:* in this tab is not directly editable. To change the name of the role, you use the *Name* input area in the *Properties* pane. The *Role description:* is a free-text description that allows you to annotate the intent behind the role. Then you have three checkboxes—Full Control (Administrator), Process database, and Read definition. These are fairly self-explanatory. You will want to create an Administrator role early on and get used to using that for administrators. By default SSAS creates a Server role behind the scenes—this role has administrator privileges, and has the members of the local administrators group as members. If you want to remove the local administrators from this role, you'll have to go into Management Studio and connect to the Analysis Server. Then right-click on the server itself and select Properties to open the server properties dialog.

At the bottom of the dialog is a checkbox for "Show Advanced (All) Properties." Check this to show a security property titled "Security \ BuiltinAdminsAreServerAdmins." If you change the value of this to "false" then built-in admins will no longer have access to the server. You can explicitly add users to the Server role on the Security tab of the Server Properties dialog.

■ **Caution** You can lock yourself out of the server if you disable the local admins from the Server role without explicitly adding users in the Security tab.

The Membership tab is pretty self-explanatory—you can add Windows users or security groups here. The Data Sources tab controls access to the data sources in the UDM for the database. The `Access` property allows users to read data from the data source directly—this is used mostly in actions or drilling through to underlying data. The `Read Definition` checkbox will allow a role to read the definition (including the connection string and any embedded passwords!) for a data source.

The Cubes tab lists the cubes in the database and the associated access rights. `Access` is `None` for no access, and `Read` to allow members of the role to read the cube (which translates to "the dimensions and measure groups that belong to the cube"). `Read/Write` enables reading, but also writeback for the cube. The `Local Cube/Drillthrough Access` option for a cube allows the role to drill-through to underlying data for a cube, while a local cube allows members of the role to pull a subset of data from the cube to create a cube on their workstation. Finally, the Process checkbox, of course, allows members of the role to process the cube. You can see here that it would be possible to set up an account that has permission only to process cubes without reading any of the data—a solid security option for unattended accounts.

The `Cell Data` tab allows you to create fine-grained access permission on the data in a cube. There are three sections—reading cube content, reading/writing cube content, and contingent cell content. I'll explain read-contingent in a moment.

For each of these sections, when you enable the permissions, the formula box will be enabled, as well as the builder button […] for editing the MDX. In each permission set you can directly define what cells the role has access to with an MDX expression. For example, you might want a certain analyst to have access only to mountain bikes that don't have a promotion going on. When we get to dimension security you'll see that the permissions are additive, so selecting the [Mountain Bike] dimension member and the [No Promotion] member would be overly inclusive. So instead we can use the Cell Data read permission and enter MDX for {[Mountain Bike], [No Promotion]}, which will be the set of those members we're looking for.

A potential problem here, and with dimension security, is that in some cases data that should be secured can be inferred. For example, we may allow a role to see Profit and Revenue data, but not Expenses. However, by seeing profits and revenues, you can calculate expenses through some simple math. Read-contingent permissions address this problem by digging into the constituent data under cell data—a role with read-contingent permissions is allowed access only to measures when that role is also allowed access to all the subordinate data.

The `Dimensions` tab is pretty basic—simply select the dimensions that the role should have access to. After selecting dimensions here, your next stop is the `Dimension Data` tab. Here you can select individual members of dimensions to make available to the role. You can either explicitly select and deselect members, or click on the Advanced tab and write MDX to indicate which members should be allowed or denied. You will also want to indicate a dynamic default member to align with the dynamic selections you're making.

■ **Caution** When you filter out a member, to the user it's as if the member doesn't exist in the cube. So be careful about hard-coding members in any MDX expressions you create (calculated measures, KPIs, etc.). If there's a hard-coded member in an MDX expression and the user is in a role that doesn't have access to that member, he or she will get an error.

Behind the scenes, Analysis Services uses a feature called *autoexists* to ensure consistent visibility of members in the hierarchies within a dimension. For example, suppose that within the Product Line hierarchy of the Products dimension you removed access to the Road member, so the role you're working on can't see any products in the Road product line. If you then looked at the Model hierarchy, you would see that no road bikes are listed. Huh? We didn't touch the Model hierarchy—so why is it filtered?

Autoexists runs against a dimension that has attribute security applied. When the cube is processed, Analysis Services will run the cross-product of all the hierarchies in the dimension, and any members wholly encompassed by a hierarchy member that is hidden from the role will also be hidden. So since all road bikes sold are in the Road product line, and the Road models in the Model hierarchy are all contained within specific members, those members are also hidden.

However, note that this cross-protection of members occurs only within the dimension where the attribute security is defined. Let's say that only road bikes were sold in 2003. If we look at sales data, we won't see the year 2003, since, with respect to the security of the role we're in, nothing we're allowed to see was sold in 2003. But that's because NONEMPTY is included in MDX queries by default. If we disable the NONEMPTY behavior, then 2003 will show up just fine (albeit with an empty cell of data).

The Dimension Data tab is also where you can set dimension members based on dynamic conditions. For example, you may want a manager to see pay data only for those employees working for him. Or you may want project managers to see only their own projects. Instead of hard-coding these selections, you can enter an MDX expression in the Allowed member set box, which will select those members dynamically based on the expression.

A more complex situation is when you want to secure a selection of members based on the logged-in user. For example, you may want users to see the sales only from the Promotions they are assigned. There are two approaches here:

- Create a many-many relationship table in the data source to connect the Promotions records to the Employee records representing the users with rights to see them. You will then have to create a fact table referencing the many-many table in order to create the relationship between the Promotions dimension and the Employees dimension in the cube. (This is referred to as a "factless" table.) Marco Russo has written an excellent white paper on working with many-many dimensions, which you can find at http://www.sqlbi.com/.

- Create a .NET assembly to be used as a stored procedure. The assembly will access some other authoritative data source to determine what Promotions the user can see (Active Directory, a SharePoint list, some other directory service) and return the appropriately formatted set via an MDX function call. The set can then be used in dimension data functions in the role.

Working with this type of detailed security is often a critical business requirement, but the architecture must be very carefully designed with consideration for maintainability (don't create additional data stores if you don't have to) and scalability (mapping user roles to millions of dimension members can get expensive quickly when processing or querying a cube). And on that note, let's take a look at performance considerations in SSAS.

Performance

Performance management is vitally important in SQL Server Analysis Services due to the tendency for cubes to encompass significantly large amounts of data. Like most development endeavors, optimizing performance takes place both at design time and once the solution is in production. Part of the design time aspect of performance management may just be noting potential trouble spots—dimensions that may get too big, or measure groups that will have to be partitioned later.

Design

The absolute first and foremost principle in designing for performance is "don't put it in the cube unless you need it." While it may seem prudent to be over-inclusive when deciding on dimensions, attributes, and measures, consider the impact that these will have on the cube performance. Be aware that extra measures can increase the size of the cube arithmetically; additional attributes will increase the number of cells in a cube geometrically; additional dimensions can increase the size of the cube logarithmically.

Dimensions

The biggest concern with respect to dimensions is the number of members. Often there are aspects to analysis we're interested in that can lead to larger dimensions. Analysis Services will scale to millions of members, but the first, biggest question is "do you need a dimension with millions of members?"

Consider the AdventureWorks cube—one of the dimensions is `Customer`, and the `Customer` attribute consists of a list of all the customers in the sales database. The most important question would be do the analysts really need to drill down to the *name* of the individual customer when they're doing analysis? Instead, consider making attributes only for the descriptive features that analysts will want to drill down on—demographics, location, etc. Then provide drill-through or reporting for customer-specific data.

This is the type of analysis you should perform on every aspect of the OLAP solution—measures, dimensions, and attributes. Is there a defined business case for it, and can that business case be solved with a smaller subset of data? I'm not trying to suggest shorting your users, just ensuring that you don't end up with a terabyte cube that will go mostly unused because it was filled with "just in case" requirements.

The other concern with large dimensions is processing. There are two methods of processing dimensions, set in the `ProcessingGroup` setting for the dimension. `ByAttribute` (the default) builds the dimension by performing a number of `SELECT DISTINCT` queries against the table for the attributes and the dimension itself. Obviously for large dimensions (from large tables) this can take a while. The alternative is the `ByTable` setting—with this setting Analysis Services loads the entire source table for the dimension into memory, and reads as necessary from there. The problem is that if you don't have enough memory to hold the entire table, this will error out.

Consider a table with five million rows and ten attributes, where each attribute averages a six-character string key. That's 300 megabytes of data. Increase any of those by a factor of ten and it's 3GB! At this size you also have to start to consider that SSAS has a 4GB limit on string storage. (Note that we're talking only key values here, not name values.)

So if you're processing `ByAttribute`, you want to be sure you have all the appropriate indexes on the underlying tables. If you're processing `ByTable`, keep the string keys short and be mindful of memory consumption during processing.

Attribute Design

In any guide to Analysis Services scalability, the top recommendation you're going to find is to ensure your attribute relationships in dimensions are properly defined. In SQL Server 2005 Analysis Services, attribute relationship design was a bit of an arcane art. However, in SSAS 2008, there is a new attribute relationship designer, as shown in Figure 13-14.

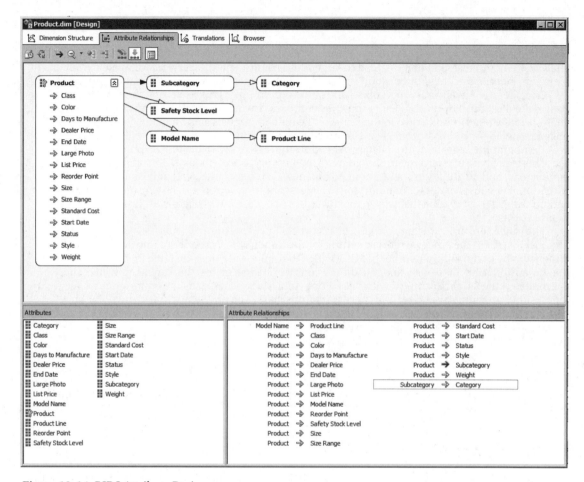

Figure 13-14. *BIDS Attribute Designer*

For maximum performance, create all your necessary attribute relationships—wherever they exist in the data or the model. It's also important to think in terms of good design practices for attributes. Attribute relationships direct Analysis Services on optimizing storage of data, are reflected in MDX queries, structure the presence of member properties, and govern referential integrity within a given dimension. Attribute relationships also make processing and queries more efficient, since a strong hierarchy results in a smaller hash table—top level attributes can be inferred instead of having to be discretely stored.

Attribute relationships are also the foundation on which you build hierarchies, which we've seen over and over again to be critical in the proper development of an OLAP solution. Hierarchies make it easier for end users to make sense of their data, and attribute relationships will ensure your hierarchy and the data structure make sense.

The first step in good attribute design is to observe natural hierarchies, for example, the standard calendar Year – Quarter – Month – Day, or geographic Country – State – County, as shown in Figure

13-15. Be sure that each step in the relationship is truly 1:n. You can create hierarchies that aren't a pyramid structure—these are referred to as "unnatural hierarchies," as shown in Figure 13-16.

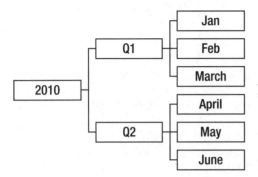

Figure 13-15. A natural hierarchy

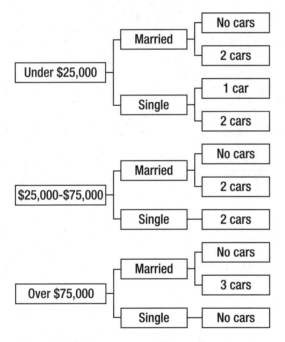

Figure 13-16. An unnatural hierarchy

This distinction between natural and unnatural hierarchies is important to be aware of because natural hierarchies will be materialized and written to disk when the dimension is processed. However, unnatural hierarchies are created on the fly when queried. Essentially, even when you build an unnatural hierarchy into the dimension, during execution Analysis Services treats it as if it were built on the fly. (This isn't to say not to build them, but rather that they don't benefit performance.)

■ **Note** See Chapter 6 for more information on Attribute design.

Partitions

We've talked about designing dimensions to provide performance and scalability. How about measure groups? The most important factor in making a cube scalable is being able to partition the data. (Note that partitions are features of cubes, which consist of measure groups; they do not apply to dimensions.)

The reasons for using multiple partitions include:

- Partitions can be added, deleted, and processed independently. The primary example of this is any measure group that's having data added on a rolling basis—you can process the "current" partition at any time and the "historical" partitions remain accessible. In addition, since there's less data to scan, processing is faster.

- Partitions can have different storage settings. Considering the preceding example, the historical data can be stored in a MOLAP mode for best performance, while the current data partition can be set to HOLAP or even ROLAP so that it processes faster and reflects more up-to-date data. In addition, you can set the storage settings on various partitions, so partitions can be stored on separate servers—keep the current data local, and partition out the historical data to a remote server.

- Partitions can be processed in parallel, maximizing usage of multiple processors.

- Queries will be isolated to the relevant partitions. So if a user is querying data that happens to be from a single partition, the cube will be more responsive as there is less data to scan.

A partition is automatically created for every measure group, with that partition group containing all the records of the measure group. You can go into the Partitions tab of the cube designer in BIDS (shown in Figure 13-17) and create additional partitions if you desire.

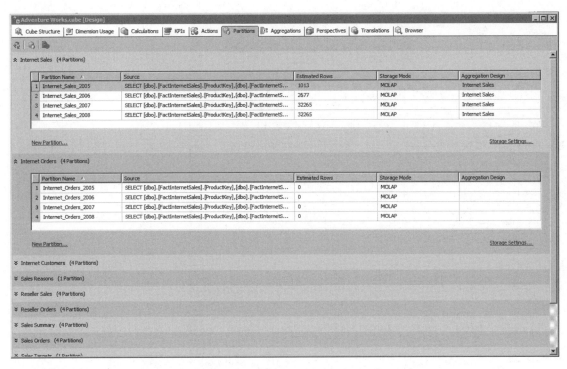

Figure 13-17. *The Partitions tab in BIDS*

Each measure group has a section listing the partitions in that measure group. You can see that each partition has a name, and the storage mode indicated. We'll talk about the Aggregation Design in the next section. The partition will also show the estimated rows of data in that partition (you should try to keep fewer than twenty million rows of data in each partition, but this is a guideline, not a hard rule). Remember that more partitions means more files, so it's a balancing act.

The actual partitioning scheme is described in the "Source" column. This is a query that acts against the underlying table to select the specific records for the partition, as shown in Figure 13-18. The Binding type has two options: Query Binding, as shown, and Table Binding. Table Binding operates on the premise that you have a data source that is already partitioned (or perhaps you're using a view), so all that needs to be done is to select the table. Another possibility is that you can create a partition for each measure in a measure group, and those measures may come from different tables. Then you can simply create a partition for each table that contains measure data.

Query Binding, on the other hand, allows you to enter a SQL Query to select the necessary records. You can see in the partition for Internet_Sales_2007 that the query is simply all the fields in the table, and a `WHERE` clause limiting the results by a date range.

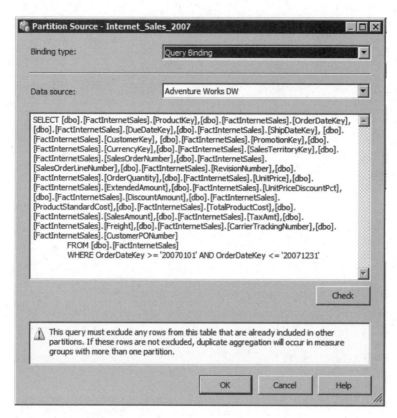

Figure 13-18. Selecting records for a partition

Two very important notes regarding the query—first, as the warning in the dialog indicates, be positive of your logic in the query, so that the record sets in each partition are truly unique and don't overlap. Also, if you're using a numeric range to divide the table, double- and triple-check your comparison operators—that if one partition uses < DATE, the next partition uses >=DATE, and so on. Leaving off that equals sign means you've dropped a day from your cube.

You can use multiple criteria to create partitions. For example, in addition to the calendar year divisions shown in the Internet Sales measure group, you might want to partition also by country. Then you could conceivably locate the partition for a country in the sales office for that country—queries against local data would be quick; global queries would travel the WAN and take more time. (In all honesty, in a global organization like that, you would probably be more interested in using replication, but I hope it illustrated the point.)

The two links under the partition list are to create a new partition, and to change the storage settings for the currently selected partition. The storage settings are the familiar ROLAP-MOLAP slider, and the custom settings for notifications. Set the storage *location* of the partition in the partition settings panel (generally on the lower right of BIDS).

The New Partition link opens the New Partition Wizard, which will walk you through creating a new partition. In the Internet Sales measure group, the group is partitioned by calendar year, so at the beginning (or soon thereafter) of each year, the DBA will have to come in, change the query on the last partition (which should be open-ended: >=0101YEAR) to add the end date for the year, and then add a new partition for the new year.

■ **Tip** Always leave your last partition open-ended, so that in the event someone doesn't create the new year's partition in the first week of January, the users aren't stumped because none of their reports are changing.

Finally, be sure to define the *slice* of the partitions you create. A property of the partition, the slice is the tuple (in MDX terms) that tells Analysis Services what data will be found in each partition. For example, the partition containing the data for the calendar year of 2007 would have a slice of [Date].[Calendar].[Calendar Year].&[2007]. If the slice you enter doesn't match the data specified in the Source, SSAS will throw an error on processing.

Aggregations

We've learned that OLAP and Analysis Services are all about aggregating data. When we process a cube, Analysis Services will pre-calculate a certain number of aggregated totals in advance. This caching of totals helps query performance. However, SSAS won't calculate *every* total in advance—the number of possible combinations generally makes that prohibitive. Also, when you consider that user queries aren't going to need every single possible combination, pre-calculating everything "just in case" is counter-productive. So SSAS does a "best guess" for the amount of advance work necessary for a given cube.

Understanding Aggregations

Consider a cube with dimensions for Geography, Products, and Time. Each of these dimensions breaks down as shown in Figure 13-19. Each dimension has the attributes shown, and the attributes are in a user-defined hierarchy in each dimension. There are other dimensions in the cube, but for this example they'll all just be at the default member. For each attribute I've indicated how many members there are.

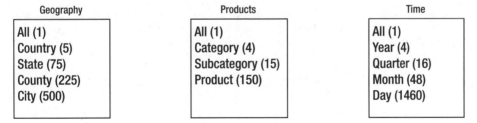

Figure 13-19. *Three dimensions in a cube*

From these dimensions, we select the attributes we want included in aggregation designs. So we make selections based on what queries we think users will make, shown in Figure 13-20. For each of these combinations, Analysis Services will calculate the values for every combination of members. So for the All-All-All aggregation, it's simply a total of every measure value, resulting in one cell. Country-Subcategory-Year is 5×15×4 = 300 values, and City-Product-Day is 500×150×1460 = 109,500,000 values.

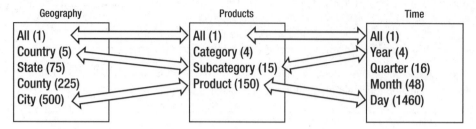

Figure 13-20. *Selecting aggregations*

Once SSAS has these values cached, then results from queries simply use the largest aggregation possible, select the necessary members, and do the math. For example, if a user wants to see data from each year for sales of Road Bikes in France, SSAS can use the Country-Subcategory-Year aggregation, select the values for Road Bikes in France, and return them. If the user made the same query, but for the Bicycles category, Analysis Services could use the same aggregation, select out the values for each of the subcategories under Bicycle, add them, and return them.

However, if a user wants to see a breakdown of Road Bike sales by Year for each of the states in the United States, SSAS cannot use the Country-Subcategory-Year aggregation, as there is no way to get to the values for each state. So it drops down to the leaf-level aggregation of City-Product-Day, queries out the values for the Road Bike sales in the United States, and performs the calculations necessary to get to a breakdown by State, Subcategory, and Year. The amount of math necessary here indicates this is going to be a longer-running query…

So aggregation design is a trade-off between the space necessary in the file system to pre-cache the aggregation values and query time. If we had tons of space and lots of time to process a cube, it seems like it would make sense to just create aggregations for every possible combination of attributes. However, when you start considering the Cartesian products of all the attribute relationships in your cube, you can see that "tons of space" quickly becomes unrealistic. So aggregation design is a trade-off between storage space/processing time and query performance.

Creating an Aggregation

Aggregations for a cube are managed on the Aggregations tab of the cube designer in BIDS, as shown in Figure 13-21. Under each measure group are listed the aggregation designs for that group. For a given measure group, you can create multiple aggregation designs, then assign the aggregation designs to various partitions. An aggregation design will contain multiple aggregations.

Figure 13-21. *The aggregation designer in BIDS*

To create a new aggregation design, right-click on the measure group and select Design Aggregations to open the Aggregation Design Wizard (Figure 13-22). The wizard will allow you to select one or more partitions and create an aggregation design for them. One reason you may want a different aggregation design for a given partition is that it is loaded under different usage scenarios. For example, take the partition design scheme we talked about earlier—partitioning the data by year. Odds are that there will be far more queries against the current year's data than previous years', so the current year partition will probably have different aggregation usage.

Figure 13-22. *The Aggregation Design Wizard*

Once you've selected one or more partitions, then the wizard will ask you to select the attributes for each dimension to be included in the aggregation design, as shown in Figure 13-23. These settings guide the aggregation designer as to whether to include an attribute when calculating aggregations.

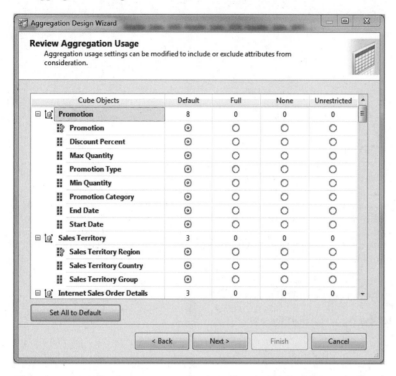

Figure 13-23. Selecting measures for the aggregation wizard

Each of these options indicates how the dimension attribute should be used in calculating aggregations. There's a list of all the dimensions in the cube, and each of the attributes in each dimension. For each attribute you indicate whether it should be included in every aggregation (Full), no aggregation (None), or "it depends" (Default or Unrestricted). Unrestricted means that you are allowing the aggregation designer to perform a cost/benefit analysis on the attribute and evaluate if it is a good candidate for an aggregation design. Attributes you anticipate to be queried often should be set to Unrestricted, while attributes you don't anticipate being queried often should be set to None.

Default implements a more complex rule depending on the attribute. The granularity attribute will be treated as unrestricted (perform cost/benefit analysis and include if suitable). All attributes that are in a natural hierarchy (1:n relationships from the top down) are considered as candidates for aggregation. Any other attribute is not considered for aggregation.

The next page in the wizard displays the counts for the measure group you're working on, as well as the attributes for all the dimensions included, as shown in Figure 13-24. You can have the wizard run a count of the current objects, or you can enter your own values if you are designing for a different scenario than what's currently defined.

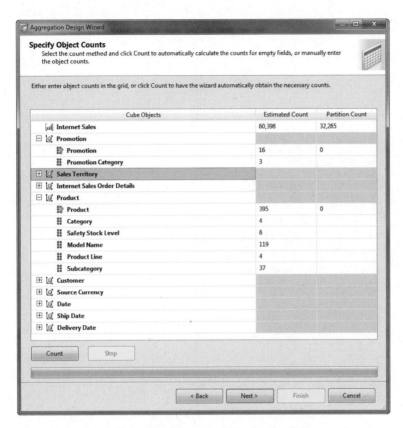

Figure 13-24. *Specifying the object counts in the database*

The next page of the wizard, shown in Figure 13-25, is the heart of the aggregation designer. Here you select how you want the designer to proceed. You can set a maximum storage (in MB or GB), a specific performance gain (estimated), or a "keep running until I click Stop" to design the aggregations. Once you select the option, click the Start button to start the designer running. You'll see a chart generating something similar to what's shown in the example. The *x* axis is the size of the aggregation storage, the *y* axis is the aggregation optimization level in percent.

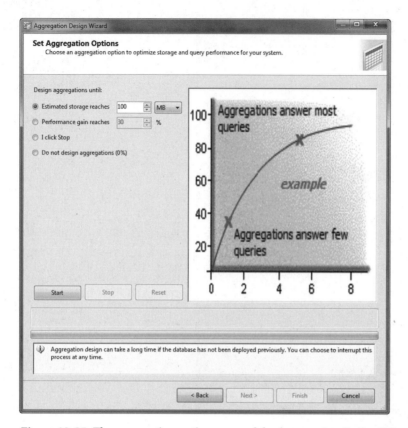

Figure 13-25. The aggregation options page of the Aggregation Design Wizard

Once the designer stops (or you stop it), you'll get a result regarding the number of aggregations designed and the optimization level. Then you can either change the settings and run it again, or click the Next button. The final page of the wizard allows you to name the Aggregation design and either save the design or deploy it and process it immediately (probably not a great idea to do in production in the middle of the workday).

Once you've completed this wizard, Analysis Services will have the initial aggregation design set for the partitions you've set up. If you want to see the results of the aggregation design, click the "Advanced View" button in the toolbar. The aggregation viewer will show which attributes were included for each aggregation, as shown in Figure 13-26. Once you select the measure group and aggregation design at the top, you'll get a grid showing a column for each aggregation, and a checked box for each dimension attribute included in the aggregation.

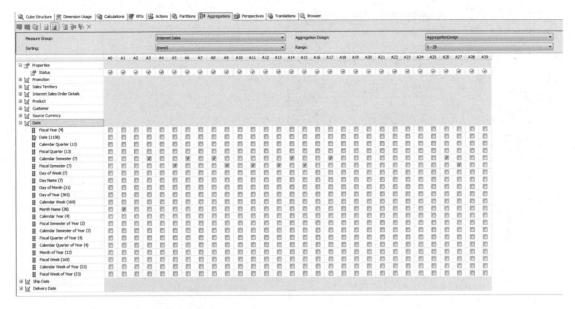

Figure 13-26. *Advanced aggregation design view*

The grid of checked boxes in Figure 13-26 is a "first best guess" by SSAS, based on a cold, clean look at the cube design. The next step is to see how the aggregation design fares once it's been tested under load. For this SSAS has a Usage Based Optimization Wizard, or UBO. The UBO will look at the query records for the partitions covered by an aggregation design and evaluate how the aggregations perform based on what queries users are actually running.

Preparing an Aggregation for Use

The first step to prepare for using the UBO is to enable the query log for SSAS. Analysis Services can run a query log to track the performance of every user query against a server. The query log is disabled by default (so unsuspecting admins don't have databases being filled up by query logs). To enable the query log, open SQL Server Management Studio and connect to the Analysis Services Server, as shown in Figure 13-27.

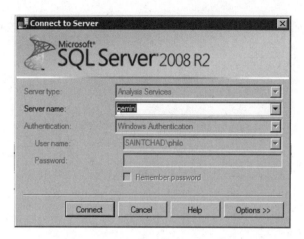

Figure 13-27. Connecting SSMS to an Analysis Server

Once you've connected, open the properties dialog for the *server* by right-clicking the server name and selecting Properties. In the Properties dialog, you're interested in the Log \ QueryLog settings. Set the QueryLogConnectionString property to connect to a SQL Server instance. The QueryLogTableName value will be the name of the log table. You can connect to an existing table (which will have to have the appropriate schema), or by setting the CreateQueryLogTable setting to True, as shown in Figure 13-28.

ForceCommit Timeout	30000	30000	30000
Log \ FlightRecorder \ Enabled	true	true	true
Log \ QueryLog \ CreateQueryLogTable	true	false	false
Log \ QueryLog \ QueryLogConnectionString	Provider=SQLN...		
Log \ QueryLog \ QueryLogSampling	10	10	10
Log \ QueryLog \ QueryLogTableName	OlapQueryLog	OlapQueryLog	OlapQueryLog
LogDir	C:\Program File...	C:\Program File...	
Memory \ HardMemoryLimit	0	0	0
Memory \ LowMemoryLimit	65	65	65

Figure 13-28. Setting server properties for query logging

After running the query logs for a while, you'll be able to run the usage-based optimization wizard. You run the wizard by right-clicking on an aggregation design and selecting "Usage Based Optimization." From this point the wizard is just like the aggregation design wizard; however, behind the scenes, in addition to evaluating the cube architecture, it is reviewing the query logs. Based on the query logs and the architecture, the UBO wizard will redesign the aggregation design optimization for the indicated usage.

■ **Note** If you haven't configured the QueryLog for the SSAS server, the UBO wizard will notify you and fail out.

Once you've established the aggregation designs and are in production, you can still run the UBO wizard periodically to review aggregation designs against cube usage, and update the designs if necessary.

Scaling

Scalability in Analysis Services depends on which kind of scaling you need to do: large numbers of end users, or large data sizes (or both). In general, the choices of how to scale are either *scale up*, meaning add additional processors, move to a more powerful CPU speed, or add more RAM; or you can *scale out*, referring to adding additional servers in parallel. Each method has benefits and drawbacks, and which direction to go also depends on the reason you need to scale.

Large Data Size

When evaluating the size of your data and where the limiting factors may be, here are some guidelines to keep in mind:

Large Dimensions: Generally the only thing you can do to deal with large dimensions is add more RAM. Remember that 32-bit servers (x86) recognize only up to 3GB. When buying any server today you really should go with 64-bit hardware (x64). As mentioned earlier in the chapter, also be sure to evaluate the dimensions and see if there is any way to reduce them (for example, shifting to drill-through reporting for granular data).

Large Fact Data: First and foremost is to evaluate whether you need all the fact data you've accumulated. After that, if your server is getting bogged down due to large measure groups, your best option is to partition the data. That alone will give you some headroom, as only the active partition will be processed, and queries that don't cross partitions will run much faster. The next step is to move partitions to alternate SSAS servers—scale out the measure groups. Adding RAM will also help with large measure groups.

Of course, cubes don't exist in a vacuum—the other pain you will feel with respect to loading and scalability is the number of users and the queries they run.

Large Number of Users

First and foremost, if you have a large number of users on an Analysis Services cube, scale out to additional servers. There are a number of architectures possible—scaling out partitions, separating query servers and processing servers, using multiple query servers for load balancing and redundancy, and so on. The architecture in Figure 13-29 provides a load-balanced queue of Query Servers that query against a read-only cube loaded on a SAN.

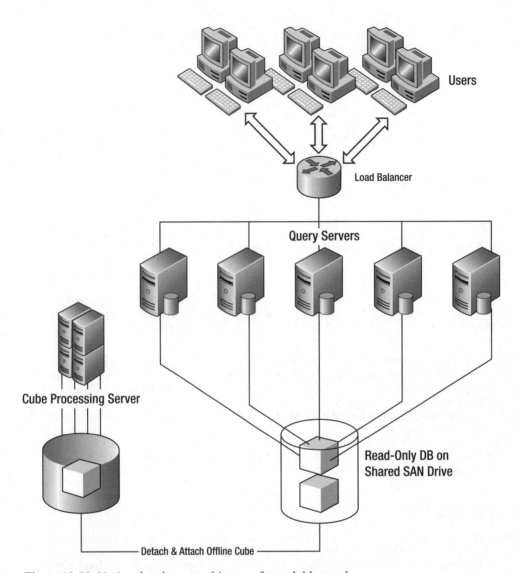

Figure 13-29. Notional scale-out architecture for scalable user base

The database is loaded and processed on a processing server; then the processed database is detached, copied to the shared SAN drive, and attached to the instance there. Since the database is read-only, no editing or writeback is possible. However, current writeback scenarios are generally much smaller-scale so you shouldn't need this approach.

■ **Tip** If you want to load-test your Analysis Services Server without having a thousand users try it and break it, check out the ASLoadSim package at `http://www.codeplex.com/SQLSrvAnalysisSrvcs`, which includes code for running load tests as well as a document on SSAS Load Testing Best Practices.

Virtualization

Until recently, Microsoft strongly discouraged running SQL Server in a virtual environment, simply because enterprise databases and database applications generally get dedicated servers on their own. So virtualizing a database server and taking the virtualization hit didn't seem to make sense.

However, with the growth of adoption of SQL Server's BI platforms (often necessitating additional dedicated servers that aren't as heavily loaded), and improvements in Hyper-V performance and integration (load testing shows a SQL Server in a Hyper-V environment takes only about a 5% hit on performance), Microsoft now supports SQL Server in a virtualized environment.

Additionally, while the new Resource Governor in the SQL Server Relational Database gives DBAs finer control over services located on the same server, there is no similar feature for the BI services. So Microsoft now recommends setting up a BI environment (Integration Services, Analysis Services, Reporting Services) in dedicated virtual servers on a single physical server. Then the individual servers can be throttled by the Hyper-V infrastructure.

Finally, a tiny note about licensing—SQL Server Enterprise Edition has an interesting quirk in the per-processor licensing. If you license all the *physical* processors on a server, then any SQL Server instance you install on the virtual guest servers is also covered. So, if you need a number of independent but lightly-loaded SQL Servers, you could buy a host server with two quad-processor CPUs, then put eight guest machines on Hyper-V, each with a dedicated CPU core. Buy two SQL Server Enterprise Edition processor licenses for the two physical procs, and all of the guest SQL Server instances are licensed as well.

SharePoint Server 2010

Since PowerPivot enables users to consume data from various sources, publish them to SharePoint, and refresh data periodically, there is a unique administrative challenge in that users may create PowerPivot reports that put excessive burdens on their servers. It's also possible that a PowerPivot report created on an ad-hoc basis turns out to be very popular, and a large number of people are running it on a regular basis, putting an extreme load on the PowerPivot server.

Given the potential for users to create significant loads on business intelligence and collaboration servers, it's very important to be able to track what's going on, and why, in the PowerPivot sandbox on your servers. With that in mind, the product group has added a management dashboard to the PowerPivot add-in for SharePoint 2010. To get to the management dashboard, open the SharePoint 2010 Central Administration application (Start ➤ All Programs ➤ Microsoft SharePoint 2010 Products). Once in Central Administration, the PowerPivot Management Dashboard is under General Application Settings. The management dashboard is shown in Figure 13-30.

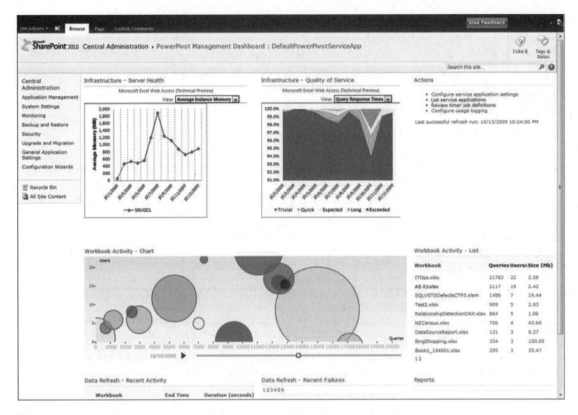

Figure 13-30. *The PowerPivot management dashboard in SharePoint 2010*

The SharePoint management dashboard provides a very useful perspective on the PowerPivot deployment from two aspects. First, you can easily identify PowerPivot workbooks that are putting a severe load on your server. Then you can analyze how heavily used that workbook is. For workbooks that are lightly used but loading the server and data sources, you may just want to contact the author and see if he or she can filter some of the data that's being loaded or take other steps to reduce the load.

The infrastructure charts at the top show Average Instance CPU, Memory, and Query Response Times. (The page is a standard SharePoint web part page, so you can add additional web parts and reports if you desire.) There are reports for Workbook Activity, Data Refresh Activity, and Data Refresh Failures.

The Workbook Activity Chart, shown in Figure 13-31, is a very useful chart. It's a bubble chart that helps to show the most active workbooks. The *y* axis is the number of users, while the *x* axis is the number of queries, and the size of the bubble is the size of the workbook. So we can interpret the bubbles—a large bubble that is low but to the far right is a large quantity of data that is heavily used by a small user base. On the other hand, a large bubble that is higher on the *y* axis is used by more people. Even if it's not as large and not as heavily used, it may be higher priority for IT review and possibly redesign.

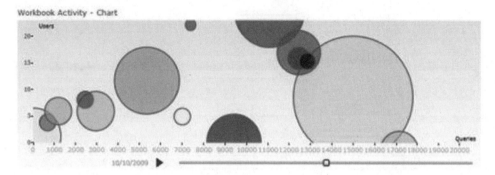

Figure 13-31. *The Workbook Activity Chart*

Workbooks that are used by a lot of analysts, however, create a different problem. Given the ad-hoc nature of PowerPivot workbooks, the loading may simply be a result of heavy usage. In that case, you will probably want to evaluate having the PowerPivot data reengineered into a formal SSAS cube that can be optimized with consideration of the techniques and practices outlined in this chapter and elsewhere. As noted in Chapter 12, there is no "built-in" way to migrate a PowerPivot cube to SQL Server Analysis Services, so the workbook will have to be rebuilt following the guidance in the workbook.

Summary

Hopefully this chapter has helped you to think in terms of enterprise management and scalability of Analysis Services solutions. SSAS is such a powerful tool, and relatively straightforward once you understand the basics, that it can be too easy to create incredibly useful tools that put excessive strains on the servers involved. However, with some forethought and investment in grooming the cubes involved, you can keep the solution responsive to users and manageable for IT.

In the next chapter we'll take a look at the front ends available for all this power we've built.

CHAPTER 14

■■■

User Interfaces

We've looked at a lot of things we can do with dimensional data, and we've picked apart how to build cubes and mine the data contained in them. However, most of our work has been in BIDS, which isn't exactly a user-friendly front end. So how do we get the value of our OLAP solutions out to "the general population"? In this chapter we'll look at the various front ends we can leverage to display the data in our cubes.

We'll start with Excel 2007, which we've used in a few exercises. I'll review connecting to an OLAP data source, and building pivot tables and pivot charts. Then I'll show how to use native Excel functions on OLAP data to create some presentable interactive reports. Also from the Office suite, I'll show how to connect Visio to SSAS for rendering data using Visio shapes.

We'll also look at publishing Excel spreadsheets to SharePoint Excel Services for other users. On the topic of SharePoint, we're also going to check out the new KPI lists feature in MOSS 2007, which enables publishing OLAP data in the form of scorecard-style green/yellow/red indicators.

For static consumption of dimensional data, we'll look at SQL Server Reporting Services. In 2008 there were significant improvements in the charting engine and some great new features in reports in general. We'll definitely leverage these improvements for our OLAP charting. We'll also look at the new Report Builder that gives average users the ability to create their own reports.

I'll spend some time on PerformancePoint and ProClarity—Microsoft's business intelligence tools, which were recently rolled into SharePoint. PerformancePoint provides the ability to build dashboards in SharePoint for tracking performance metrics. ProClarity is a great tool for ad-hoc analysis.

Finally we'll look at some ways to work with Analysis Services from code, using both Analysis Management Objects (manipulating cubes via an object model) and ADOMD.NET (running queries against multidimensional data sources).

Excel 2007

In earlier chapters we used Excel to view cube data and as a data mining client. In this chapter we'll dig a bit deeper into the capabilities of Excel as a front end for Analysis Services. Aside from the basic usefulness as a browser similar to the browser in BIDS, you also have all the formatting and analysis features of Excel, truly blurring the line between the back-end OLAP engine and the desktop tools. The end result can be very powerful and informative reports, as shown in Figure 14-1.

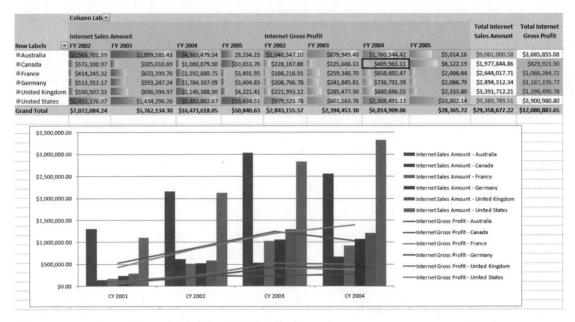

Row Labels	Internet Sales Amount				Internet Gross Profit				Total Internet Sales Amount	Total Internet Gross Profit
Column Labels	FY 2002	FY 2003	FY 2004	FY 2005	FY 2002	FY 2003	FY 2004	FY 2005		
⊞Australia	$2,568,701.39	$2,099,585.43	$4,383,479.54	$9,234.23	$1,040,547.10	$879,949.40	$1,760,344.42	$5,014.16	$9,061,000.58	$3,685,855.08
⊞Canada	$573,100.97	$305,010.69	$1,088,879.50	$10,853.70	$228,167.88	$125,668.33	$469,963.11	$6,122.19	$1,977,844.86	$829,921.50
⊞France	$414,245.32	$633,399.70	$1,592,880.75	$3,491.95	$166,216.91	$259,348.70	$658,692.47	$2,006.64	$2,644,017.71	$1,086,264.72
⊞Germany	$513,353.17	$593,247.24	$1,784,107.09	$3,604.83	$206,706.78	$241,845.81	$736,731.39	$2,086.79	$2,894,312.34	$1,187,370.77
⊞United Kingdom	$550,507.33	$696,594.97	$2,140,388.50	$4,221.41	$221,993.12	$285,477.30	$880,686.55	$2,333.80	$3,391,712.21	$1,390,490.78
⊞United States	$2,452,176.07	$1,434,296.26	$5,483,882.67	$19,434.51	$979,523.78	$602,163.76	$2,308,491.13	$10,802.14	$9,389,789.51	$3,900,980.80
Grand Total	$7,072,084.24	$5,762,134.30	$16,473,618.05	$50,840.63	$2,843,155.57	$2,394,453.30	$6,814,909.06	$28,365.72	$29,358,677.22	$12,080,883.65

Figure 14-1. A pivot table and pivot chart in Excel 2007

■ **Note** This section is wholly about Excel 2007. While Excel 2003 does have a plug-in available that gives similar capabilities, we'll be sticking with the latest version of Excel for this discussion.

Data Source Connections

The starting point for bringing OLAP data into Excel is the Data tab in the ribbon. The Data tab has a number of features related to working with tables of numbers in Excel. However, we're going to start with the section on the far left—`Get External Data`, highlighted in Figure 14-2. Note the listed options— importing data from Access, web-based tables (while we're not covering this here, it's very cool—check it out), and text (importing CSV files).

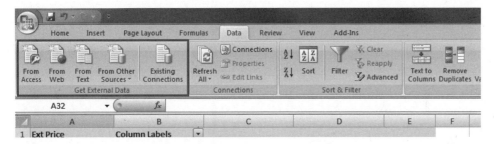

Figure 14-2. *Importing external data into Excel*

Those three options, with the options under "From Other Sources" (which we'll look at in a moment), are to create new data connections, which will be stored in the local file system. If you already have the data connection you need, then you can click the Existing Connections button to browse for it. This will allow you to look for connections in either the local file system or a SharePoint data connection library. However, since we don't have a connection already, we want to check options available under the From Other Sources drop-down, shown in Figure 14-3.

Figure 14-3. *Importing data from SQL Server Analysis Services*

When we select the option to import from Analysis Services, Excel will present us with a dialog for the server name and instance, as shown in Figure 14-4. If you enter the server name by itself, the connection will link to the default instance (remember that you can install multiple SSAS instances on a single server). If you need to indicate a specific instance, then use a backslash: `server\instance`. On this dialog you can also indicate whether to use Windows integrated authentication or a specific user name and password.

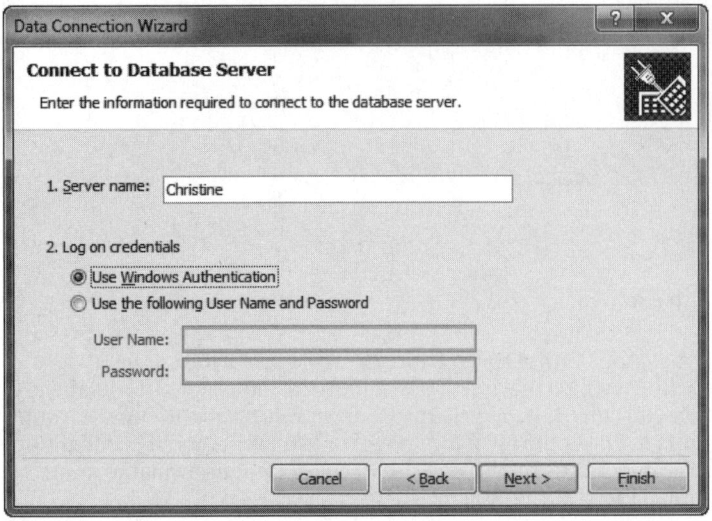

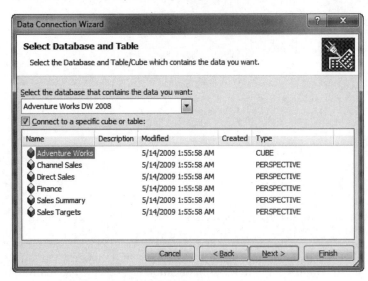

Figure 14-4. *Selecting the server in the data connection wizard*

The next page in the wizard lists the databases on the server in a drop-down at the top, as shown in Figure 14-5. Once you've selected a database, it lists the cubes and perspectives in the database. Then you can select a cube or perspective for the connection. (You can also uncheck "Connect to a specific cube or table"—then you'll be prompted to select the cube or perspective when you use the connection.)

Figure 14-5. *Selecting a cube*

The final page of the wizard, as shown in Figure 14-6, allows you to define the connection file. This includes the file name and location, a description, and the friendly name to be displayed for the connection file. You can also add metadata keywords to be used when indexing the data connection file. We'll cover the Excel Services Authentication Settings later in the chapter.

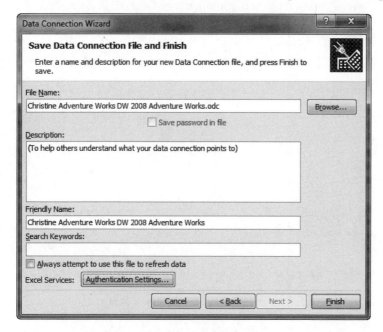

Figure 14-6. *Finishing the data connection wizard*

Once you finish the data connection wizard, Excel will use the connection you created, and show the Import Data dialog, which you can see in Figure 14-7. Here you can select whether to create a PivotTable, a PivotChart, both, or neither. You can also choose whether to place the created items in the current worksheet or to create a new worksheet for them.

Figure 14-7. *The import data wizard*

The Properties button at the bottom of the Import Data dialog will give you options governing the connection in the workbook, as shown in Figure 14-8. The dialog will show you when the data was last refreshed (remember that in Excel you work on a static snapshot of the data until it's refreshed). You can also set properties on when to refresh the data—on a regular basis, or when the file is opened. The connection can always be refreshed manually using the Refresh button on the Data tab in the ribbon.

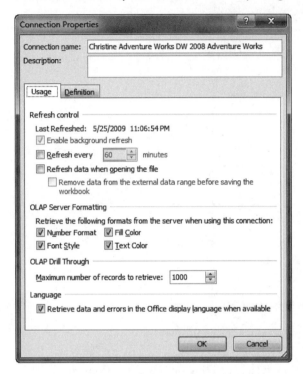

Figure 14-8. The Connection Properties dialog

■ **Note** Be wary of setting connections to automatically refresh. Most business users expect to have control over when the data refreshes. Many financial analysts will store a file as a financial report, and it can cause unhappiness if the numbers change when the file is reopened.

The OLAP Server Formatting section allows you to select which aspects of formatting Excel will retrieve from the cube and automatically apply. Finally, the Definition tab is as it sounds—the definition of the connection is stored here, and you can edit it if necessary.

Once you've created the connection, Excel will create the pivot table and pivot charts for you in the workbook. Each of them has a corresponding task pane that's active when they are (click in the pivot table, the pivot table task bar displays; click in the pivot chart, the task bars for the pivot table and pivot chart display). The pivot chart does require a pivot table—if you create a pivot chart on its own (from the Insert tab), Excel will also create an attached pivot table.

Pivot Tables

When you create a data connection in Excel 2007, Excel creates the pivot table constructor for you, as shown in Figure 14-9. The constructor consists of the placeholder in the worksheet and the task pane on the right-hand side (the task pane can be moved if you choose to.) The task pane has a list of available measures and dimensions in the upper section and the table detail sections in the lower section.

■ **Tip** For more information on dealing with task panes, see `http://office.microsoft.com/en-us/help/HA010785861033.aspx`.

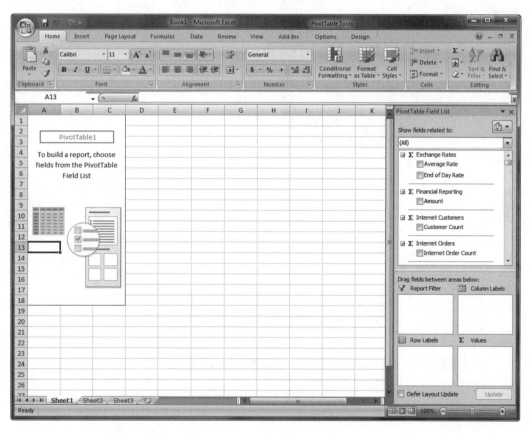

Figure 14-9. A pivot table in Excel 2007

Pivot Table Task Pane

You can change the layout of the task pane by using the drop-down button at the top, as shown in Figure 14-10. Under the layout selector is a drop-down list that lets you filter the measures and dimensions by dimension group. At the bottom of the task pane is a checkbox labeled Defer Layout Update. If you're working with a large cube, some updates may take a few seconds (or longer…). The cube will be more responsive as you pare down the amount of data you're working with, so you may want to check this to hold off on updating the table while you're setting up the initial filters. You can then click the Update button to update the table when you want to, or uncheck the box when you've got a smaller data set.

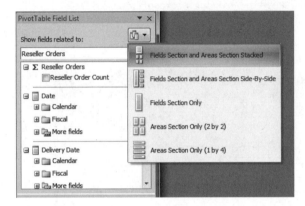

Figure 14-10. *Changing the layout of the task pane*

The four areas at the bottom of the task pane are Report Filter, Column Labels, Row Labels, and Values. These correspond to the table, as shown in Figure 14-11. You can either check the check box for measures to show those in the pivot table, or drag them to the Values area. You'll have to drag dimension hierarchies to the other areas as needed.

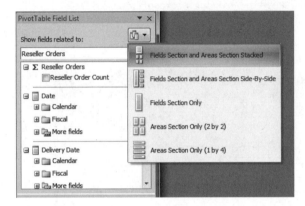

Figure 14-11. *Mapping the task pane areas to the pivot chart*

Another aspect of assembling a pivot table that's worth noting is the ability to nest dimensions. Simply dragging another hierarchy to a row or column will nest them in the pivot table, as shown in Figure 14-12. Note that you can create the hierarchy in whichever order you choose—group by Sales Reasons first, and then by Product Categories if you need to.

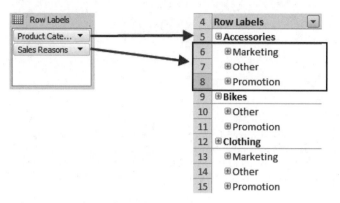

Figure 14-12. *Nesting hierarchies*

Pivot Table Ribbons

The easiest way to work with a pivot table is with the PivotTable Tools ribbons, which appear when you click inside the pivot table. There are two ribbons—one for general options and one for design, as shown in Figures 14-13 and 14-14. The ribbons are fairly self-explanatory, but I strongly recommend browsing through them as you start working with pivot tables, as it can be easy to forget they're there and get frustrated trying to work with the pivot table itself. For example, selecting a pivot table to move it in a workbook is nontrivial, but straightforward with the Move PivotTable button on the Options ribbon.

Figure 14-13. *The pivot table options ribbon*

The most notable function on the Options tab is the ability to rename the pivot table. If you work with Excel named ranges a lot, you may find yourself a bit baffled as to how to name the pivot table by selecting a range on the worksheet. The easy answer is to just click inside the table and rename it on the Options tab. You can also select fields (column and row headers, value fields) and work with their settings, and expand or collapse them.

With the Group Selection and Ungroup buttons you can select members from a hierarchy and create a custom grouping. For example, if you have all 50 states listed, you may want to group them into geographic regions. To do so, just select the states you want to group together and click the Group Selection button. Repeat to create additional groups—as you go, any ungrouped members will be swept into a group called Other.

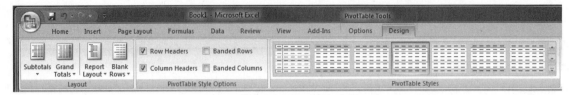

Figure 14-14. *The pivot table design ribbon*

The pivot table design ribbon is pretty descriptive—it provides the tools to change the appearance and layout of your pivot table. The drop-down buttons in the Layout section allow you to select whether to show grand totals and subtotals (for rows and/or columns), and whether to insert blank rows between items. The Report Layout selector offers compact, outline, or tabular layouts, as shown in Figure 14-15.

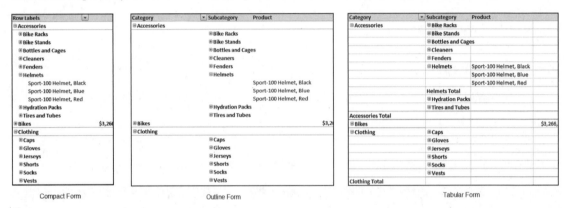

Figure 14-15. *Pivot table formats*

The checkboxes in the PivotTable Style Options section let you switch the styles for row and column headers on and off, and select whether to shade alternating rows or columns. Finally, the PivotTable Styles gallery provides a number of preformatted styles for pivot tables. You can preview how the style will affect your table by mousing over the styles without clicking on them. Also, you can drop down the gallery with the lower down-arrow to the right of the display, as shown in Figure 14-16.

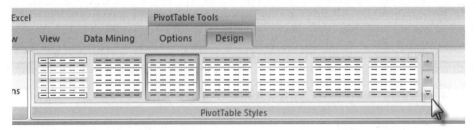

Figure 14-16. *Opening the style gallery*

Formatting a Pivot Table

Formatting the contents of a pivot table is as easy as highlighting the cells you want to change the format on and using the controls in the ribbon or the context menu. Note that you won't be able to alter the structure of the pivot table—for example, trying to merge cells in a pivot table will raise an error prompting you with the options for what you're trying to do. You can also use the Cell Styles drop-down, shown in Figure 14-17, for quick access to some standard styles and number formats. (I just discovered that there's a quick style on there for Currency with no decimals. I've been doing that by hand....)

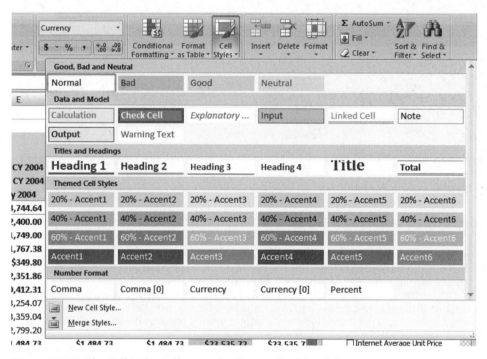

Figure 14-17. The Cell Styles gallery on the Home tab of the ribbon

Another powerful option in Excel 2007 is to use conditional formatting for values. With conditional formatting, you can apply a specific style to cells based on the specific numeric value or the value in relation to other values, or even if it contains a specific text or is a date between certain dates. Conditional formatting is under the Conditional Formatting drop-down gallery in the Styles section of the Home tab on the ribbon. Figure 14-18 shows the Conditional Formatting gallery opened to the Highlight Cells Rules selection.

Figure 14-18. *Conditional Formatting selectors*

Selecting New Rule near the bottom of the drop-down opens a dialog that provides a more fine-grained rule-editing capability. Ultimately you have the option of entering a formula to select which cells to format, and a custom format using any of the Excel cell formatting styles.

One of my favorite format schemes is to apply data bars to a selection of cells, shown in Figure 14-19. The bars behind the numbers give a quick visual indication of the relative magnitude of the values. It is important, however, to be careful not to include any subtotal or total rows, as those will obviously be far larger and dominate the layout.

⊞ **Accessories**	**$173,550.63**
⊟ **Bikes**	**$4,024,025.47**
⊞ **Mountain Bikes**	**$1,658,240.43**
⊟ **Road Bikes**	**$1,280,739.34**
Road-250 Black, 44	$75,743.85
Road-250 Black, 48	$78,187.20
Road-250 Black, 52	$75,743.85
Road-250 Black, 58	$73,300.50
Road-250 Red, 58	$90,403.95
Road-350-W Yellow, 40	$113,966.33
Road-350-W Yellow, 42	$112,265.34
Road-350-W Yellow, 44	$96,956.43
Road-350-W Yellow, 48	$95,255.44
Road-550-W Yellow, 38	$60,506.46
Road-550-W Yellow, 40	$52,663.03
Road-550-W Yellow, 42	$59,385.97
Road-550-W Yellow, 44	$50,422.05
Road-550-W Yellow, 48	$51,542.54
Road-750 Black, 44	$46,439.14
Road-750 Black, 48	$49,139.09
Road-750 Black, 52	$53,459.01
Road-750 Black, 58	$45,359.16
⊞ **Touring Bikes**	**$1,085,045.70**
⊞ **Clothing**	**$86,053.86**

Figure 14-19. Data bars in a pivot table

Now that we understand how to create and format a pivot table, we also want to be able to use these values in other calculations (that's a major reason we're using Excel, right?). So let's take a look at the ins and outs of using pivot table values in formulas.

Formulas Using Pivot Table Values

Let's say you wanted to create a column next to your pivot table showing what the numbers would look like with 10% growth added on top. So you click in a cell, type an equal sign, and then click on a cell in the pivot table to reference it. You'll end up with something looking like this:

```
=GETPIVOTDATA("[Measures].[Internet Sales Amount]",$A$3,
"[Product].[Product Categories]",
"[Product].[Product Categories].[Product].&[374]")
```

This is the function Excel puts in place to fetch the pivot table value, and it's the reference you'll get. This will work well enough—type "*110%" afterwards and you'll see it calculates fine. The problem is when you copy that formula and try to paste it for all the values in the column. You'll get the same value for every cell. Why is that?

The `GETPIVOTDATA` function in this case has four arguments. These are the data field (Internet Sales Amount), a descriptor for the pivot table (`$A$3`, the cell reference for the table anchor), the field, and then the unique name for the specific cell. So when we copy and paste the cell reference, it keeps the unique name and we get the same value.

Couldn't we just type a cell reference for the value? Well, that would work, except that if either rows or columns contain a hierarchy, the user can expand and collapse the members of that hierarchy, giving us the result shown in Figure 14-20. Since the `GETPIVOTDATA` formula uses unique names to identify cells, then so long as the pointer to the pivot table is valid, the value will still work.

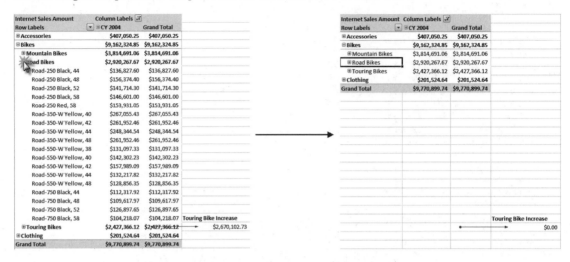

Figure 14-20. *The dangers of using cell references with pivot tables*

■ **Note** One problem you can still run into is if the value you're referencing gets hidden. For example, in Figure 14-20, if you were referencing a specific bike model and you then closed up the subcategory, the reference would break as well (but it would be valid again if the user opened that category again).

Now that we have a feeling for putting our data together and formatting it in a pivot table, let's create some visual displays of our data with pivot charts.

Pivot Charts

Being able to visualize data is every bit as important as actually viewing the data itself. In this section we're going to create pivot charts in Excel to give a graphical view of the data in the pivot table. Figure 14-21 shows an example of such a chart.

It's important to remember that a pivot chart is always bound to a pivot table. When we talk about ProClarity and SQL Server Reporting Services charts later in the chapter, we'll look at some very powerful charting capabilities as the result of linking directly to SSAS cubes. However, Excel pivot charts are limited to what's in the pivot table.

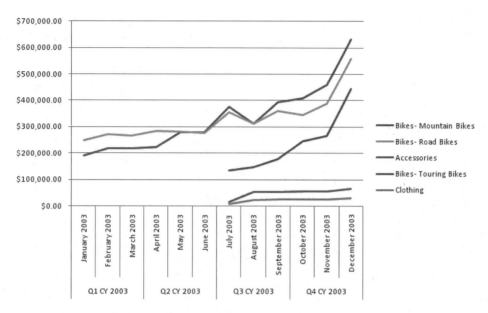

Figure 14-21. *A pivot chart showing sales by product category over twelve months*

There are three ways to create a pivot chart in Excel:

- From the Insert tab on the ribbon, drop-down the PivotTable button and select PivotChart. You will be prompted to select either a table or range, or to use an external data source. Excel will automatically create a pivot table linked to the data source, and the pivot chart linked to the pivot table.

- When you create a data connection, one of the options is to create a pivot chart with the pivot table.

- If you've already created a pivot table, then while the table has focus, the Options tab under PivotTable Tools in the ribbon has a PivotChart button.

There are many types of charts available in Excel, including bar charts, line charts, pie charts, and area charts. For each chart type there are various stylistic subtypes, as shown in Figure 14-22 for bar charts. Be careful not to spend too much time worrying about which type of chart to use—remember that the goal is to convey information, not win a beauty contest. For a more in-depth discussion about charting and chart types, I recommend *Show Me the Numbers* by Stephen Few (2004, Analytics Press).

Figure 14-22. *Selecting a bar chart type*

Each of the subtypes is generally a minor stylistic variation on the main chart type, but there are some significant standouts, most notably the stacked variations (where values are added together or presented as percentages).

■ **Note** When creating a chart from pivot table data, you can't use an XY (scatter) chart, bubble chart, or stock chart, since these chart types require additional dimensions of data. Unfortunately, they're not disabled in the chart-type selector, so you'll get an error if you try to use them.

As with the pivot table, when you create a pivot chart, Excel presents you with a number of new tabs on the ribbon to manage the chart, shown in Figure 14-23. This time we have four tabs:

- **Design:** Here you can change the fundamentals of the chart—the chart type, data, rough layout (title and legend), and style.

- **Layout:** This tab offers more specific layout options. There are selectors for the chart and axis titles, legend, data table (if there is one), axes, gridlines, plot area, and specific parts of the chart itself. You can also set the chart name here and insert objects into the chart.

- **Format:** More finely-grained formatting for the chart objects. By selecting a chart item, e.g., the title or data labels, you can use the format galleries in the Format tab to apply a specific font, outline, fill, or other formatting.

- **Analyze:** Here you'll find the toggles for the field list and filters, as well as a button to refresh the pivot chart data and expand or collapse hierarchies, if used.

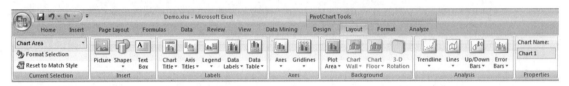

Figure 14-23. The PivotChart Tools tabs in Excel

Pivot charts really are fairly straightforward—since they're bound to the underlying pivot table, most of your work will be done ensuring the data in that table is in the proper format for the chart. In Exercise 14-1 we're going to use the AdventureWorks cube to create a pivot table and pivot chart so we can get a feel for how they work together.

Exercise 14-1. Create an Excel PivotTable and PivotChart

In this exercise we'll walk through using Excel as a front end for an Analysis Services cube, including connecting to the cube, building a pivot table, configuring the pivot chart, and applying some Excel formatting.

    1.   Open Excel 2007.

2. Click the Data tab in the ribbon.

3. Click the From Other Sources button to drop-down the selector, and then click From Analysis Services, as shown in Figure 14-24.

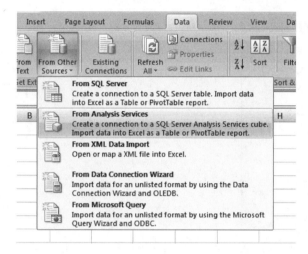

Figure 14-24. Creating an Analysis Services connection

4. This will open the Data Connection wizard, shown in Figure 14-25. Enter the name of your SSAS server ("localhost" if it's on the same machine).

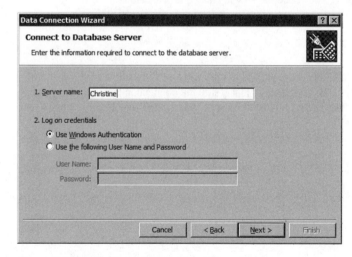

Figure 14-25. The Data Connection Wizard

5. Click the Next button.

6. The next page allows you to select the database and cube that you want for the pivot table. Select Adventure Works DW 2008 (or whatever you named your AdventureWorks database) and then the Adventure Works cube, as shown in Figure 14-26.

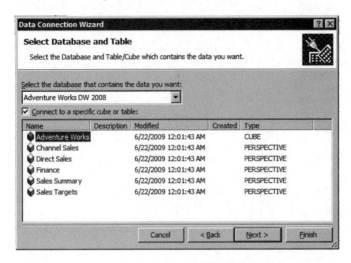

Figure 14-26. *Selecting the SSAS database and cube*

7. On the "Save Data Connection File and Finish" page, you can leave the defaults or edit the file name, friendly name, and description if you choose to. This step creates the data connection file in your file system (by default in $DOCUMENTS\My Data Sources\).

8. Click the Finish button.

9. Now you should see the Import Data dialog, which prompts whether you want to create a PivotTable, PivotTable and PivotChart, or just create the connection file (Figure 14-27). Select PivotTable Report and leave "Existing worksheet" selected for the location.

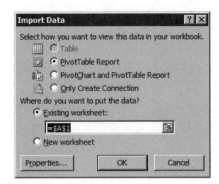

Figure 14-27. Inserting the PivotTable and PivotChart

10. Click the OK button.

11. You'll have a PivotTable placeholder and the PivotTable Field List task pane open, as shown in Figure 14-28. (If you don't see the task pane, click inside the PivotTable placeholder, select the PivotTable Tools – Options tab in the ribbon, and then click Field List in the Show/Hide section to the far right.)

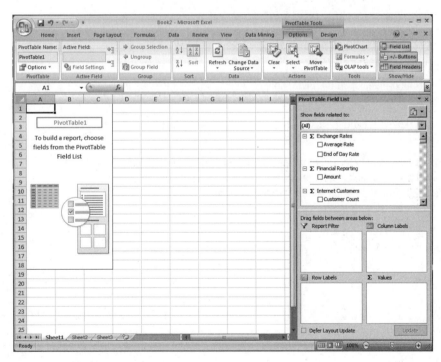

Figure 14-28. Excel with a pivot table embedded

12. Now let's add some data to our pivot table. In the Field List on the right, check the boxes for `Internet Sales Amount` and `Internet Gross Profit Margin`. This will add the two values to the pivot table. With no dimensions or filters, they represent their respective values for the entire cube. (Also note that the Column Labels section now has an entry for Values to indicate the values dimension added with the two measure members.)

13. Let's break these measures down by product. In the "Show fields related to" drop-down at the top of the PivotTable Field List task pane, select Internet Sales.

14. Scroll down in the list of fields—under Product check the box next to Product Categories. Note the categories now in the left-hand column, as shown in Figure 14-29.

	A	B	C
1		Values	
2	Row Labels ▼	Internet Sales Amount	Internet Gross Profit Margin
3	⊞ Accessories	$700,759.96	62.60%
4	⊞ Bikes	$28,318,144.65	40.63%
5	⊞ Clothing	$339,772.61	40.15%
6	Grand Total	$29,358,677.22	41.15%

Figure 14-29. Adding product categories to the pivot table

15. Click on the [+] next to Bikes to see the category open and display the subcategories underneath.

16. Scroll to the Date dimension in the PivotTable Field List. Click [+] to open the Calendar folder.

17. Check the box next to Calendar Year. This moves the Calendar Year hierarchy of the Date Dimension to the Column Labels area.

18. We don't have a lot of data for Calendar Years 2001 and 2002, so let's leave them off for now. In the pivot table click the drop-down arrow next to Column Labels and uncheck 2001 and 2002, as shown in Figure 14-30.

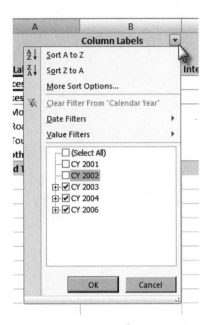

Figure 14-30. *Unselecting CY 2001 and 2002*

19. Click the OK button.

20. We now have a pivot table showing Internet sales and profit margin by Category/Subcategory/Product and by calendar year.

21. Open up the subcategory Road Bikes by clicking the [+] symbol next to it.

22. Select the cells with Road-250 bikes by clicking and dragging. Then right-click and select Group, as shown in Figure 14-31.

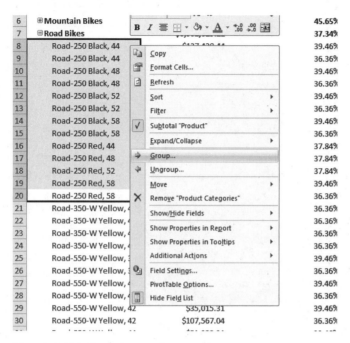

Figure 14-31. *Grouping products*

23. The group will be named Group1 by default—rename it Road-250.

24. Do the same for the Road-350, Road-550, Road-650, and Road-750 bikes.

25. Now we can look at numbers by product line. (The cube actually has a hierarchy for Product Lines, but you get the idea.)

26. "Internet Gross Profit Margin" is a bit verbose, so let's trim it a bit. Right-click on one of the column headers, and then select Value Field Settings.

27. For Custom Name, type Internet PM. Note the other options available here for manipulating the display of the value field.

28. Now we have Sales and Profit Margin broken down by calendar year and product, as shown in Figure 14-32. The Fields area in the sidebar should look like Figure 14-33.

	CY 2003		CY 2004		Total Internet Sales Amount	Total Internet PM
Row Labels ▾	Internet Sales Amount	Internet PM	Internet Sales Amount	Internet PM		
⊞ Accessories	$293,709.71	62.60%	$407,050.25	62.60%	$700,759.96	62.60%
⊟ Bikes	$9,359,102.62	40.96%	$9,162,324.85	40.54%	$18,521,427.47	40.75%
⊞ Mountain Bikes	$3,989,638.48	45.65%	$3,814,691.06	45.45%	$7,804,329.54	45.55%
⊟ Road Bikes	$3,952,029.21	37.34%	$2,920,267.67	36.36%	$6,872,296.88	36.92%
⊞ Road 250	$2,144,214.15	37.67%	$735,448.35	36.36%	$2,879,662.50	37.34%
⊞ Road 350	$540,914.82	36.36%	$1,039,304.89	36.36%	$1,580,219.71	36.36%
⊞ Road 550	$676,095.66	37.33%	$692,462.82	36.36%	$1,368,558.48	36.84%
⊞ Road 650	$264,650.62	37.84%			$264,650.62	37.84%
⊞ Road 750	$326,153.96	36.36%	$453,051.61	36.36%	$779,205.57	36.36%
⊞ Touring Bikes	$1,417,434.93	37.84%	$2,427,366.12	37.84%	$3,844,801.05	37.84%
⊞ Clothing	$138,247.97	40.16%	$201,524.64	40.14%	$339,772.61	40.15%
Grand Total	$9,791,060.30	41.60%	$9,770,899.74	41.45%	$19,561,960.04	41.52%

Figure 14-32. Breaking down sales and profit margins by calendar year and product

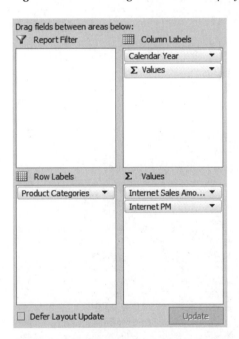

Figure 14-33. The field area in the sidebar task pane

29. If we're focused on our Road Bike models, having the other categories and subcategories displayed can be a bit distracting—let's hide them. Click on the drop-down arrow next to Row Labels.

30. In the "Select field" selector, uncheck Accessories, Clothing, and Components. Then click the [+] next to Bikes, and uncheck Mountain Bikes and Touring Bikes, as shown in Figure 14-34.

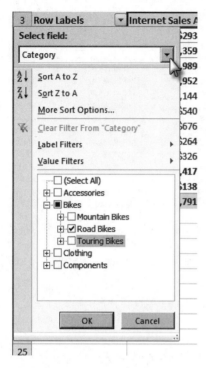

Figure 14-34. *Filtering out other products*

31. Click the OK button and you should be left with a cleaner display, as shown in Figure 14-35.

	Column Labels					
	CY 2003		CY 2004		Total Internet Sales Amount	Total Internet PM
Row Labels	Internet Sales Amount	Internet PM	Internet Sales Amount	Internet PM		
⊟ Bikes	$3,952,029.21	37.34%	$2,920,267.67	36.36%	$6,872,296.88	36.92%
⊟ Road Bikes	$3,952,029.21	37.34%	$2,920,267.67	36.36%	$6,872,296.88	36.92%
⊞ Road 250	$2,144,214.15	37.67%	$735,448.35	36.36%	$2,879,662.50	37.34%
⊞ Road 350	$540,914.82	36.36%	$1,039,304.89	36.36%	$1,580,219.71	36.36%
⊞ Road 550	$676,095.66	37.33%	$692,462.82	36.36%	$1,368,558.48	36.84%
⊞ Road 650	$264,650.62	37.84%			$264,650.62	37.84%
⊞ Road 750	$326,153.96	36.36%	$453,051.61	36.36%	$779,205.57	36.36%
Grand Total	$3,952,029.21	37.34%	$2,920,267.67	36.36%	$6,872,296.88	36.92%

Figure 14-35. *Breaking down sales and profit margin for all road bikes*

32. For CY 2003, the Road 250 stands out with $2.1M out of almost $4M in sales, but it takes some staring and thinking to figure out how the other models rank. Let's add a visual cue.

33. Click and drag to highlight the sales amounts for CY 2003 for the five Road models.

34. Click the Conditional Formatting button in the ribbon (on the Home tab), and then select Data Bars and choose one of the color schemes, as shown in Figure 14-36.

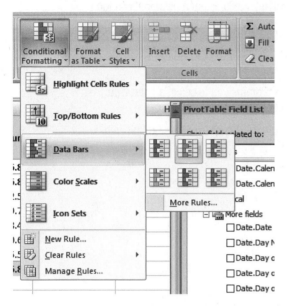

Figure 14-36. Applying color bar conditional formatting

35. You should see each of the values, now accented with a color bar showing their relative values, as shown in Figure 14-37.

	CY 2003	
Row Labels	Internet Sales Amount	Internet PM
⊟Bikes	$3,952,029.21	37.34%
⊟Road Bikes	$3,952,029.21	37.34%
⊞Road 250	$2,144,214.15	37.67%
⊞Road 350	$540,914.82	36.36%
⊞Road 550	$676,095.66	37.33%
⊞Road 650	$264,650.62	37.84%
⊞Road 750	$326,153.96	36.36%
Grand Total	$3,952,029.21	37.34%

Figure 14-37. Color bars showing relative values

■ **Note** The color bars themselves shouldn't be read like a chart; they just show relative values.

Now that we've seen how to create a pivot chart in Excel, let's quickly put a pivot table on top.

1. Click a cell inside the pivot table to bring up the task pane and PivotTable Tools.

2. In the PivotTable Tools/Options tab, click the PivotChart button in the Tools section.

3. This will open the Insert Chart dialog, which prompts you to select a chart type. Leave the default selection and click the OK button.

4. You should see a chart similar to Figure 14-38. Note that Excel has created series for both the year and the measures.

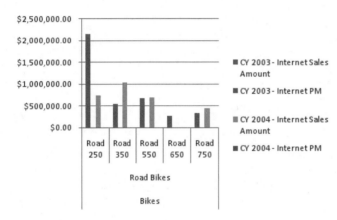

Figure 14-38. *The default pivot chart*

5. If you look at the PivotChart Filter Pane, you'll see that you can filter the fields down, but not much else.

6. Experiment with right-clicking on various parts of the chart to understand the options available.

7. Save the workbook—we'll use this later with Excel Services.

In my experience, Excel 2007 is awesome for pivot tables—if you need to do tabular analysis on SSAS data, Excel is a top-tier client. Unfortunately, the fact that the pivot chart is bound to a pivot table, combined with the lack of ability to manipulate data in the chart, makes it pretty limited. Later in this chapter we'll look at Reporting Services and ProClarity for more powerful charting tools. For now let's look at the other member of the Office family that understands Analysis Services—Visio.

Visio 2007

With Visio 2003, Microsoft started introducing more data-binding capabilities into the product. We won't dive too deeply into this, other than to look at the SSAS integration. Visio 2007 includes a construct called a *pivot diagram,* which we can use to map cube data. An example of a breakdown of Internet sales data is shown in Figure 14-39.

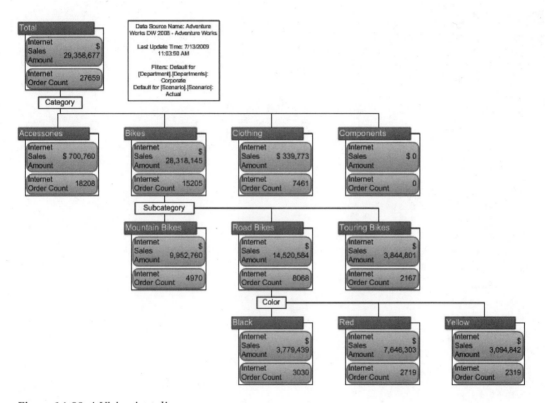

Figure 14-39. *A Visio pivot diagram*

■ **Note** I read a comment online that I have to agree with—Visio is generally used for designing, not reporting. There are some publication capabilities built into Visio, but they're not used as extensively as Excel Services, and certainly nothing like Reporting Services. So pivot diagrams are fairly niche; but if you have the need to bring Analysis Services data into Visio, there it is.

Creating a pivot diagram is pretty easy to do—in a new Visio drawing, select `PivotDiagram → Insert PivotDiagram`. This will start the PivotDiagram wizard, which is essentially a dialog box to either select an existing connection or create a new connection (which will launch the New Connection wizard we're familiar with).

Once you've created the connection, Visio will add a shape representing the default measure from the measure group, and add a task pane listing measures ("Totals") and dimension hierarchies ("Categories") in the selected cube, as shown in Figure 14-40. Checking a Total adds it to the diagram. If you select a pivot node (the shapes in the pivot diagram) and select a Category, Visio will add a breakdown of that hierarchy under the selected node (replacing an existing hierarchy if one is in place).

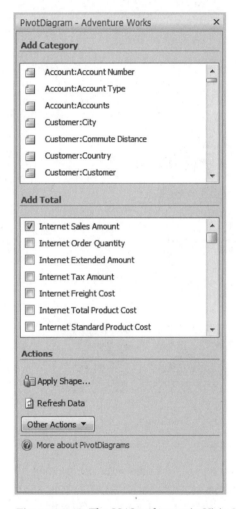

Figure 14-40. *The SSAS task pane in Visio 2007*

This operates in much the same way as the ProClarity Decomposition Tree, which we'll look at later in the chapter. There are two significant differences:

- ProClarity is web-based and enables publishing charts to the web, while Visio is client-based.

- ProClarity charts are similar to all charts in that once you've designed a chart, it's relatively inflexible. On the other hand, Visio pivot diagrams are composed of Visio shapes, so in addition to manipulating nodes with data, you can simply drag them, edit them, or add additional data.

We won't walk through an exercise, because when considered with the other front ends we're going to look at, the task of creating a pivot diagram is pretty straightforward. Now let's look at some ways we can publish SSAS data for broad consumption, starting with SQL Server Reporting Services.

SQL Server Reporting Services

Another one of the BI services in the SQL Server platform is SQL Server Reporting Services. Reporting Services is a web-based service for hosting and publishing reports. Reports are natively published in HTML format, but can also be rendered as XML, in Excel, PDF, TIFF, or Word. While Reporting Services runs as a SQL Server service, it can render reports from a variety of data sources, though the data source we're most interested in is, of course, SQL Server Analysis Services.

■ **Note** Remember that SQL Server services can be run "a la carte." Even if you have a significant investment in SQL Server 2005 as your data storage, you can still use SQL Server 2008 Reporting Services for reporting, and given the improvements in SSRS 2008, I highly recommend doing so.

Let's take a look at some of the features of SSRS—paired with Analysis Services, this is an incredibly powerful platform for reporting on business data. We're going to look at reports in general, and then a new feature in SSRS 2008 called Tablix, which combines the best features of tables and matrices. We'll examine the new charting engine in Reporting Services, and finally take a quick look at the new Report Builder and how it enables self-service reporting.

■ **Note** This is going to be a fairly lightweight overview of SQL Server Reporting Services. For a more in-depth examination of the subject, I recommend *Microsoft SQL Server 2008 Reporting Services* by Brian Larson (McGraw-Hill, 2008) or *Pro SQL Server 2008 Reporting Services* by Landrum, McGehee, and Voytek (APress, 2008).

Reports

Between the new tablix control and advanced charting engine, Reporting Services is more powerful than ever. Formatting reports is more intuitive, and far more WYSIWYG than previously.

The report designer is once again BIDS—in this case you can create either a new project with a wizard or just an empty project. Reports use data connections—either shared connections deployed to the server, or connections embedded in the report. (The benefit of a shared connection is that it's easy to modify it in either BIDS or SSMS—to point it at a different server, change a password, etc.)

Once you've created a report project, report design is fairly straightforward—create datasets for your needs, and then create regions on the report using the datasets to populate them. You have three types of data regions: tablix regions, charts, and gauges, as shown in Figure 14-41.

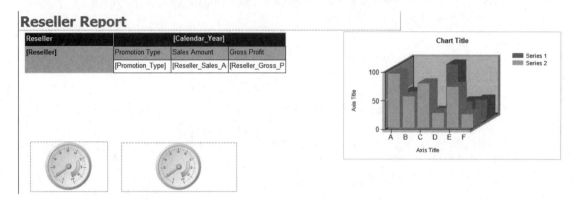

Figure 14-41. A report in design mode with tablix, a chart, and two gauges

Let's take a look at each of these data containers and how they can relate to visually presenting Analysis Services data.

Tablix

In SQL Server 2005 Reporting Services, you had the option of creating a report with a table, as shown in Figure 14-42, or a matrix, as shown in Figure 14-43. Tables were great for straight tabular data—record after record (think green-bar reports). Matrices were used for aggregating data (a topic that should be near and dear to our hearts now)—generally think of them like pivot tables.

Order Table

Order Date	Sales Order Number	Purchase Order Number	Sub Total	Tax Amt	Freight	Total Due	Order Qty	Unit Price	Line Total	Name
7/1/2001 12:00:00 AM	SO43659	PO522145787	24643.9362	1971.5149	616.0984	27231.5495	1	2024.9940	2024.994000	Mountain-100 Black, 42
7/1/2001 12:00:00 AM	SO43659	PO522145787	24643.9362	1971.5149	616.0984	27231.5495	3	2024.9940	6074.982000	Mountain-100 Black, 44
7/1/2001 12:00:00 AM	SO43659	PO522145787	24643.9362	1971.5149	616.0984	27231.5495	1	2024.9940	2024.994000	Mountain-100 Black, 48
7/1/2001 12:00:00 AM	SO43659	PO522145787	24643.9362	1971.5149	616.0984	27231.5495	1	2039.9940	2039.994000	Mountain-100 Silver, 38
7/1/2001 12:00:00 AM	SO43659	PO522145787	24643.9362	1971.5149	616.0984	27231.5495	1	2039.9940	2039.994000	Mountain-100 Silver, 42
7/1/2001 12:00:00 AM	SO43659	PO522145787	24643.9362	1971.5149	616.0984	27231.5495	2	2039.9940	4079.988000	Mountain-100 Silver, 44
7/1/2001 12:00:00 AM	SO43659	PO522145787	24643.9362	1971.5149	616.0984	27231.5495	1	2039.9940	2039.994000	Mountain-100 Silver, 48
7/1/2001 12:00:00 AM	SO43659	PO522145787	24643.9362	1971.5149	616.0984	27231.5495	3	28.8404	86.521200	Long-Sleeve Logo Jersey, M
7/1/2001 12:00:00 AM	SO43659	PO522145787	24643.9362	1971.5149	616.0984	27231.5495	1	28.8404	28.840400	Long-Sleeve Logo Jersey, XL
7/1/2001 12:00:00 AM	SO43659	PO522145787	24643.9362	1971.5149	616.0984	27231.5495	6	5.7000	34.200000	Mountain Bike Socks, M
7/1/2001 12:00:00 AM	SO43659	PO522145787	24643.9362	1971.5149	616.0984	27231.5495	2	5.1865	10.373000	AWC Logo Cap
7/1/2001 12:00:00 AM	SO43659	PO522145787	24643.9362	1971.5149	616.0984	27231.5495	4	20.1865	80.746000	Sport-100 Helmet, Blue
7/1/2001	SO43660	PO1885012750	1553.1025	124.2482	38.8276	1716.1704	1	419.4589	419.458000	Road-650 Red

Figure 14-42. A table in SQL Server Reporting Services

Matrix Report

	FY 2002	FY 2003	FY 2004
Australia			$1,594,335
Canada	$3,079,807	$5,615,169	$5,682,950
France		$1,428,020	$3,179,518
Germany			$1,983,988
United Kingdom		$1,406,492	$2,872,517
United States	$13,208,635	$19,471,989	$20,927,177

Figure 14-43. A matrix in SQL Server Reporting Services

These two options were "good enough" for most purposes; however, most report developers would eventually learn that many reports don't fall squarely into being either tabular or a matrix, but rather a hybrid between the two. For example, a report user may want the report in 14-43 to have some summary information as well. Consider the report in Figure 14-44, which shows the profit margins by country. Veterans of SQL Server 2005 Reporting Services know this is a nontrivial task to accomplish.

	FY 2002	FY 2003	FY 2004	Profit Margin
Australia			$1,594,335	-6.82%
Canada	$3,079,807	$5,615,169	$5,682,950	2.60%
France		$1,428,020	$3,179,518	0.78%
Germany			$1,983,988	-5.61%
United Kingdom		$1,406,492	$2,872,517	2.95%
United States	$13,208,635	$19,471,989	$20,927,177	2.96%

Figure 14-44. Matrix with a summary column

Adding columns outside a grouping, splitting cells in the report, and adding other forms of data outside the main paradigm of "table" or "matrix" have generally meant either doing gymnastics in SQL to get a data set that roughly conforms to the required report, or writing code to generate the data values. As a report designer, we don't have the option to tell our user "your requirements don't match what the software can do. Please adjust your expectations accordingly."

In response to the limitations of the table/matrix approach in SSRS 2005, Microsoft has added a component called *tablix* to Reporting Services. The tablix data region combines the capabilities of the matrix and table—you can add pivot-type features to a table, table-type features to a matrix, and "mix and match." The end result is that reports like the one shown in Figure 14-45 are possible.

Products by *Quarter & Trend*	2003			2004			Y / Y %	Monthly Trend Jan03 - Jun04	Total Product Sales	% of Total Sales in Top 3 Regions		
	Q1	Q2	Year Total	Q1	Q2	Year Total				US	CA	AU
Total	$6,682,510	$8,365,317	$42,308,575	$11,405,246	$14,413,899	$25,819,146	-64 %		$68,127,721	52 %	14 %	11 %
Bikes	$8,101,755	$7,028,317	$35,199,346	$10,279,653	$12,336,326	$22,615,980	-56 %		$57,815,326	52 %	13 %	12 %
Mountain Bikes	$2,517,500	$2,908,659	$12,851,826	$3,473,750	$4,268,134	$7,741,884	-66 %		$20,593,710	55 %	14 %	10 %
Road Bikes	$3,584,255	$4,119,659	$15,282,929	$3,391,876	$4,001,737	$7,393,614	-107 %		$22,676,542	55 %	13 %	11 %
Touring Bikes			$7,064,592	$3,414,027	$4,066,455	$7,480,482	6 %		$14,545,074	43 %	10 %	15 %
Components	$459,086	$1,111,521	$5,489,741	$657,742	$1,433,809	$2,091,551	-162 %		$7,581,292	58 %	18 %	3 %
Bottom Brackets			$30,793	$7,727	$13,307	$21,034	-46 %		$51,826	54 %	16 %	8 %
Brakes			$45,231	$8,115	$12,716	$20,831	-117 %		$66,062	52 %	15 %	8 %

Figure 14-45. A complex report using tablix in SQL Server Reporting Services 2008

Note that we have a matrix (products broken down by calendar quarter), but there's also a year-over-year percentage and trend, a sparkline chart, total sales, and breakdown by top three regions. Before 2008, a chart request like this would have meant a significant amount of code behind the scenes. Tablix makes it easy.

If you've cracked open BIDS and looked for Tablix in the toolbox, you won't find it. You'll see Table and Matrix items, but no Tablix. Where is it? Well, you're looking at it. Both the table and matrix toolbox items are simply special cases of the tablix control—adding either of these gives you the same capabilities in the long run. This was an excellent design decision on Microsoft's part, as you won't ever run into a situation where you'll need a feature in a table and think "I can do this in a matrix…"

So you start with a table or matrix, and then the tablix features start to reveal themselves. For example, in our matrix with the summary columns we just have to right-click on the rightmost column to see the context menu shown in Figure 14-46. By having the ability to add columns (either inside or outside the current grouping) we've got a lot more flexibility.

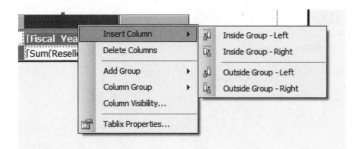

Figure 14-46. Manipulating a tablix item

■ **Note** A tablix data region is still limited to being tied to a single dataset, so keep that in mind. You can add subreports to nest datasets for more complex reports.

Let's take a moment to build a quick report using the tablix control—this will be an easy introduction to Reporting Services as well as leveraging the tablix capabilities.

Exercise 14-2. Build a Report with Tablix

This exercise will show you how to build a basic matrix report on SSAS data, as well as introduce the tablix features. We won't cover deployment of the report or more advanced SSRS features.

1. Open BIDS, create a new project (File → New → Project).

2. Select Report Server Project Wizard and give the project a name.

3. Click the OK button to open the Report wizard.

4. Ensure "New data source" is selected. Name it SSAS DW.

5. Select Microsoft SQL Server Analysis Services as the data source type.

6. Click the Edit button. In the Connection Properties dialog enter the server name for your SSAS server, and then select Adventure Works DW 2008 for the database. Click the OK button.

7. Check the box for "Make this a shared data source," as shown in Figure 14-47.

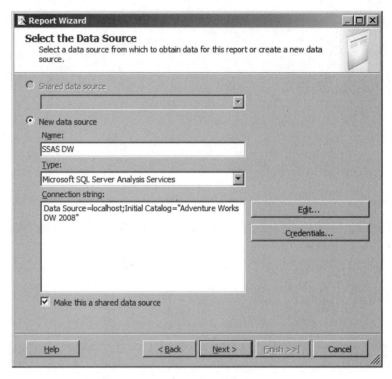

Figure 14-47. *Configuring the report data source*

8. Click the Next button.

9. In the Design the Query page, click the Query Builder button. This will open a dimensional data editor that should look pretty familiar by now.

10. Open the Measures group, and then Reseller Sales. Drag Reseller Sales Amount over to the designer.

11. Under Geography, drag Country to the designer.

12. Under Date, open Fiscal and drag Date.Fiscal Year to the designer.

■ **Note** The designer query results will show a tabular view of the data. This can be a bit unnerving when we plan to design a matrix, but it will all work out—you'll see.

---

13. Drag Reseller Gross Profit Margin over to the designer; it should look like Figure 14-48.

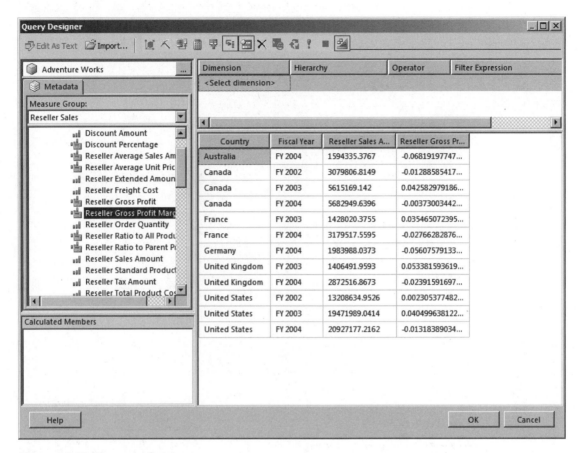

Figure 14-48. The query designer

14. Click the OK button.

15. Click the Next > button.

16. In the Report Type page, select Matrix, and then click the Next > button.

17. In the designer, select Fiscal Year and click the Columns button.

18. Select Country and click the Rows button.

19. Select Reseller_Sales_Amount and click the Details button.

20. Leave Reseller_Gross_Profit_Margin in the left pane, as shown in Figure 14-49.

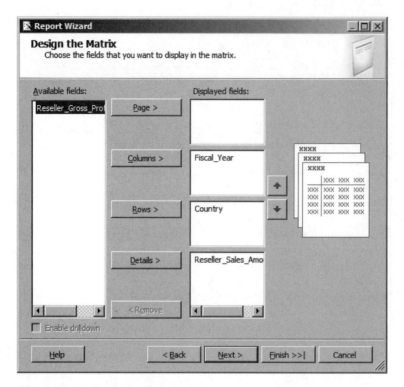

Figure 14-49. Designing the matrix

21. Click the Next button.

22. Select a style and click the Next button.

23. Leave the defaults for the Report Server and folder—we won't be deploying the report.

24. Click the Next button.

25. Name the report Profit Margin Matrix and click the Finish button.

26. The report should end up as shown in Figure 14-50.

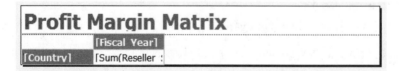

Figure 14-50. The matrix report

27. If you click on the Preview tab at the top of the designer, you can see how this looks now. Let's fix the formatting—right-click on the cell where it says [Sum(Reseller … and select Text Box Properties.

28. In the Text Box Properties dialog, select Number in the left-hand pane. For Category, select Currency. Set the decimal places to 0 and check the box for "Use 1000 separator." Click the OK button.

29. Now let's add the Profit Margin to our table. With a matrix in SSRS 2005, the only option would have been to add the new measure at the same granularity as the existing measure.

30. If you click in the detail cell, you should see the tablix headers appear, as shown in Figure 14-51.

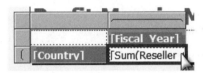

Figure 14-51. Tablix showing the headers

31. Right-click on the column header above the Fiscal Year label, opening the context menu shown in Figure 14-52.

32. Select Insert Column, and then Outside Group – Right to indicate we're adding the column outside the grouping signified by the Fiscal Year.

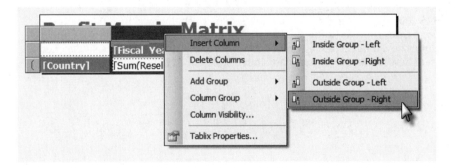

Figure 14-52. Adding a column to the tablix

33. Now click inside the newly created textbox—you should get a drop-down showing the fields in the dataset. Select Reseller_Gross_Profit_Margin. Note that BIDS has automatically inserted a SUM function on the field.

34. In the text box properties, change the formatting for the profit margin text box to Percentage.

35. Click the Preview tab—you should see a report similar to Figure 14-53.

Profit Margin Matrix

	FY 2002	FY 2003	FY 2004	Reseller Gross Profit Margin
Australia			$1,594,335	-6.82%
Canada	$3,079,807	$5,615,169	$5,682,950	2.60%
France		$1,428,020	$3,179,518	0.78%
Germany			$1,983,988	-5.61%
United Kingdom		$1,406,492	$2,872,517	2.95%
United States	$13,208,635	$19,471,989	$20,927,177	2.96%

Figure 14-53. The finished matrix with profit margin

36. Note that the profit margin isn't broken down by fiscal year—it's a summary value on the end.

37. Save the project.

For more information about tablix, a good place to start is "Understanding the Tablix Data Region" at http://msdn.microsoft.com/en-us/library/bb677552.aspx.

Now that we've looked at tabular results of numerical data, let's take a look at a more visual representation using the new charting engine in SSRS 2008.

Charts

The charting engine in SSRS 2005 wasn't too bad, but it had its share of challenges. In 2007, Microsoft licensed the reporting engine from Dundas Software. The result of this effort is the charting engine in SSRS 2008. SSRS 2008 includes new chart types, advanced support for multiple series in a single chart, secondary axes, and a host of new layout features. An advanced SSRS 2008 chart is shown in Figure 14-54.

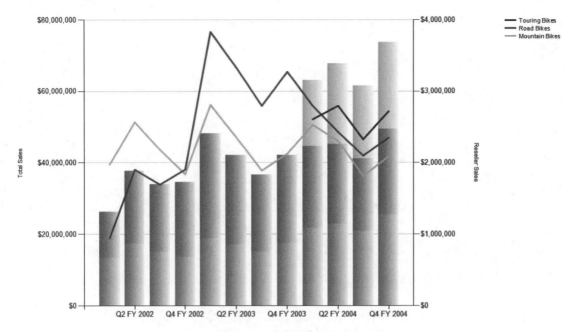

Figure 14-54. *A chart in SQL Server Reporting Services 2008*

To me, one of the best enhancements is access to properties. In SSRS 2005, all charting properties were accessed through a single properties dialog, as shown in Figure 14-55. This was limiting to say the least—any little change or tweak to an axis or chart meant digging through pages of properties and hoping you had the right one—lots of trial and error.

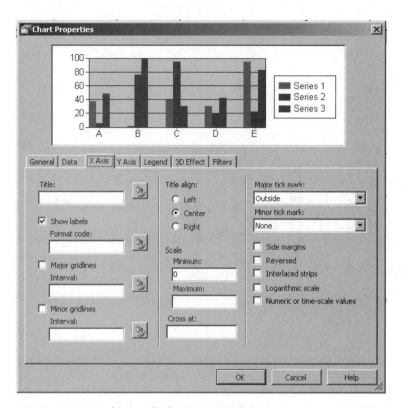

Figure 14-55. Working with charts in SSRS 2005

In SSRS 2008, you can format each chart component by simply right-clicking on it and selecting Properties. For example, to format the *y* axis in SSRS 2005, you have to right-click on the chart, select Properties, select the *y*-axis tab, and then enter the format code for the numerical format you want (yes, you have to know the format codes…). In SSRS 2008, you simply right-click on the *y* axis, and select Axis Properties to open the Value Axis Properties dialog shown in Figure 14-56. Also note that as you make changes in SSRS 2008, they are shown dynamically on the chart, though the dialog may be covering the part of the chart you're changing…

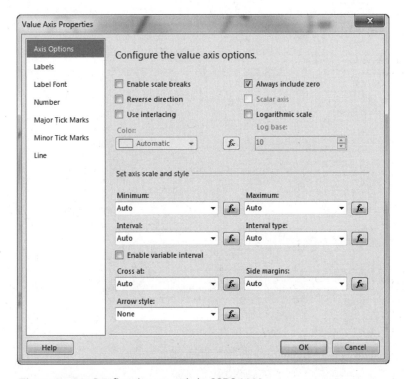

Figure 14-56. Configuring an axis in SSRS 2008

The best way to understand working with charts in Reporting Services is to build one. In Exercise 14-3 we'll create a chart from dimensional data, including two different data series.

Exercise 14-3. Create a Chart in SSRS 2008

In this exercise we'll use the AdventureWorks data to create a two-axis chart like the one in Figure 14-54.

1. Open BIDS to the project you created in Exercise 14-2.

2. Right-click on Reports in the Solution Explorer, select Add…, and then New Item.

3. This will open the Add New Item dialog, as shown in Figure 14-57.

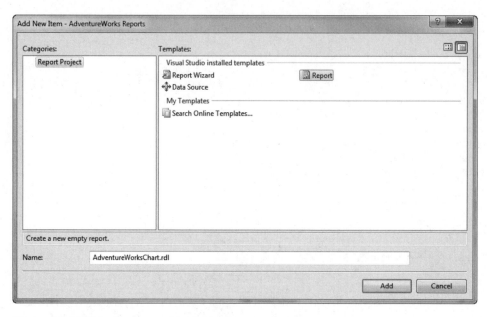

Figure 14-57. *Creating a new, blank report*

4. Select Report and give it the name AdventureWorksChart. Then click the Add button.

5. In the Report Data pane on the right for the new report, click the New button and select Data Source.

6. In the Data Source Properties dialog, select "Use shared data source reference" and select the AdvWorks data source in the drop-down list box.

7. Click the OK button.

8. Click the New button in the Report Data pane and select Dataset to open the Dataset Properties dialog.

9. Ensure DataSource1 is selected in the Data source drop-down. Click the Query designer button to open the multidimensional query designer.

10. Drag Reseller\Reseller_Sales_Amount and Sales Summary\Sales_Amount from the measure groups to the design pane.

11. Drag Product\Category to the filter area at the top of the query designer. Scroll to the right and check the box under Parameter.

12. Drag Product\Subcategory and Date\Fiscal\Date.Fiscal\Fiscal Quarter to the design pane, so that it looks as shown in Figure 14-58.

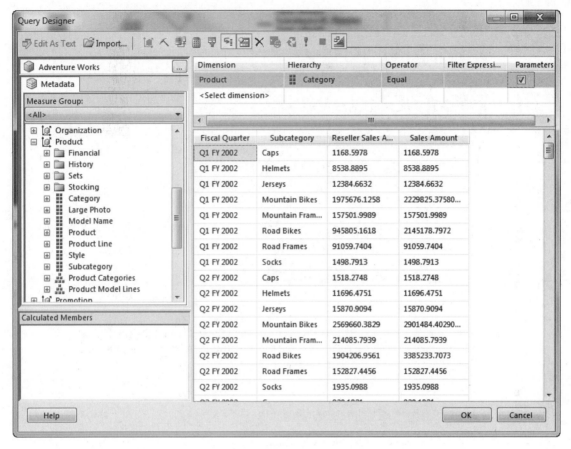

Figure 14-58. The query for the chart data source

13. Click the OK button.

14. Click the OK button on the Dataset Properties dialog.

15. Now open the Toolbox in BIDS (View → Toolbox).

16. Click on the Chart item in the toolbox and click and drag to fill the report canvas.

17. In the Select Chart Type dialog, select the default Column chart, as shown in Figure 14-59.

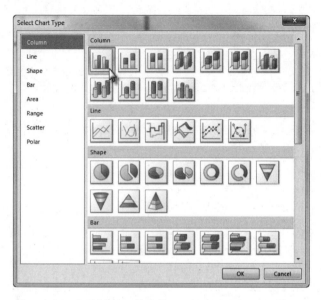

Figure 14-59. Selecting the chart type

18. Switch back to the Report Data tab (or View → Report Data). Click on the chart to select it, and then click the chart area. You should see areas appear around the chart to drop data fields, as shown in Figure 14-60.

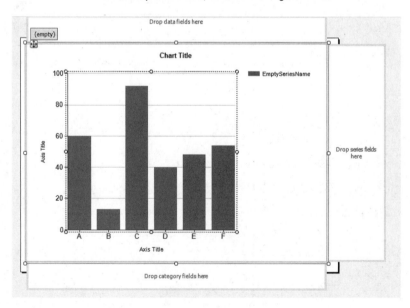

Figure 14-60. Data areas on an SSRS chart

19. Drag the Reseller_Sales_Amount and Sales_Amount fields from the Report Data pane on the left to the area above the chart labeled "Drop data fields here."

20. Drag the Subcategory field to the area labeled "Drop series fields here."

21. Drag the Fiscal_Quarter field to the area below the chart labeled "Drop category fields here."

22. If you preview the chart now, you should see something similar to Figure 14-61, which isn't very useful at all.

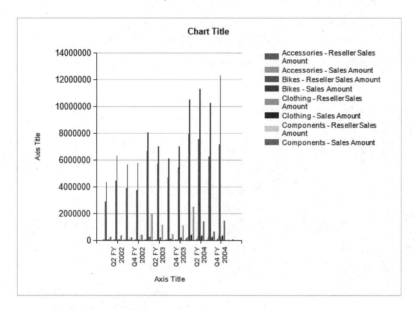

Figure 14-61. A not-very-helpful chart

23. To make the chart more readable, let's change the Reseller Sales data to a line chart—right-click on the data field that reads [Sum(Reseller_Sales_Amount)] and select Change Chart Type.

24. Select the second line chart, and then click the OK button.

25. Now let's add some formatting to the *y* axis—right-click on it, and then select Axis Properties.

26. In the Value Axis Properties dialog, select the Number page, and then Currency. Check "Use 1000 separator" and set Decimal Places to 0.

27. Click the OK button.

28. Finally, let's break up the legends. Right-click on the bars in the chart, select Chart → Add New Legend. This will create a second (empty) legend in the Chart area.

29. Right-click on the bars again, and select Series Properties. Select the Legend page, and then under "Show in legend:" select ChartLegend1.

30. Click the OK button and you'll see the sales amount data legend is now in the new legend area. You can drag the legend areas around to format the chart to your taste. When you're done, preview the chart.

31. You'll have to select a category from the parameter at the top of the preview page—the category Bikes usually works well. Then click the View Report button.

32. You should end up with a chart similar to that shown in Figure 14-62.

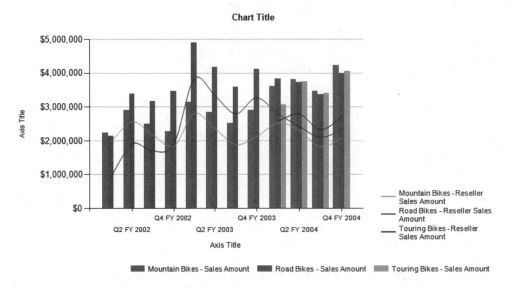

Figure 14-62. *Our finished chart with two series of data*

Report Builder 2.0

SQL Server 2005 Reporting Services introduced a client-side ad-hoc report designer, named Report Builder. It was fairly good, but suffered from the limitation that reports could be built based only on predesigned report models (basically a lightweight cube residing in the report server).

SQL Server 2008 Reporting Services brings us Report Builder 2.0, a free download from Microsoft.com. (You can find the download link at `http://www.microsoft.com/sqlserver/2008/en/us/report-builder.aspx`). Report Builder 2.0 allows end users to create ad-hoc reports from any number of data sources, as shown in the default connection type list in Figure 14-63.

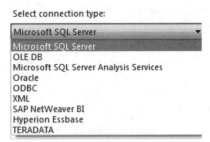

Figure 14-63. *The connection types available in Report Builder 2.0*

When you create a new report, you can either create the report by hand, or use one of the two wizards (Table/Matrix or Chart). The wizards will walk you through the familiar process of creating a data connection, and building a query. If you selected Table/Matrix, then the first layout screen will be a little different, as shown in Figure 14-64, but the rest is pretty straightforward.

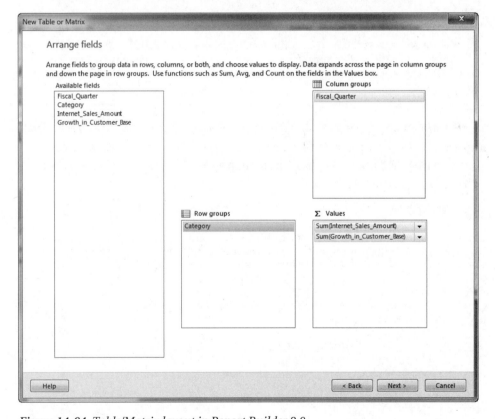

Figure 14-64. *Table/Matrix layout in Report Builder 2.0*

For the most part, Report Builder works exactly like BIDS for designing reports (at least in basic data access, layout, and formatting). Once a user finishes a report design, the user can save it back to the Report Server or SharePoint library where reports are published.

■ **Note** Report Builder 1.0 was a "click-once" deployment application, and when Reporting Services was run in SharePoint integrated mode, Report Builder could be automatically run from the reports library. Report Builder 2.0 takes a little more work. It must be installed on the client, and the reports library has to be "tweaked" to activate the builder.

That finishes up the quick tour of Reporting Services and Report Builder. Let's take a look at our next stop for displaying dimensional data—Microsoft Office SharePoint Server (MOSS) 2007.

MOSS 2007

Microsoft Office SharePoint Server (MOSS) 2007 introduced a number of new features that make it a great platform for business applications. For our purposes, we're deeply interested in two of these new features—KPI lists and Excel Services.

■ **Note** Another new feature in MOSS 2007 is InfoPath Forms Server, which gives you a way to easily create and publish web-based forms. This is important because often a major part of the problem with a data warehouse is getting data in. InfoPath can solve the "data entry by hand" problems pretty easily. For more information about InfoPath, see my book *Pro InfoPath 2007* (APress, 2007).

MOSS makes it very easy to assemble a dashboard with KPI lists and related content, specifically focusing on Excel Services spreadsheets for reporting. Before January 2009, SharePoint dashboards were somewhat problematic, as the KPI capabilities were fairly limited from a business intelligence perspective. However, after Microsoft moved PerformancePoint from being a separate product to MOSS Enterprise with KPI lists, Excel Services, and InfoPath Services, MOSS is a very respectable BI platform. Figure 14-65 shows a dashboard in MOSS using a KPI list, Excel Services, and SQL Server Reporting Services report (we'll add PerformancePoint in later).

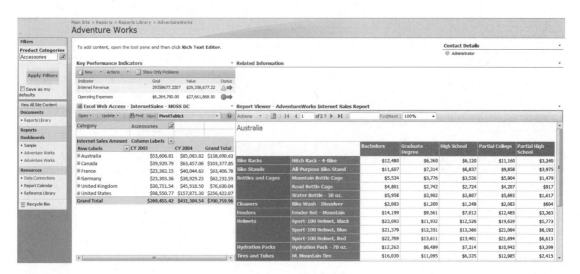

Figure 14-65. *A MOSS dashboard*

KPI Lists are a way to build ad-hoc dashboards from data available to SharePoint, including SharePoint lists, SQL Server Analysis Services, Excel spreadsheets, and manually-entered data. A KPI list is a standard SharePoint list, designed with some extra logic for representing a KPI-type indicator. You can see a MOSS KPI list in Figure 14-66.

Key Performance Indicators

Indicator	Goal	Value	Status
Internet Revenue	29358677.2207	$29,358,677.22	△ ➡
Operating Expenses	$6,264,750.00	$27,661,868.50	● ➡

Figure 14-66. *A KPI list in MOSS*

The final native feature in MOSS that we're interested in is Excel Services. Excel Services allows you to publish a spreadsheet from Excel 2007 to a thin client display on a SharePoint page, while still maintaining all data connections, formulas, etc. This allows Excel power users to create "reports" in the platform they are most familiar with, but then publish those reports in a full-fidelity display.

KPI Lists

I've always found KPI lists to be a little odd. They seem very powerful at first—a web-based scorecard key performance indicator feature that can be hooked to any number of data items (including SharePoint list data). However, when you've worked with KPIs in other products, the KPIs in MOSS come up very short.

MOSS KPIs can use data from several sources, as shown in Figure 14-67. Once you select a data source, you'll be given a definition page appropriate for the data source.

Figure 14-67. KPI List data sources

The definition page for SSAS data requires you to select a data connection from a library on the SharePoint server. Once you select a data connection to an Analysis Server, MOSS will provide a list of KPIs on the server, as shown in Figure 14-68. You can then give the indicator a name, and a URL for the KPI to use as a link to details.

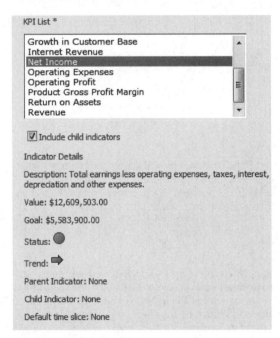

Figure 14-68. Selecting a KPI in MOSS

That's it—everything about the KPI must come from the definition in SSAS. This is, of course, a blessing and a curse. It follows our "one version of the truth" philosophy to the letter. However, any changes we need in a KPI (for example, tweaking a display format) require making changes to the cube. KPIs are then stored in a SharePoint list, and dashboard presentation is about configuring a view for the KPI list.

As I mentioned, it is an odd little beast. It does exactly one thing. I feel it would be a good tool for SSAS KPIs as a lightweight indicator, or for creating KPIs from data stored in SharePoint lists. But they're not useful for much else.

Excel Services

Excel Services is a very robust platform—the server enables processing large, complex Excel spreadsheets on server hardware, and also publishing Excel spreadsheets to a thin client. This enables consumption of Excel spreadsheets as analytics reports.

■ **Note** The spreadsheets in the web-based display are not interactive as spreadsheets (you can't click in a cell and start typing). However, they can be parameterized, and end users can change parameters to drive the display.

Excel Services is a pretty robust subject, and the configuration alone is fairly complex. For that reason I'm going to leave covering the topic to the experts, and recommend *Professional Excel Services* by Shahar Prish (Wrox, 2007) if you want to dig more deeply into using Excel Services.

PerformancePoint

PerformancePoint was launched in 2007 as a standalone product (PerformancePoint Server 2007). PerformancePoint Server offered advanced business intelligence capabilities, including scorecards, strategy maps (using Visio), analytic charts, and a planning engine. In January 2009, Microsoft announced that they were moving the capabilities of PerformancePoint into MOSS 2007, with the exception of the planning engine, which was being retired.

All the features of PerformancePoint work with a broad array of data sources, but scorecards and analytics really shine with an Analysis Services data source. The scorecards, key performance indicators, charts, and dashboards are all designed to leverage dimensions and members. A PerformancePoint Server scorecard is shown in Figure 14-69.

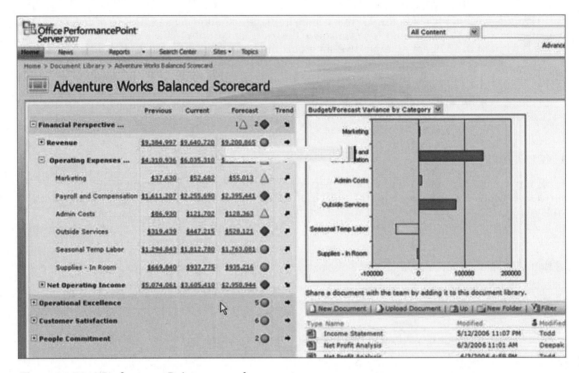

Figure 14-69. A PerformancePoint scorecard

The PerformancePoint designer is a click-once application executed from a web page on the server. The Dashboard Designer is a WYSIWYG designer that allows you to create data sources, KPIs, reports, and scorecards, as shown in Figure 14-70. Once you've designed a dashboard (consisting of one or more scorecards and associated analytics reports), you can publish it to SharePoint, where it will generate a page to display the components of the dashboard.

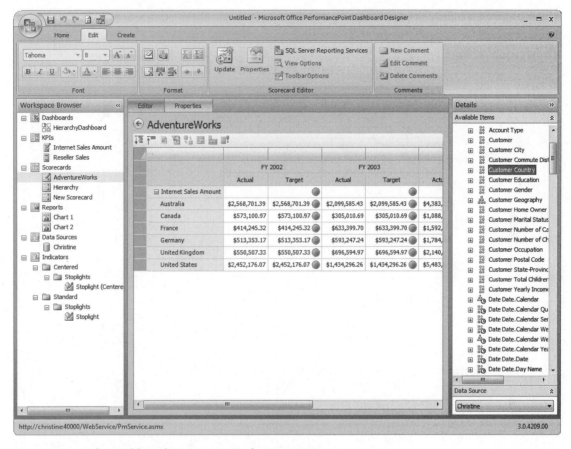

Figure 14-70. The Dashboard Designer in PerformancePoint

Let's run through a quick exercise to build a scorecard in PerformancePoint 2007. Follow along with Exercise 14-4.

Exercise 14-4. Build a Scorecard in PPS 2007

In this exercise you'll learn to build a scorecard using data from Analysis Services that you load into PerformancePoint 2007.

1. Open a browser, and navigate to `http://[server]:40000/` where [server] is the name of your PerformancePoint Server.

2. This will open the Dashboard Designer site. Click the Run button next to Download Dashboard Designer to open the Designer.

3. Once the Designer is open, right-click on Data Sources in the Workspace Browser on the left, and select New Data Source.

4. In the Select a Data Source Template dialog, select Multidimensional, and Analysis Services, and then click the OK button, as shown in Figure 14-71.

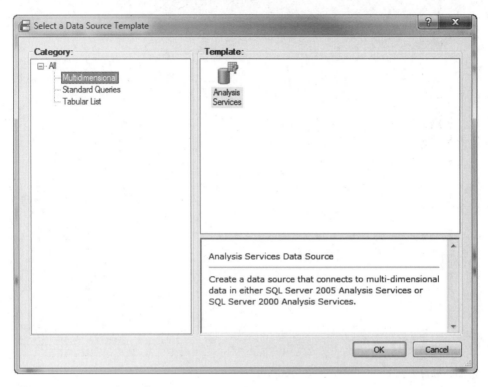

Figure 14-71. *Creating a data source*

5. Name the data source AdventureWorks, and then click the Finish button.

6. In the configuration editor, enter the name of the server, and then select the AdventureWorks DW 2008 database.

7. Select the AdventureWorks cube, as shown in Figure 14-72.

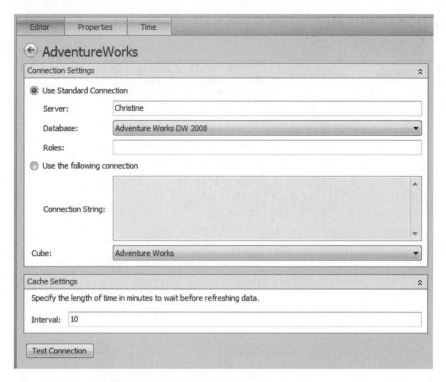

Figure 14-72. Configuring the data connection

8. Right-click on the new data source and select Publish.

9. Now right-click on Scorecards in the Workspace Browser and select New Scorecard.

10. Select Microsoft, and then Analysis Services. Click the OK button.

11. Name the scorecard AdventureWorks in the Create an Analysis Services Scorecard wizard. Click Next.

12. Select the AdventureWorks data source you just created. Click Next.

13. Ensure "Create KPIs from SQL Server Analysis Services measures" is selected. Click Next.

14. Click Add KPI.

15. We'll accept the defaults of Internet Sales Amount, for now. Click Next.

16. Don't make any changes on Add Measure Filters. Click Next.

17. Check Add Column Members, and then click the Select Dimension button.

18. Select Date.Date.Fiscal and click OK.

19. Click Select Members, and then right-click on All Periods and select Check Children, which should select all the Fiscal Years (Figure 14-73.)

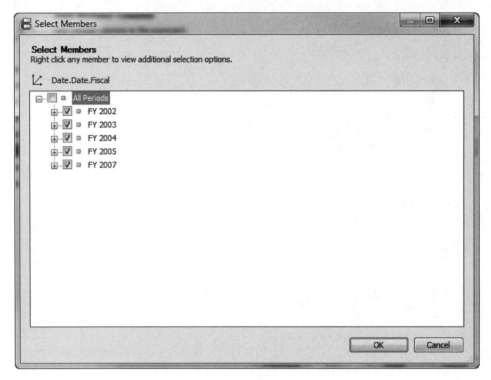

Figure 14-73. *Selecting Fiscal Year members*

20. Click OK, Finish, and then Close.

21. You should now have a scorecard that shows the Internet Sales Amount for each FY, as shown in Figure 14-74.

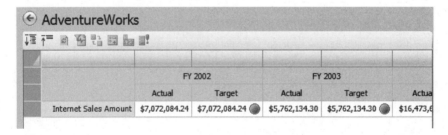

Figure 14-74. *Scorecard, almost*

22. Let's drill down a bit. In the Details pane on the right, click and drag Customer Country until it's on the right border of where Internet Sales Amount is, as shown in Figure 14-75.

Figure 14-75. *Adding countries to the scorecard*

23. Right-click on All Customers in the dialog and select Check Children, and then click the OK button.

24. Now we have the countries lined up underneath our KPI. If you click Update on the Edit tab, it should populate, as shown in Figure 14-76.

	FY 2002		FY 2003		FY 2004		FY 2005		FY 2007	
	Actual	Target	Actual	Target	Actual	Target	Actual	Target	Actual	Target
⊟ Internet Sales Amount	●		●		●		●			
Australia	$2,568,701.39	$2,568,701.39 ●	$2,099,585.43	$2,099,585.43 ●	$4,383,479.54	$4,383,479.54 ●	$9,234.23	$9,234.23 ●		◇
Canada	$573,100.97	$573,100.97 ●	$305,010.69	$305,010.69 ●	$1,088,879.50	$1,088,879.50 ●	$10,853.70	$10,853.70 ●		◇
France	$414,245.32	$414,245.32 ●	$633,399.70	$633,399.70 ●	$1,592,880.75	$1,592,880.75 ●	$3,491.95	$3,491.95 ●		◇
Germany	$513,353.17	$513,353.17 ●	$593,247.24	$593,247.24 ●	$1,784,107.09	$1,784,107.09 ●	$3,604.83	$3,604.83 ●		◇
United Kingdom	$550,507.33	$550,507.33 ●	$696,594.97	$696,594.97 ●	$2,140,388.50	$2,140,388.50 ●	$4,221.41	$4,221.41 ●		◇
United States	$2,452,176.07	$2,452,176.07 ●	$1,434,296.26	$1,434,296.26 ●	$5,483,882.67	$5,483,882.67 ●	$19,434.51	$19,434.51 ●		◇

Figure 14-76. *Finished scorecard*

That's the "quick and dirty" on creating a scorecard in PerformancePoint Services. To dig much deeper into the subject, I humbly recommend my book *Pro PerformancePoint Server 2007*, published by Apress in 2007.

Summary

In this chapter, I introduced you to several front-end tools to display your cube data. You used Excel 2007 pivot tables to attach to an SSAS cube to retrieve and display data. You also worked with formulas and pivot charts. After giving you a quick look into using Visio 2007, I guided you through the new Tablix component added to SSRS. Finally, you discovered the new KPI list and Excel Services features in MOSS 2007.

You did it! This chapter concludes your exploration of Pro SQL Server 2008 Analysis Services. The authors sincerely hope that you have gained valuable experience and insights into SSAS by choosing to spend your time with us. Thank you.

APPENDIX A

■ ■ ■

Setting Up Adventure Works

Through most of this book, I work with a demonstration data set called *AdventureWorks*. It's a multidimensional data warehouse and Analysis Services project based on a fictional bicycle company, which has both retail and Internet sales.

I really enjoy working with this data set. It's true that it's a bicycle store, and as I've often commented, "just about wholly unrelated to anything anyone does." However, although it may not be *related* to the business of various SSAS users, it is something that just about everyone can *understand*.

If you've downloaded sample databases from Microsoft before, you've gotten them from Microsoft.com. A few years ago, Microsoft moved all the samples to www.codeplex.com, their open source community, so they would be covered by the open source licensing there.

The sample databases are located at the following site:

www.codeplex.com/MSFTDBProdSamples

Click the Downloads link in the top-right corner. Then download SQL2008.AdventureWorks_All_Databases—either x64 or x86, depending on the architecture you're running. (There's a .zip file that has the same contents.) Then run the MSI on your SQL Server (Figure A-1).

■ **Note** CodePlex uses a Silverlight applet for the download prompt. So if you try to download from a locked-down server, the link won't do anything. To enable it, add *.codeplex.com to your trusted sites. Alternatively, download the file from a desktop PC and copy it to the server.

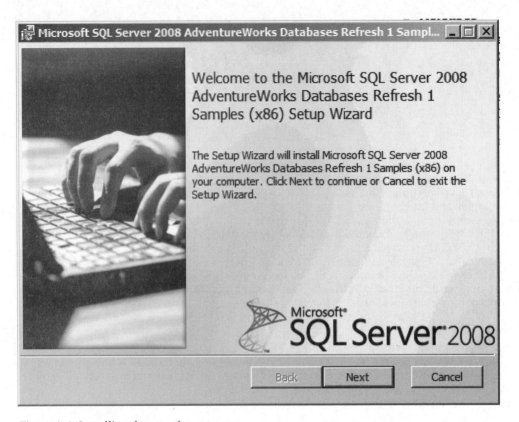

Figure A-1. Installing the samples

The wizard prompts you for the location to install the files (Figure A-2). The default installation location is C:\Program Files\Microsoft SQL Server\100\. The Program Menu Shortcuts option will add shortcuts to the Start menu. The Create AdventureWorks DBs option will generate the databases in the server you select in the next step.

■ **Note** You must have Full Text Search and FileStream installed on the server to create the databases, or the wizard will error out.

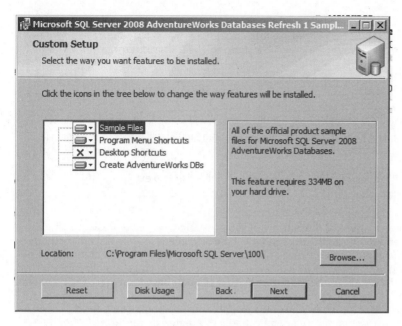

Figure A-2. Selecting options and the location to install the sample files

The final step is to select the SQL Server instance to install the databases into (Figure A-3).

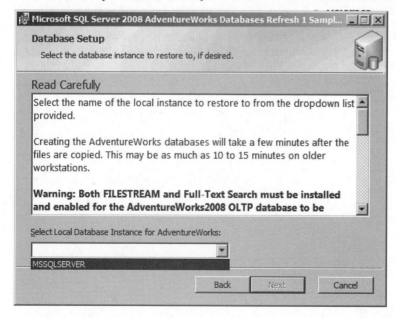

Figure A-3. Selecting the database instance

If you don't use the wizard, you can always install the databases yourself with the script files installed. To do this, open the Tools\Samples\AdventureWorks 2008 Data Warehouse folder from where you installed it. (If you installed the shortcuts with the MSI, there will be a shortcut to the folder in the Start Menu.)

■ **Note** If you use the wizard to install the databases, do not run the scripts afterward. The first thing the script will do is drop the existing database, and if you haven't set up the script properly, it will error out immediately afterward. It's a very confusing state to be left in, because it will look like the script created all the databases but one, when in fact the databases were already created and the script dropped that one.

Open the instawdwdb.sql file. If you have the SQL client tools installed, double-clicking the file should open SQL Server Management Studio. If you don't have the SQL client tools installed, you can copy the scripts to a PC where the client tools are available, or you can use the SQLCMD command-line parameter to run the scripts. If you installed the files into the default location, you should be able to run this query file, and it will create the database and populate it. After the database is created, verify that it's in place on the server.

To build the cubes necessary for reviewing some of the examples in the book, you'll need to open the OLAP projects and build the cubes (these were not automatically created by the wizard). Open Tools\Samples\AdventureWorks Analysis Services Project. You'll see two folders, enterprise and standard. Choose the folder that matches the edition of SQL Server Analysis Services you have installed. In that folder, you'll find an Adventure Works.sln file; double-click that to open the solution in BIDS.

After you have the solution open in BIDS, you'll need to make two changes. First, in the Solution Explorer, double-click the AdventureWorks data source, and enter the server name for where you installed the AdventureWorks DW database. Next, right-click on the solution (the topmost *Adventure Works* in bold), and select Properties. Click Deployment in the list on the left. Find the Server entry under Target and change that to the server name you want to deploy the cubes to when you process them. Click OK.

Deploy the project, process the cubes (see Chapter 8), and you're all set!

APPENDIX B

∎∎∎

Data-Mining Resources

Books

Data Mining with SQL Server 2005, by ZhaoHui Tang and Jamie MacLennan (Wiley, 2005)

Data Warehousing, Data Mining, & OLAP, by Alex Berson and Larry Dubov (McGraw-Hill, 2007)

Delivering Business Intelligence with Microsoft SQL Server 2005, by Brian Larson (McGraw-Hill, 2006)

Foundations of SQL Server 2005 Business Intelligence, by Lynn Langit (Apress, 2007)

Web Sites

SQLServerDataMining.com:
www.sqlserverdatamining.com

Microsoft SQL Server 2008 Data Mining page:
www.microsoft.com/sqlserver/2008/en/us/data-mining.aspx

Microsoft SQL Server forums:
http://social.msdn.microsoft.com/Forums/en-US/sqldatamining/threads

SQL Server Books Online:
http://msdn.microsoft.com/en-us/library/ms130214.aspx

About.com: "Data Mining: An Introduction":
http://databases.about.com/od/datamining/a/datamining.htm

Analysis Services Resource Hub:
www.ssas-info.com

SQL Server TechCenter:
http://technet.microsoft.com/en-us/sqlserver/default.aspx

Sample Databases for Microsoft SQL Server 2008: Samples Refresh 4:
http://msftdbprodsamples.codeplex.com/releases/view/37109

Index

■ C

■ D

■ E

■ O

■ P

■ Q

■ R

You Need the Companion eBook

Apress®
THE EXPERT'S VOICE™

eBookshop

233 Spring Street, New York, NY 10013